NONLINEAR TIME SERIES ANALYSIS OF ECONOMIC AND FINANCIAL DATA

Dynamic Modeling and Econometrics in Economics and Finance

VOLUME 1

NONLINEAR TIME SERIES ANALYSIS OF ECONOMIC AND FINANCIAL DATA

edited by

Philip Rothman
East Carolina University

KLUWER ACADEMIC PUBLISHERS
BOSTON / DORDRECHT / LONDON

Distributors for North, Central and South America:
Kluwer Academic Publishers
101 Philip Drive
Assinippi Park
Norwell, Massachusetts 02061 USA
Telephone (781) 871-6600
Fax (781) 871-6528
E-Mail <kluwer@wkap.com>

Distributors for all other countries:
Kluwer Academic Publishers Group
Distribution Centre
Post Office Box 322
3300 AH Dordrecht, THE NETHERLANDS
Telephone 31 78 6392 392
Fax 31 78 6546 474
E-Mail <orderdept@wkap.nl>

Electronic Services <http://www.wkap.nl>

Library of Congress Cataloging-in-Publication Data

Nonlinear time series analysis of economic and financial data / edited by Philip Rothman.
p. cm. — (Dynamic modeling and econometrics in economics and finance ; v. 1)
Includes bibliographical references and index.
ISBN 0-7923-8379-6
1. Econometric models. 2. Finance—Econometric models. 3. Time-series analysis. 4. Chaotic behavior in systems. 5. Nonlinear systems. I. Rothman, Philip. II. Series.
HB141.N658 1999
330'.01'5195—dc21 98-45191
CIP

Printed on acid-free paper.

Printed in the United States of America.

This book is dedicated to the memory of my father, Murray Rothman, who would have received a great deal of pleasure in seeing its publication.

Philip Rothman

Contents

Contributing Authors

Heather M. Anderson
Texas A&M University

Richard Ashley
Virginia Polytechnic Institute and State University

Chris Brooks
University of Reading

Patrick De Fontnouvelle
General Electric

John Driffill
University of Southampton

Alvaro Escribano
Universidad Carlos III de Madrid

Andrew J. Filardo
Federal Reserve Bank of Kansas City

Philip Hans Franses
Erasmus University, Rotterdam

Marie-Josée Godbout
Bank of Canada

Stephen F. Gordon
Université Laval

Melvin J. Hinich
University of Texas at Austin

Oscar Jordá
University of California, Davis

Junsoo Lee
Vanderbilt University and University of Central Florida

Panos Michael,
University of Liverpool

Bruce Mizrach
Rutgers University

Kiseok Nam
Texas A&M University

A. Robert Nobay
London School of Economics

David A. Peel
Cardiff University

James B. Ramsey
New York University

Morten O. Ravn
Universitat Pompeu Fabra, CEPR, and University of Aarhus

Philip Rothman
East Carolina University

Huntley Schaller
Carleton University

Michael J. Smith
University of Reading

Martin Sola
University of London, Universidad Torcuato Di Tella, and London Business School

Norman R. Swanson
Pennsylvania State University

Hui Boon Tan
Universitii Putra Malaysia

David J. Thomson
Bell Laboratories, Lucent Technologies

Farshid Vahid
Texas A&M University

Simon van Norden
École Des Hautes Études Commercial

James Watkins
American Express

INTRODUCTION

Philip Rothman
East Carolina University

1. Introduction

This volume presents a collection of papers from some of the world's leading researchers in the field of economic and financial nonlinear time series analysis. There is today increasing interest amongst practitioners in nonlinear time series methods. This behavior stems from several important developments in the literature.

Roughly fifteen years ago, econometricians began a "search for chaos" in economic and financial data. This line of inquiry was motivated along two fronts. First, physicists had shown that certain classes of chaotic attractors generate time series with an intriguing property, i.e., deterministic white noise. This raised the possibility that the white noise processes often observed (e.g., stock returns or residuals from fitted models) were actually chaotic. Second, growth theorists had shown that, for certain parameter values, conventional models could generate chaotic dynamics. This seemed to provide some theoretical justification for the wide-spread testing for chaos that was about to take place; see Boldrin (1994), however, for an argument that this justification was rather weak.

The subsequent empirical literature produced on the whole negative results on the question of whether economic time series are generated by low-dimensional chaotic attractors; see, e.g., Ramsey, *et al.* (1990). By the start of this decade interest had shifted from detecting chaos per se to testing for "neglected nonlinearity;" see, e.g., Lee, *et al.* (1993). Researchers began to focus on the extent to which the algorithm of obtaining white noise residuals through the standard linear second-order analysis could mask the existence of a stochastic nonlinear data generating process.

An important parallel development in the literature was the introduction of tractable nonlinear time series models. Foremost amongst these have been the AutoRegressive Conditional Heteroskedasticity (ARCH) model and its many extensions, the Threshold AutoRegressive (TAR) model, the Markov-switching time series model, and the class of Smooth Transition AutoRegressive (STAR) models; see Engle (1982), Tong (1993), Hamilton (1989), and Luukkonen, *et al.* (1988), respectively. Through use of such models we now have an arguably improved characterization of the dynamical properties of many important economic and financial time series.

The papers in this book strongly reflect these trends. Following Hamilton's (1989) seminal contribution, five papers in this volume utilize Markov-switching time series methods. Filardo and Gordon compare a set of non-nested empirical business cycle models. The linear models considered are a VAR and Stock and Watson's (1991) unobserved components model. The nonlinear models studied are a time-varying transition probability Markov-switching model and an integration of the Markov-switching model with the Stock and Watson model. Filardo and Gordon focus on how well these alternative approaches model the NBER-identified turning point in the U.S. economy in 1990.

In their chapter Mizrach and Watkins examine the relative merits of two standard numerical procedures, the DFP and EM algorithms, used to estimate the parameters of Markov-switching time series models. Their analysis shows that while the EM algorithm is more robust than the DFP method with respect to poor likelihood surfaces, it is computationally far more demanding. Their simulations also demonstrate that the computational burden of the EM algorithm can be decreased significantly through use of an approximation technique.

Using time-varying transition probability Markov-switching models, in their paper Ravn and Sola carry out an exhaustive analysis of the extent to which one can identify factors which, based on postwar quarterly data for the U.S., can be said to cause business cycle shifts between recessions and expansions. They concentrate on those macroeconomic variables which appear to have predictive power in modelling the time variation in the transition probabilities of the estimated Markov-switching time series models. The framework these authors employ also allows them to examine whether there are asymmetric effects between the macrovariables considered and output. Ravn and Sola's empirical results and discussion provide very useful insights into the nature of the transmission mechanism for the U.S. economy.

Driffill and Sola introduce regime-switching into a VAR in their study of the present value hypothesis. In their paper they extend Campbell and Shiller's (1987) well-known analysis in two important directions. In contrast to Campbell and Shiller (1987), with their approach Driffill and Sola show that the expectations present value model appears to be inconsistent with U.S. data. They argue that Campbell and Shiller's (1987) inability to reject the present value model stems from their use of a misspecified VAR.

van Norden and Schaller study the link between regime-switching and stock market crashes in their chapter. They begin with a switching-regression model in which stock returns switch between survival and collapse regimes. The authors then extend the model to allow for regime-switching in economic fundamentals, which allows them to examine the extent to which stock market crashes can be linked to such regime-switching. van Norden and Schaller's approach is the first in the literature which explores this particular source of stock markets panics.

The chapter by Godbout and van Norden does not utilize Markov-switching time series methods directly. It does, however, provide a very useful cautionary note for practitioners of such techniques when analyzing the behavior of excess returns in financial markets. For example, Evans and Lewis (1993, 1995) argue that results suggesting that excess returns to holding foreign exchange are I(1) should be interpreted as evidence of the importance of regime-switching in asset prices. The purpose of Godbout and van Norden's analysis is to call into question such cointegration-based claims that excess returns are I(1). They assert that these claims are based to some extent on inadequately constructed Monte Carlo experimental designs.

Three papers in this volume are concerned with STAR-type modelling. Michael, Nobay, and Peel reconsider the well-known persistence in deviations from equilibrium money demand in light of the existence of lumpy costs of adjustment. Such non-convex costs imply that a linear adjustment model is inappropriate. Accordingly, in their chapter these authors test the null hypothesis of linear adjustment to equilibrium money demand against the alternative of a STAR adjustment process using two sets of quarterly U.S. data. They strongly reject the linear null hypothesis and their estimated STAR models imply no adjustment for small deviations, but rapid adjustment for large deviations from equilibrium.

In their chapter Escribano and Jordá extend some of their previous work and introduce new *LM* specification procedures to identify Logistic and Exponential Smooth Transition Regression (STR) models. They argue that their specification algorithms are simpler, consistent, and more powerful than those previously available in the literature; see Teräsvirta (1994). The authors provide extensive Monte-Carlo evidence on the performance of their proposed tests and also apply their new procedures in a study of the monthly U.S. postwar unemployment rate.

Anderson, Nam, and Vahid introduce a class of asymmetric nonlinear smooth transition GARCH models. They argue that such models are useful for studying the asymmetric response of volatility shocks of different magnitudes and different signs. This chapter departs from previous approaches which utilize time-varying GARCH parameter models in the study of stock market data by: (1) allowing a less restrictive switching setup; and (2) allowing the volatility response to price shocks to depend on the current volatility state. In their empirical application to an international stock market dataset, Anderson, Nam, and Vahid show that their estimated models dominate specifications that only allow for time variation in the ARCH parameters.

The remainder of the papers in this volume do not fit neatly into either the Markov-switching or STAR paradigm. But they do address many important issues of interest to analysts concerned with improved modelling, relative to the standard linear approach, of economic and financial time series data. In their selective survey, Swanson and Franses focus on two important recent developments in the nonlinear time series literature: (1) stochastic unit root models; and (2) artificial neural networks. These authors have made important contributions to the literature on both of these topics, so it is useful to see displayed both their technical expertise and their reflective comments on general modelling issues.

In their chapter Tan and Ashley argue that frequency dependent relationships are inherently nonlinear. Their analysis shows that when the true relationship between a set of variables is frequency dependent, the coefficient in the linear model is unstable across frequencies due to model misspecification. This implies that there is a fundamental problem in using the usual linear cross spectral analysis to study bivariate relationships subject to frequency dependence. In the empirical part of their paper Tan and Ashley produce evidence suggesting that consideration of nonlinear dynamics is necessary to properly model the relationship between U.S. time series data for the CPI and PPI.

Nonlinear cross-dependence is also considered by Brooks, Hinich, and Smith in their study of high frequency stock returns and volume data for stocks traded on the London Stock Exchange Automated Quotation System. Through use of higher-order spectral and cross-spectral analysis, these authors produce very strong evidence of nonlinear dependence and within and between the series. However, the strength of these nonlinear dependencies is short-lived, since the market appears to evolve rapidly. This leads Brooks, Hinich, and Smith to be skeptical about the efficacy of applying tests and estimating models over very long spans of stock market data, since the associated assumption of a fixed data generating process is unlikely to hold.

Ramsey and Thomson re-explore the second-order spectral properties of the growth rates for a set of long-run real and financial monthly time series data. In their paper they employ recent innovations in the estimation of sample spectra that provide much greater efficiency than current techniques; these new procedures are especially important in light of the ever present problem of nonstationarity. In their analysis of industrial production indices, they find that after allowing for a set of main frequencies, there still remains a residual predictable, yet complex, component. Ramsey and Thomson conclude that these data are inconsistent with a linear data generating process. They also find that the stock market is efficient at seasonal frequencies in growth rate terms, but has significant levels of coherence with the real data at business cycle frequencies.

Following Perron's (1989) pioneering contribution, Lee's chapter is concerned with stationarity testing in the presence of multiple endogenized structural breaks. In particular, he considers the converse of the problem studied by, e.g. Bai and Perron (1998). Bai and Perron (1998) are concerned with testing a unit root null hypothesis in the presence of multiple structural breaks with unknown break points. In contrast, Lee considers the problems of testing the null hypothesis of stationarity around an unknown number of breaks at unknown break points, against a unit root alternative hypothesis. Lee employs the Bayes information criterion to determine both the number of breaks and their placements endogenously from the data. He shows that the proposed testing procedure performs reasonably well under the stationarity null hypothesis, when the magnitude of the structural breaks is big enough. The endogenized stationarity tests also have decent power under the alternative hypothesis of a unit root.

In his chapter De Fontnouvelle examines the structural foundations of ARCH behavior in stock market data in the context of the Mixture of Distributions model. His test of this model relies upon the notion of time reversibility, an important time series property exploited by Ramsey and Rothman (1996) in their study of business cycle asymmetry. De Fontnouvelle establishes a set of important relationships between the reversibility properties of the information arrival process and the time reversibility/irreversibility for volatility and trading volume. In this paper De Fontnouvelle introduces a test of time reversibility which he argues is more general than the *TR* test of Ramsey and Rothman (1996). He provides reversibility testing results which he interprets as statistical evidence against the Mixture of Distributions financial model.

Time reversibility testing is also the concern of Rothman's chapter in this volume. He carries out a set of Monte Carlo simulations to explore the power properties of Ramsey and Rothman's (1996) *TR* test against a set of simple bilinear and threshold AutoRegressive models. The results show that the power of the test unambiguously increases as an index of time irreversibility increases. In addition, Rothman establishes a set of patterns revealed by the *TR* test statistics which in principle can help in model identification of these classes of nonlinear time series models.

Many people are due thanks in bringing about the production of this book volume. First and foremost I wish to thank the many contributors who were generous and considerate enough to participate in this project. I would like to also acknowledge the suggestions of Clive Granger and Buz Brock in helping me assemble this particular set of contributors.

This book was delivered to the publisher in "camera-ready" form. The labor requirements behind this effort were more than considerable. Special thanks go once again to all of the book's contributors for their patience and willingness to cooperate in this terribly tedious task; in particular, the efforts of Andy Filardo, Bruce Mizrach, and David Thomson were truly above the call of duty. The ECU Economics Department secretary, Laurie Dixon, played a crucial role in typesetting the book. Finally, I wish to thank Ranak Jasani and Yana Lambert, my contact people at Kluwer, for all of the help and guidance they have given me in this process.

References

Bai, J. and P. Perron (1998). "Estimating and Testing Linear Models with Multiple Structural Changes," *Econometrica*, 66, 47-78.

Boldrin, M. (1994). "Competitive Markets and Endogenous Cycles: An Evaluation," in Semmler, W. (ed.) *Business Cycles: Theory and Empirical Methods*. Boston: Kluwer Academic Publishers, 53-71.

Campbell, J. Y. and R. J. Shiller (1987). "Cointegration and Tests of Present Value Models," *Journal of Political Economy*, 95, 1062-1088.

Engle, R. F. (1982). "AutoRegressive Conditional Heteroskedasticity with Estimates of the Variance of UK Inflation," *Econometrica*, 50, 251-276.

Evans, M. D. D. and K. Lewis (1993). "Trends in Excess Returns in Currency and Bond Markets," *European Economic Review*, 37, 1005-1019.

Evans, M. D. D. and K. Lewis (1995). "Do Long-Term Swings in the Dollar Affect Estimates of the Risk Premia?" *Review of Financial Studies*, 8, 709-742.

Hamilton, J. D. (1989). "A New Approach to the Economic Analysis of Nonstationary Time Series and the Business Cycle," *Econometrica*, 57, 357-384.

Lee, T.-H., H. White, and C. W. J. Granger (1993). "Testing for Neglected Nonlinearity in Time Series Models: A Comparison of Neural Network Methods and Alternative Tests," *Journal of Econometrics*, 56, 269-290.

Luukkonen, R., P. Saikkonen, and T. Teräsvirta (1988). "Testing Linearity in Univariate Time Series Models," *Scandinavian Journal of Statistics*, 15, 161-175.

Perron, P. (1989). "The Great Crash, the Oil Price Shock, and the Unit Root Hypothesis," *Econometrica*, 57, 1361-1401.

Ramsey, J. B. and P. Rothman (1996). "Time Irreversibility and Business Cycle Asymmetry," *Journal of Money, Credit, and Banking*, 28, 1-21.

Ramsey, J. B., C. L. Sayers, and P. Rothman (1990). "The Statistical Properties of Dimension Calculations Using Small Data Sets: Some Economic Applications," *International Economic Review*, 31, 991-1020.

Stock, J. H. and M. Watson (1991). "A Probability Model of the Coincident Economic Indicators," in Lahiri, K. and G. H. Moore (eds.) *Leading Economic Indicators: New Approaches and Forecasting Records*. New York: Cambridge University Press, 63-85.

Teräsvirta, T. (1994). "Specification, Estimation and Evaluation of Smooth Transition Autoregressive Models," *Journal of the American Statistical Association*, 89, 208-218.

Tong, H. (1983). *Threshold Models in Non-Linear Time Series Analysis. Lecture Notes in Statistics, Vol. 11.* New York: Springer-Verlag.

1 BUSINESS CYCLE TURNING POINTS: TWO EMPIRICAL BUSINESS CYCLE MODEL APPROACHES

Andrew J. Filardo
Federal Reserve Bank of Kansas City
Stephen F. Gordon
Université Laval

1. Introduction

In the last decade, two business cycle models have dominated discussions and research of empirical business cycles. The first model is the Markov switching model of business cycle phases pioneered by Hamilton (1989). In this nonlinear business cycle model, cycles are composed of expansionary and contractionary phases. Empirically, this model is consistent with the classic research methodology of Burns and Mitchell (1946). The second model is the unobserved dynamic model of Stock and Watson (1989, 1991, 1993). In this linear model, cycles arise from the internal propagation mechanism of the system of equations. Econometrically, this model is consistent with the Frischian view of business cycles. Even though the model is not constructed to model business cycle expansions and contractions, the model is helpful in predicting downturns and upturns in the economy.

Recently, some researchers have tried to reconcile these two fundamentally different characterizations of business cycles. In particular, Diebold and Rudebusch (1996) and Chauvet (1994) have embedded a Markov switching model within a dynamic linear unobserved components model. Using a particular functional form and maximum likelihood methods, they find that the unobserved component appears to switch. Our paper extends their analyses by focusing on the information content of the models around turning points in the business cycle. Moreover, we offer an alternative estimation procedure to more tractably estimate the model and exploit new methods to compare the two competing models.

Econometrically, the model comparison strategy is made tractable with the Gibbs sampler and data augmentation. We use techniques developed by Albert and Chib (1993a) and McCulloch and Tsay (1994) for Markov switching models, a multivariate time-varying transition probability model (Filardo and Gordon 1994) and model comparison methods of Geweke (1994). Issues of identification, estimation, and model comparison are highlighted.

This paper estimates a number of empirical business cycle models and compares their predictive accuracy. The next section describes the data used in the investigation. The third section describes the alternative linear and nonlinear models. The fourth section outlines the model comparison strategy. The fifth section outlines the estimation procedure. In particular, the steps of the Gibbs sampler for each model are listed. The sixth section reports results and the last section draws conclusions and highlights issues that warrant further study.

2. Data Description

Monthly data for the employees on nonagricultural payrolls (L_t) are from the Bureau of Labor Statistics. Personal income less transfer payments (I_t) in 1987 dollars, manufacturing and trade sales (M_t) in 1987 dollars, and the composite index of the eleven leading indicators (Z_t) are from the Bureau of Economic Analysis. The industrial production data (Y_t) are from the Federal Reserve Board. The employment, income, sales, and industrial production series correspond to the component series in the Department of Commerce's coincident index of economic activity. All variables are transformed into growth rates by log first differences. The data run from January 1960 to January 1995.

3. Empirical Business Cycle Models

Two empirical business cycle approaches have come to dominate the literature—one linear and the other nonlinear. Each approach has certain benefits and costs. The chief benefits from linear models arise from the ease of estimation and wide success in capturing certain features of the business cycle data—such as means, variances, and time series dynamics. The benefits of the nonlinear models arise from the ability to account for nonlinear features of the data—such as turning points. Complexity in the models and difficulty in estimation derives from the nature of unobserved variables and from nonlinearities.

3.1 VAR Model

The multiequation linear autoregressive model has been a workhorse model of modern econometrics. This paper uses the simple, yet informative, vector autoregression (VAR) to model economic fluctuations. While historically quite important and empirically informative, the VAR model only serves as a benchmark in this paper.[1] As a benchmark, this model will help us understand the relative contributions, if any, of two more sophisticated macroeconomic models.

The structure of the VAR is standard. Let x_t be a $(k \times 1)$ vector of growth rates $(x_1,...,x_k)'_t$ at time t. The model is a pth order Gaussian vector autoregression

$$\begin{aligned} x_t &= \beta_{VAR} + \Phi_1 x_{t-1} + \cdots + \Phi_p x_{t-p} + e_t \\ &= \beta_{VAR} + \Phi_{VAR}(L) x_{t-1} + e_t \end{aligned} \tag{1}$$

where $\Phi(L) = 1 + \Phi_1 x_{t-1} + \cdots + \Phi_p x_{t-p}$, and e_t is a normally distributed $N\left(0, \Sigma_e\right)$ error vector.

3.2 Stock-Watson Unobserved Components Model

The unobserved components model of Stock and Watson (SW) is a multiequation autoregressive model which serves an alternative to the standard VAR. The model can capture two important features of business cycle fluctuations. First, business cycle behavior is common across the data series in x_t. Second, the business cycle component is unobservable, but inferable.

The model has a dynamic latent variable structure,

$$\begin{aligned} x_t &= \beta_{SW} + \Phi_{SW}(L) x_{t-1} + D^{-1}(L)\gamma \Delta C_t + \xi_t \\ \psi(L)\Delta C_t &= \delta + \eta_t \end{aligned} \tag{2}$$

where $\xi \sim N(0, \Sigma_\xi)$. $\psi(L)$ is a lag polynomial and $D(L)$ is a matrix lag polynomial of lags q and r, respectively. ΔC_t is the unobserved component and, without loss of generality, η_t is distributed $N(0,1)$.[2] With the assumption that $(\xi_1,..., \xi_T, \Delta C_t)$ are mutually uncorrelated at all leads and lags, ΔC_t is identified as the common component. In addition, the variance of η_t is assumed to be 1 without loss of generality. Linearity of the model is apparent. Stock and Watson point out that the identification restrictions imply that the extracted common component is a linear function of the observable variables, i.e., $\Delta C_{t|t} = W(L) X_t$ where $W(L)$ is a lag polynomial.

3.3 Time-Varying Markov Switching Model

This paper considers two nonlinear models of the Markov switching type. The first is a multivariate version of Hamilton's (1989) Markov switching model with time-varying transition probabilities. As in the SW unobserved components model, an unobserved state variable measures the common business cycle across the series. In contrast to the SW model, the state variable in the TVTP model, S_t, is assumed to take on one of two values. This binary variable can pick up business cycle phases that correspond to notions of expansions and recessions. Even though this binary state variable takes on fewer values than the continuous random variable ΔC_t, S_t can capture sudden shifts in data series; such shifts may be empirically important around business cycle turning points.

Structurally, the TVTP model has phase-dependent means within an otherwise linear dynamic system,

$$x_t = \mu(S_t) + \Phi_{TVTP}(L)[x_{t-1} - \mu(S_{t-1})] + \varepsilon_t$$

or

$$\begin{pmatrix} x_{t,1} \\ \vdots \\ x_{t,k} \end{pmatrix} = \begin{pmatrix} \mu_1(S_t) \\ \vdots \\ \mu_k(S_t) \end{pmatrix} + \begin{pmatrix} \Phi_{11}(L) & \cdots & \Phi_{1k}(L) \\ \vdots & \ddots & \vdots \\ \Phi_{k1}(L) & \cdots & \Phi_{kk}(L) \end{pmatrix} \begin{pmatrix} x_{t-1,1} - \mu_1(S_{t-1}) \\ \vdots \\ x_{t-1,k} - \mu_k(S_{t-1}) \end{pmatrix} + \begin{pmatrix} \varepsilon_{t,1} \\ \vdots \\ \varepsilon_{t,k} \end{pmatrix} \quad (3)$$

where the mean is $\mu_i(S_t) = \mu_{il} + \mu_{iH} S_t, \Phi_{ij}(L) = \phi_{ij,1} + \phi_{ij,2} L^1 + \ldots + \phi_{ij,r} L^{r-1}$ and the error $(\varepsilon_{t,1}, \ldots, \varepsilon_{t,k})' \sim N(0, \Sigma_\varepsilon)$.

The unobserved binary state variable is assumed to have a stochastic process governed by time-varying Markovian transition probabilities. Time-varying transition probabilities gauge the persistence of the business cycle phases of expansions and recessions. Moreover, the time-varying transition probabilities allow the probability of exiting a particular phase to vary with economic information

$$P(S_t = s_t \mid S_{t-1} = s_t, \mathrm{z}_t) = \begin{bmatrix} q(\mathrm{z}_t) & 1 - P(\mathrm{z}_t) \\ 1 - q(\mathrm{z}_t) & P(\mathrm{z}_t) \end{bmatrix} \quad (4)$$

This paper uses a latent variable version of the probit model to measure the transition probability matrix at each t,

$$P(S_t = 1) = P(S_t^* \geq 0) \quad (5)$$

where S_t^* is a latent variable defined by

$$S_t^* = \gamma_0 + \gamma_z z_t + \gamma_s s_{t-1} + u_t = [1, z_t, s_{t-1}]\Gamma_{TVTP} + u_t \tag{6}$$

and u_t is distributed as an independent standard normal variable; without loss of generality, setting the variance equal to unity allows identification of the model. Moreover, this distributional assumption simplifies the calculation of the transition probabilities. Calculating transition probabilities at each t is done by evaluating the conditional cumulative distribution function for u. Let the function $\Phi_{U|Z}(u(z))$ represent a conditional $N(0,1)$ CDF such that

$$\int_{-\infty}^{\bar{u}(z)} \frac{1}{\sqrt{2\pi}} \exp(-\frac{1}{2}x^2)dx$$

where $\bar{u}(z)$ is the upper limit of the integration and is determined by equations (5) and (6). Consequently, p_t and q_t are

$$\begin{aligned} p_t &\equiv P(S_t = 1 \mid s_{t-1} = 1) = P(U_t \geq -\gamma_0 - \gamma_z z_t - \gamma_s) \equiv 1 - \Phi_{U|Z}(\gamma_0 - \gamma_z z_t - \gamma_s) \\ q_t &\equiv P(S_t = 0 \mid s_{t-1} = 0) = P(U_t < -\gamma_0 - \gamma_z z_t) \qquad \equiv \Phi_{U|Z}(-\gamma_0 - \gamma_z z_t) \end{aligned} \tag{7}$$

In this model, the unobserved state variable, S_t, is a nonlinear function of the observable variables. The nonlinearity reflects the abruptness with which the economy switches from expansion to recession and from recession to expansion.

3.4 Nonlinear Stock-Watson Model

Diebold and Rudebusch (1996) and Chauvet (1994) investigate an empirical business cycle model that combines business cycle features of the Stock-Watson unobserved component model and the TVTP model. They extend the SW model to include a Markovian state variable in the unobserved component equation. The second line in equation (2) can be recast as

$$\delta_t = \mu_0 + \mu_1 S_t\,. \tag{8}$$

With equation (8), the resulting model is nonlinear. This SW-TVTP model incorporates the dynamic structure of the unobserved component ($D(L)\Delta C_t$) in the SW model with the abruptness of the phase-dependent means μ_{s_t} in the TVTP model. The stochastic process on S_t is the same as in Section 3.3.

4. Estimation

This paper estimates and compares models in a Bayesian framework with Gibbs sampling methods. The Gibbs sampling methods not only simplify estimation of the unobserved components and TVTP models, but also facilitate non-nested model comparisons. Geweke (1994) has proposed model comparison methods that are a natural byproduct of Markov chain Monte Carlo methods such as Gibbs sampling methods. Both marginalized likelihoods and Bayes factors are used to compare the models described in Section 3.

4.1 Gibbs Sampling Methods

Gibbs sampling methods are simulation-based estimation methods that avoid the need to directly estimate the joint distribution of parameters in a model with unobserved variables; direct estimation is both cumbersome and unnecessary. Because each model provides conditional distributions that are easily derived and simulated, Gibbs sampling methods are particularly well suited to estimate the models in this paper. This paper adopts a Bayesian methodology and also draws on data augmentation techniques that treats all unobserved elements of the model as parameters to be estimated (including the unobserved random variables).

In the standard VAR in equation (1), the model parameters of interest are

$$\theta_{VAR} = \{\Phi_{VAR}, \Sigma_e\}. \tag{9}$$

In the unobserved components model in equation (2), the parameters of interest are

$$\theta_{SW} = \left\{\beta_{SW}, \Phi_{SW}, D, \gamma, \Sigma_\xi, \psi, \delta, \{\Delta C_t\}_{-q}^{T}\right\}. \tag{10}$$

In the TVTP model in equations (3), (4), (6), and (7) the parameters of interest are

$$\theta_{TVTP} = \left\{\mu, \Phi_{TVTP}, \Sigma_\varepsilon, \Gamma_{TVTP}, \{s_t\}_0^T, \{s_t^*\}_1^T, \{p_t\}_1^T, \{q_t\}_1^T\right\}. \tag{11}$$

In the SW-TVTP model in equations (2), (6), (7), and (8), the parameters of interest are

$$\theta_{SW-TVTP} = \left\{\beta, \gamma, \delta_0, \delta_1, \Phi, D, \psi, \Sigma_\xi, \{\Delta C_t\}_{-q}^{T}, \Gamma_{TVTP}, \{s_t\}_0^T, \{s_t^*\}_1^T, \{p_t\}_1^T, \{q_t\}_1^T\right\}. \tag{12}$$

5. Model Comparison

To compare the linear and nonlinear models of Section 3, this paper adopts the strategy of Geweke (1994). Geweke proposes the use of the marginalized likelihood and Bayes factors to compare models. The chief benefit of his strategy is that model comparison can be done without a significant computational burden above calculating the parameter estimates. This strategy is particularly useful in the business cycle comparisons for this paper, because this method provides a convenient metric to measure the performance of these models observation by observation.

5.1 Methodology

To understand how the marginalized likelihood and the Bayes factor are easily calculated with Gibbs sampling methods, the link between the marginal likelihood and predictive likelihood needs to be outlined. For a given set of data $\{x_t\}_{t=1}^{T}$ and a model *J* denoted by θ_j, the partial likelihood function is

$$L_{jt}(\theta_j, X_t,) = \prod_{s=1}^{t} f_{js}(x_s \mid Y_{s-1}, \theta_j)$$

with a posterior conditional distribution

$$p_{jt}(\theta_j \mid X_t) = \frac{f_{j0}(\theta_j) L_{jt}(\theta_j, X_t)}{\int_{\Theta_j} f_{j0}(\theta_j) L_{jt}(\theta_j, X_t) d\theta_j} \tag{13}$$

where $f_{jt}(x_t \mid Y_{t-1}, \theta_j)$ is the conditional density of x_t, given the previous data and the model *j*. Corresponding to this partial likelihood function is the marginal likelihood function. This function is a mixture of the partial likelihood functions and the mixing weights of the prior distribution, $f_0(\theta_j)$, for model θ_j:

$$M_{jt} = \int_{\Theta_j} f_{j0}(\theta_j) L_{jt}(\theta_j, Y_t) d\theta_j .$$

The Bayes factor is calculated as $B_{j|kt} = \dfrac{M_{jt}}{M_{kt}}$.

Geweke shows that the predictive likelihood can be written as a linked predictive likelihood. This decomposition is important in order to compare models. The predictive likelihood can be written as

$$\hat{p}^{t}_{jt} = \int_{\Theta_j} p_{js_0}(\theta_j \mid X_{s_0}) \prod_{s=s_0+1}^{s_q} f_{js}(x_s \mid X_{s-1}, \theta_j) d\theta_j = \prod_{t=1}^{q} \hat{p}^{s_t}_{js_{t-1}}$$

where

$$\hat{p}^{s_t}_{js_{t-1}} = \frac{\int_{\Theta_j} p_{js_0}(\theta_j) \prod_{s=s_0+1}^{s_t} f_{js}(x_s \mid X_{s-1}, \theta_j) d\theta_j}{\int_{\Theta_j} p_{js_0}(\theta_j) \prod_{s=s_0+1}^{s_{t-1}} f_{js}(x_s \mid X_{s-1}, \theta_j) d\theta_j}.$$

By interpreting equation (13) as the importance sampling density for the posterior density $p_{j,t+1}(\theta_j \mid X_{t=1})$, the appropriate weights are

$$\frac{L_{j,t+1}(\theta_j, X_{t+1})}{L_{j,t}(\theta_j, X_t)} = f_{j,t+1}(x_{t+1} \mid X_t, \theta_j).$$

This observation by Geweke means that evaluation of the posterior density $p_{j,t+1}(\theta_j \mid X_{t=1})$ is a natural byproduct of the drawn sample $\{\theta_j^{\{m\}}\}$. In particular, for each draw of $\{\theta_j^{\{m\}}\}$, $\{f_{j,t+1}(x_{t+i} \mid X_{t+i-1}, \theta_j)\}_{i=1}^{t}$ is computed recursively. Notice that this computation does not require a Monte Carlo sample of the posterior distribution of the predictive likelihoods.[3]

The predictive Bayes factor has a similar decomposition

$$\hat{B}^{s_q}_{j|k,s_0} = \prod_{l=1}^{q} \left(\hat{p}^{l}_{ju} / \hat{p}^{l}_{ku}\right) = \prod_{l=1}^{q} \hat{B}^{s_l}_{j|k,s_0}. \tag{14}$$

This measure helps to gauge whether an observation or a group of observations are more likely to occur in one model versus another.

In the SW unobserved components model and the TVTP model, evaluation of the predictive density will employ simulation methods. The simulated predictive density is simpler than numerically integrating the predictive density. For the models in this paper and given the observations $\{1, \ldots, u\}$, the simulated predictive density for observations $\{u+1, \ldots, t\}$ and the unobserved state variables $\{S_{u+1}, \ldots, S_t\}$ is

$$L(\theta \mid x) = \int \cdots \int L(\theta, S \mid x) dS_1, \ldots, dS_t = (2\pi)^{-\frac{(t-u)n}{2}} \left|\Sigma^{-1}\right|^{\frac{t-u}{2}}$$

$$\int \cdots \int \exp\left(\sum_{s=u+1}^{t} -\varepsilon(S)'\Sigma^{-1}\varepsilon(S)\right) P(S_{u+1}, \ldots, S_t \mid S_1, \ldots, S_u, \theta) d\Delta S_{u+1} \cdots d\Delta S_t.$$

5.2 Applications

In the models of Section 3, the predictive density can be computed with the following formulae. Given the observations from $\{1,\ldots,u\}$, each model's predictive density $\hat{p}^{t}_{ju} = \prod_{t=1}^{q} \hat{p}^{s_t}_{js_{t-1}}$ for $\{u+1,\ldots,t\}$ is easy to compute. First, for the VAR model, the predictive density is

$$\hat{p}^{t}_{VAR,u} = (2\pi)^{-\frac{(t-u)n}{2}} \left|\Sigma_e^{-1}\right|^{\frac{t-u}{2}} \exp\left(\sum_{s=u+1}^{t} -(1/2)\, e_s'\Sigma_e e_s\right).$$

Second, for the unobserved component model, the predictive density is

$$\hat{p}^{t}_{SW,u} = (2\pi)^{-\frac{(t-u)n}{2}} \left|\Sigma_\xi^{-1}\right|^{\frac{t-u}{2}}$$

$$\int\cdots\int \exp\left(\sum_{s=u+1}^{t} -(1/2)\, \xi_s'\Sigma_\xi \xi_s\right) p(\Delta C_{u+1},\ldots,\Delta C_t)\, d\Delta C_{u+1}\cdots d\Delta C_t.$$

Third, for the TVTP model, the predictive density is

$$\hat{p}^{t}_{TVTP,u} = (2\pi)^{-\frac{(t-u)n}{2}} \left|\Sigma_\varepsilon^{-1}\right|^{\frac{t-u}{2}}$$

$$\int_{[0,1]}\cdots\int_{[0,1]} \exp\left(\sum_{s=u+1}^{t} -(1/2)\, \varepsilon_s'\Sigma_\varepsilon \varepsilon_s\right) p(S_{u+1},\ldots,S_t)\, dS_{u+1}\cdots dS_t.$$

Fourth, for the SW-TVTP model, the predictive density is

$$\hat{p}^{t}_{TVTP-SW,u} = (2\pi)^{-\frac{(t-u)n}{2}} \left|\Sigma_\xi^{-1}\right|^{\frac{t-u}{2}} \int_{-\infty}^{\infty}\cdots\int_{-\infty}^{\infty} \int_{[0,1]}^{\cdots}\int_{[0,1]} \exp\left(\sum_{s=u+1}^{t} -(1/2)\, \xi_s'\Sigma_\xi \xi_s\right)$$

$$\times p(S_{u+1},\ldots,S_t)\, p(\Delta C_{u+1},\ldots,\Delta C_t)\, dS_{u+1}\cdots dS_t\, d\Delta C_{u+1}\cdots d\Delta C_t.$$

For the last three models, the simulated values of the predicted likelihoods are formed by exploiting the time series dynamics of the unobserved variables. In the unobserved state model, for each draw of $\{\theta_j^{\{m\}}\}$, R sets of $\{\Delta C_{u+1},\ldots,\Delta C_t\}$ can be drawn from $\psi(L)\Delta C_t = \delta + \eta_t$ in equation (2). For the TVTP model, the vector $\{S_{u+1},\ldots,S_t\}$ can be simulated from $s_t = (1-q_t) + (-1+p_t+q_t)s_{t-1} + v_t$. The conditional distribution of v_t is

$$v_t = \begin{cases} \left.\begin{array}{l} 1-p_t \text{ with probability } p_t \\ -p_t \text{ with probability } 1-p_t \end{array}\right\} & \text{for } S_t = 1 \\ \left.\begin{array}{l} -(1-q_t) \text{ with probability } q_t \\ q_t \text{ with probability } 1-q_t \end{array}\right\} & \text{for } S_{t-1} = 0 \end{cases} .$$

6. Gibbs Samplers

This section develops several Gibbs samplers to estimate the posterior distributions of the parameters for each empirical business cycle model discussed in Section 3. Bayesian methods link prior distributions and conditional likelihoods together to yield the posterior distributions of interest.

6.1 VAR Model Sampler

The VAR sampler generates a posterior distribution of the VAR parameters in equation (1) and summarized in equation (9).

6.1.1 Prior Distributions

First, prior beliefs for $\Gamma_{VAR} = \{\beta_{VAR}, \Phi_{VAR}\}$ are assumed to be represented by a multivariate normal distribution:

$$P(\Gamma_{VAR}) = N(\hat{\Gamma}_{VAR}, \hat{A}_{\Gamma_{VAR}}).$$

Second, the prior distribution for Σ_e is of inverse-Wishart form:

$$P(\Sigma_e^{-1}) = W(\hat{S}_e, \hat{v}_e).$$

6.1.2 Conditional and Posterior Distributions

Step 1: $\{\Gamma_{VAR}\}$ Consider Γ_{VAR} and Σ_e. The VAR model in equation (1) can be rewritten in a vectorized form to simplify the estimation,

$$\begin{aligned} x_t &= \beta_{VAR} + \Phi_1 x_{t-1} + \cdots + \Phi_p x_{t-p} + e_t \\ \mathbf{x}_t &= \mathbf{X}_{t-1} \hat{\Gamma}_{VAR} + \mathbf{e} \end{aligned} \tag{15}$$

where $\tilde{\Gamma}_{VAR} = [\beta_1, \phi_{11,1}, \ldots, \phi_{kk,p}]'$ is a column vector of the coefficients of Γ_{VAR}.[4] Given the data and Σ_e, the posterior distribution of $\tilde{\Gamma}_{VAR}$ is multivariate normal:

$$P(\overline{\Gamma}_{VAR} \mid \Sigma_e, \mathbf{x}, \mathbf{X}) = MVN(\overline{\Gamma}, \overline{A}_\Gamma) \tag{16}$$

where

$$\begin{aligned}\overline{\Gamma} &= \overline{A}_\Gamma(\hat{A}_\Gamma^{-1}\hat{\Gamma} + \mathbf{X}'_{t\text{-}1}(\Sigma_e^{-1} \otimes I)\mathbf{x}_t \\ \overline{A}_\Gamma &= (\hat{A}_\Gamma^{-1} + \mathbf{X}'_{t\text{-}1}(\Sigma_e^{-1} \otimes I)\mathbf{X}_{t\text{-}1})^{-1}.\end{aligned}$$

Step 2: $\{\Sigma_e\}$ If Γ_{VAR} is assumed to be known, equation (15) delivers e_t which is the values of the $MVN(0, \Sigma_e)$ error vector at each t. Given these observations and an inverse-Wishart prior, the posterior distribution of Σ_e has an inverse-Wishart form:

$$P(\Sigma_e{}^{-1} \mid \Gamma, \mathbf{x}, \mathbf{X}) = W(\overline{S}_e, \overline{\nu}_e) \tag{17}$$

where

$$\begin{aligned}\overline{S}_e &= \left\{\hat{S}_e^{-1} + \sum e_t e_t'\right\}^{-1} . \\ \overline{\nu}_e &= T + \hat{\nu}_e\end{aligned}$$

6.2 SW Model Sampler

The SW sampler generates a posterior distribution of the parameters from the Stock and Watson unobserved components model in equation (2) and summarized in equation (10).

6.2.1 Prior Distributions

The Gibbs sampler for the SW model would be very similar to the VAR model except for the presence of the unobserved component ΔC_t. In order to simplify the sampler, however, Stock and Watson's original notation will be used. The prior beliefs over the parameter vector $\theta_{SW} = \{\beta_{SW}, \gamma_{SW}, D, \Sigma_\xi, \psi, \delta, \{\Delta C_t\}_{-q}^{T}\}$ can be summarized by the prior distribution over a subset of the parameters, $\{\beta_{SW}, \gamma_{SW}, D, \Sigma_\xi, \psi, \delta\}$ from the model

$$\begin{aligned}x_t &= \beta_{SW} + \gamma\Delta C_t + u_t \\ \psi(L)\Delta C_t &= \delta + \eta_t \\ D(L)u_t &= \xi_t\end{aligned} \quad . \tag{18}$$

The prior beliefs of $\Gamma_{SW}=[\beta_{SW},\gamma_{SW}]$, D, $\Psi=[\delta,\psi_1,\ldots,\psi_q]$ and Σ_ξ are represented with standard conjugate priors: $P(\Gamma_{SW})=N(\hat{\Gamma}_{SW},\hat{A}_{\Gamma_{SW}})$, $P(D)=N(\hat{D},\hat{A}_D)$, $P(\Sigma_\xi)=W(\hat{S}_\xi,\hat{v}_\xi)$, and $P(\Psi)=N(\hat{\Psi},\hat{A}_\Psi)$.

6.3.2 Conditional and Posterior Distributions

Step 1: $\{\Gamma_{SW}\}$. Given the parameters $\{\{\Delta C_t\}_{-q}^{T},\{u_t\}_{-r}^{T}\}$, the conditional distribution is normal for Γ_{SW}. Define $\ddot{x}_t=x+(D(L)-I)u_{t-1}$ and $\mathrm{X}_{t-1}=[1,\Delta C_t]$. The first line in equation (18) can be rewritten as

$$\ddot{\mathrm{x}}_t=\mathrm{X}_{t-1}\Gamma_{SW}+\xi_t.$$

The posterior distribution of Γ_{SW} is

$$P(\Gamma_{SW,A}\mid\{\Delta C_t\}_{-q}^{T},\{u_t\}_{-r}^{T},\mathrm{x},\mathrm{X})=MVN(\overline{\Gamma},\overline{A}_\Gamma)$$

where

$$\begin{aligned}\overline{\Gamma}&=\overline{A}_\Gamma(\hat{A}_\Gamma^{-1}\hat{\Gamma}+\mathrm{X}_{t-1}'(\Sigma_\xi^{-1}\otimes I)\ddot{\mathrm{x}}_t)\\ \overline{A}_\Gamma&=(\hat{A}_\Gamma^{-1}+\mathrm{X}_{t-1}'(\Sigma_\xi^{-1}\otimes I)\mathrm{X}_{t-1})^{-1}\end{aligned}.$$

Step 2: $\{D\}$. By transforming the second line in equation (18), the posterior distributions of D can be derived. Given $\{\Gamma_{SW},\{\Delta C_t\}_{-q}^{T}\}$, u_t can be calculated as $x_t-\beta_{SW}-\gamma\Delta C_t$. Let $U_{t-1}=[u_{t-1},\ldots,u_{t-r}]$, and set the $u_t, t\leq 0$ to their unconditional mean of 0. The posterior distribution of D becomes

$$P(D\mid\Gamma_{SW},\{\Delta C_t\}_{-q}^{T})=MVN(\overline{D},\overline{A}_D)$$

where

$$\begin{aligned}\overline{D}&=\overline{A}_D(\hat{A}_D^{-1}\hat{D}+\mathrm{U}_{t-1}'(\Sigma_\xi^{-1}\otimes I)\mathbf{u})\\ \overline{A}_D&=(\hat{A}_D^{-1}\hat{D}+\mathrm{U}_{t-1}'(\Sigma_\xi^{-1}\otimes I)\mathrm{U}_{t-1})^{-1}\end{aligned}.$$

Step 3: $\{\Sigma_\xi\}$. Given an inverse-Wishart prior, the posterior distribution of Σ_ξ also has an inverse-Wishart form:

$$P(\Sigma_\xi^{-1}\mid\Gamma_{SW},D,\{\Delta C_t\}_1^{T})=W(\overline{S}_\xi,\overline{v}_\xi)$$

where

$$\bar{S}_\xi = \left\{\hat{S}_\xi^{-1} + \sum \xi_t \xi_t'\right\}^{-1}$$
$$\bar{v}_\xi = T + \hat{v}_\xi$$

Step 4: $\{\Psi\}$. Given $\{\Delta C_t\}_{-q}^{T}$, the posterior distribution for Ψ is

$$P(\Psi \mid \{\Delta C_t\}_{-q}^{T}) = MVN(\bar{\Psi}, \bar{A}_\Psi)$$

where

$$\begin{aligned} \bar{\Psi} &= \bar{A}_\Psi(\hat{A}_\Psi^{-1}\hat{\Psi} + \Delta C_{-1}'\Delta C_{-1}) \\ \bar{A}_\Psi &= (\hat{A}_\Psi^{-1} + \Delta C_{-1}'\Delta C_{-1})^{-1} \\ \Delta C_{-1} &= [1, \Delta C_{t-1}, \ldots, \Delta C_{t-q}] \end{aligned}$$

Step 5: $\{\{\Delta C_t\}_{-q}^{T}\}$. Given $\{\Psi\}$, the unobserved component can be simulated recursively. The initial conditions, ΔC_t for $t \le 0$ are set to their unconditional mean, $\psi^{-1}(1)\delta$. Then each ΔC_t is simulated from draws of $P(\Delta C_t \mid \{\Delta C_{t-j}\}_{t=1}^{t+q}, \Psi) = N(C_t, 1)$ where

$$\bar{C}_t = -\left(\frac{\psi(L) - 1}{L}\right)\Delta C_{t-1} + \delta .$$

6.3 TVTP Model Sampler

The TVTP sampler generates the posterior distribution of the parameters vector of interest: the output equation parameters in equation (3), the multinomial probit coefficients in equation (6), and the transition probabilities parameters in equation (7):

$$\theta_{TVTP} = \left\{\mu, \Phi_{TVTP}, \Sigma_e, \Gamma_{TVTP}, \{s_t\}_0^T, \{s_t^*\}_1^T, \{p_t\}_1^T, \{q_t\}_1^T\right\}. \tag{19}$$

6.3.1 Prior Distributions

There is no need to specify prior distributions over all the parameters of interest in the model because many of the parameters of interest are functions of a much smaller subset. In particular, prior information about the model parameters in the vector θ_{TVTP} can be summarized by the prior distribution over $(\mu, \Phi, \Sigma_\varepsilon, \Gamma, \text{and } s_0)$. First, prior beliefs for $\tilde{\Phi}$ are assumed to be represented by a multivariate normal distribution:

$$P(\widetilde{\Phi}) = N(\hat{\Phi}, \hat{A}_{\Phi}).$$

Second, the prior distribution for μ is assumed to be a multivariate normal distribution truncated so that the state $S_t = 1$ is identified as representing the high-growth rate state:

$$P(\mu) = TN(\hat{\mu}, \hat{A}_{\mu}).$$

Third, the prior distribution for Σ_e is of the inverse-Wishart form:

$$P(\Sigma_{\varepsilon}^{-1}) = W(\hat{S}_{\varepsilon}, \hat{v}_{\varepsilon}).$$

Fourth, the transition probability equation from the probit model depends on the parameters, Γ. For computational convenience, the prior distribution over Γ is a multivariate normal:

$$P(\Gamma) = MVN(\hat{\Gamma}, \hat{A}_{\Gamma}).$$

Finally, the initial state, s_0, is assumed to have a binomial distribution, $P(s_0 = S_i) = \hat{\pi}_i$, $i = (1,2)$. The hyperparameters $\{\hat{\Phi}, \hat{A}_{\Phi}, \hat{\mu}, \hat{A}_{\mu}, \hat{S}_{\varepsilon}, v_{\varepsilon}, \hat{\Gamma}, \hat{A}_{\Gamma}, \hat{\pi}_1\}$ for all these prior distributions are assumed known. In this application, we use values that are associated with very diffuse prior information, so that the prior plays almost no role in determining the form of the posterior.

6.3.2 Conditional and Posterior Distributions

Step 1: $\{\Phi, \Sigma_{\varepsilon}, \mu\}$. The conditional distribution for $\{\Phi, \Sigma_{\varepsilon}, \mu\}$ is relatively complicated when conditioning on $\{\{s_t\}_0^T\}_i$ because of the nonlinear relationship between Φ and μ. The formulation of the conditional distributions is simplified by proceeding in two sub-steps. First, consider Φ and Σ_{ε}, given values of each series' state and state dependent mean, $\{s_t\}_0^T$ and $\mu_i(s_t)$. Define $X_t^* = X_t - M^{s_t}$ and $\widetilde{\mathbf{X}}_{t-1} = [X_{t-1}^*, \ldots, X_{t-r}^*]$. Equation (3) can then be rewritten as:

$$\begin{aligned} X_t^* &= \Phi(L) X_{t-1}^* + \varepsilon_t \\ &= \widetilde{\mathbf{X}}_{t-1} \widetilde{\Phi} + \varepsilon_t \end{aligned} \tag{20}$$

where $\widetilde{\Phi} = [\phi_{11,1}, \ldots, \phi_{kk,r}]'$ is a column vector of the coefficients in Φ. Given Σ_{ε} and the data, the posterior distribution of Φ is multivariate normal:

$$P(\Phi \mid \Sigma_{\varepsilon}, \mu, \mathbf{X}, \{s_t\}_0^T) = MVN(\overline{\Phi}, \overline{A}_{\Phi}) \tag{21}$$

where

$$\begin{aligned}\overline{\Phi} &= \overline{A}_{\Phi}(\hat{A}_{\Phi}^{-1}\hat{\Phi} + \widetilde{\mathbf{X}}'_{t-1}(\Sigma_{\varepsilon}^{-1}\otimes I)X_t^*) \\ \overline{A}_{\Phi} &= (\hat{A}^{-1} + \widetilde{\mathbf{X}}'_{t\text{-}1}(\Sigma_{\varepsilon}^{-1}\otimes I)\widetilde{\mathbf{X}}_{t\text{-}1})^{-1}\end{aligned}.$$

If Φ is known, equation (20) delivers the realized values of the $MVN(0,\Sigma_{\varepsilon})$ error vector, ε_t. Given these observations and an inverse-Wishart prior, the posterior distribution of Σ_{ε} also has an inverse-Wishart form:

$$P(\Sigma_{\varepsilon}^{-1} \mid \Phi, \mu, \mathbf{X}, \{s_t\}_0^T) = W(\overline{S}_{\varepsilon}, \hat{v}_{\varepsilon}) \tag{22}$$

where

$$\begin{aligned}\overline{S}_{\varepsilon} &= \left\{\hat{S}_{\varepsilon}^{-1} + \sum_{t=1}^{T}\varepsilon_t\varepsilon_t'\right\}^{-1} \\ \overline{v}_{\varepsilon} &= T + \hat{v}_{\varepsilon}\end{aligned}.$$

Given Σ_{ε} and Φ, the output model can be rewritten in terms of μ (for details see Appendix A in Filardo and Gordon, 1994).

$$\ddot{X}_t = \ddot{X}_{t-1}\mu + \varepsilon_t\,. \tag{23}$$

Since equation (23) is also of the linear multivariate normal form, the posterior distribution for μ is also normally distributed:

$$P(\mu \mid \Phi, \Sigma_{\varepsilon,} X, \{s_t\}_0^T) = I_{\{M_L<M_H\}}MVN(\overline{\mu}, \overline{A}_{\mu}) \tag{24}$$

where

$$\begin{aligned}\overline{\mu} &= \overline{A}_{\Phi}(\hat{A}_{\Phi}^{-1}\hat{\Phi} + \widetilde{X}'_{t-1}(\Sigma_{\varepsilon}^{-1}\otimes I)X_t^*) \\ \overline{A}_{\mu} &= (\hat{A}^{-1} + \widetilde{X}'_{t\text{-}1}(\Sigma_{\varepsilon}^{-1}\otimes I)\widetilde{X}_{t\text{-}1})^{-1}\end{aligned}.$$

Identification of the low- and high-growth rate states restricts the form of the posterior distribution. $P(\mu \mid \Phi, \Sigma_{\varepsilon}, X, \{s_t\}_0^T)$ is a truncated multivariate normal distribution where the truncation guarantees that the low-growth state has a lower mean growth rate than does the high-growth state.

Step 2: $\{s_t\}_{t=0}^T$. Values for the unobserved states can be drawn from the full conditional distribution of s_t given $\{X_t, S_{-t}, \mu, \Phi, \Sigma_{\varepsilon}, \Gamma\}$. The procedure for simulating s_t is a straight forward extension of the procedure outlined in Albert and Chib (1993a) and McCulloch and Tsay (1994). The full conditional is adapted from Albert and Chib (1993a):

$$P(s_t \mid \mathbf{X}, S_{-t}) \propto \begin{cases} P(s_t \mid s_{t-1})P(s_{t+1} \mid s_t)P(X_t,\ldots,X_\tau \mid \mathrm{X}_{t-1},S_\tau)\prod_{k=r+1}^{t+r} f(X_k \mid \mathrm{X}_{k-1},S_k),\ t \leq r \\ P(s_t \mid s_{t-1})P(s_{t+1} \mid s_t)\prod_{k=t}^{t+r} P(y_k \mid X_{k-1},S_k),\ r+1 \leq t \leq T-r+1 \\ P(s_t \mid s_{t-1})P(s_{t+1} \mid s_t)\prod_{k=t}^{T} P(y_k \mid X_{k-1},S_k),\ T-r \leq t \leq T \end{cases} \tag{25}$$

Working backwards from $t = T$, samples of s_t can be simulated by a series of discrete distributions using the probabilities in equation (25).

Given a draw of S_1, the transition probability matrix $P(S_1 = s_1 \mid S_0 = s_0, \mathrm{z}_1)$, and prior probability $\hat{\pi}(S_0 = s_0, \mathrm{z}_1)$, the discrete posterior distribution of S_0 is

$$P(S_0 = s_0 \mid S_{-0}) = \frac{P(S_1 = s_1 \mid S_0 = s_0)\hat{\pi}(S_0 = s_0)}{\sum_{s_0 \in \{0,1\}} P(S_1 = s_1 \mid S_0 = s_0)\hat{\pi}(S_0 = s_0)}. \tag{26}$$

Step 3: $\left\{\{s_t^*\}_{t=0}^T, \Gamma, \{p_t, q_t\}_{t=1}^T\right\}$. Given the values of s_t, the leading indicator data, and Γ, the latent variable s^* can be simulated using a truncated multivariate normal distribution based on equation (6). The truncation ensures that $s_t = 1(=0)$, then $s_t^* \geq 0(<0)$.

Given values for s^*, equation (6) becomes a linear normal model, and the techniques used to draw values for the parameters of equation (3) can be applied here as well. Given s^* and the covariates W in equation (6), the posterior distribution for the transition probability coefficients, Γ, is a multivariate normal:

$$P(\Gamma \mid \Sigma, s, W) = MVN(\Gamma, A_\Gamma) \tag{27}$$

where

$$\begin{aligned} \overline{\Gamma} &= \overline{A}_\Gamma(\hat{A}_\Gamma^{-1}\hat{\Gamma} + W'S^*) \\ \overline{A}_\Gamma &= (\hat{A}_\Gamma^{-1} + W'W)^{-1} \end{aligned}.$$

6.4 SW-TVTP Model Sampler

This sampler simply combines the steps of the SW sampler and the TVTP sampler. The combined sampler yields the posterior distribution of the parameters in Equation (12).

6.5 Running the Gibbs Sampler

The sampler is initiated with a set of arbitrary starting values, and the parameter values generated by each pass of the steps are collected. It is well known (see, for example, Gelfand and Smith 1990) that this sequence of random draws of the identified parameters converges in distribution to the joint posterior distribution of the parameters. In practice, the first M of the N draws are discarded in order to minimize the effect of not being able to draw the starting value directly from the joint distribution. As interest is in the marginal distribution of the parameters (and functions of the parameters), the Gibbs sampler can deliver the marginal distribution of the means and functions of $\theta_i \in \Theta$ with density estimators:

$$E(\theta_i \mid \cdot) = \frac{1}{N-M}\sum_{j=1}^{N-M}\theta_i \quad \text{and} \quad E(g(\theta_i) \mid \cdot) = \frac{1}{N-M}\sum_{j=1}^{N-M} g(\theta_i) \qquad (28)$$

where $g(\cdot)$ is a measurable function.

7. Results

The four models in the previous section were estimated using three and four variables. The employment series was ultimately dropped from the results because of the fragility of estimation.[5] Despite the exclusion of the employment series, the common component that was estimated looked similar to Stock and Watson's reported estimation results. The correlation of this unobserved component with the Department of Commerce's composite coincident index was 0.9. The models were run from January 1952 to January 1995.

The Gibbs sampling procedures outlined in the previous section were implemented successfully. The parameter estimates of the models tended to settle down to a stable posterior distribution. The numerical standard errors of the simulation provided evidence that the parameter estimates (not reported) were all less than 0.01. One exception to the relatively quick convergence was the nonlinear Stock-Watson model. Relatively tight priors have to be imposed to estimate the model, and the transition probabilities are assumed to equal their unconditional means (fixed transition probabilities). The state variable in the Markov switching regressions were highly correlated with the NBER business cycle dates as has been found by Filardo and Gordon (1998).

The results from the prediction exercise are reported in Figures 1-10. These figures display the log marginalized predictive likelihood (top panel), the cumulative log Bayes factor (middle panel), and the observations (bottom panel). The (log) marginalized predictive likelihood for each model measures the (log) model probabilities for the observations. (Note that the dashed line in the log marginalized predictive likelihood panel (top panel) corresponds to the first variable in the title of the figure.)

The cumulative log Bayes factor measures the relative prediction accuracy of the various business cycle models for the given set of observations.[6] The cumulative log Bayes factor indicates the probability that observations are more probable under the one model relative to another. In a sense, the panel provides a convenient way to compare the empirical business cycle models described above in a simple horse race. The horse race is run pairwise, pitting two alternative models against each other in each figure. Interpretation of the figures is fairly straight forward. A change in the cumulative log Bayes factor signals a revision of model probabilities in a certain direction. If the cumulative log Bayes factor rises, the observations favor the model represented in the numerator of the Bayes factor; if it declines, the observations favor the model represented in the denominator. (For simplicity in reporting the results, all figures except Figures 4 and 9 use the results from the TVTP model in the numerator of the cumulative log Bayes factor. In these figures a rise in the cumulative log Bayes factor indicates favorable support for the TVTP model.)

Figures 1-5 and Figures 6-10 display the same model comparison over two different time periods. The first set of figures serves as a control set of results, that is, when the economy was not at a turning point. The starting date is January 1990 and the ending point is June 1990. The latter set of figures covers a NBER recognized turning point from expansion to recession. The starting point in these figures is July 1990 and the ending point is December 1990.

Figures 1 and 6 compare the TVTP and VAR models; Figures 2 and 7 compare the TVTP and SW models; Figures 3 and 8 compare the TVTP and SW-TVTP models; Figures 4 and 8 compare the SW and SW-TVTP models; and Figures 5 and 10 compare the TVTP and constant mean models. The constant mean model represents a naïve forecasting model where the forecast is simply the unconditional mean of each series.

Comparisons of the figures offer several insights about the importance of nonlinearity in business cycle modeling and the ability of the empirical business cycle models to capture turning points. First, in Figures 1-5, all the models appear to perform equally well in the January-June period (non-turning point period). The cumulative log Bayes factors are uniformly low and hover close to zero. This suggests that during the first half of 1990, the nonlinear and linear distinctions are not important.

Second, in the latter half of 1990, however, certain models outperform others. The TVTP model performs comparatively well. This might not be unexpected since this model, if correctly specified, is constructed to capture abrupt changes in the economy. Somewhat surprising was the superior performance of the linear VAR model. The VAR appears to finish second in the comparisons, while the linear SW and nonlinear SW-TVTP models finish third and fourth.

Third, somewhat surprisingly, the linear and nonlinear SW models perform equally well, or poorly, depending on your priors. The ability of the nonlinear SW to capture abrupt changes in economic activity suggests that it would perform better

Figure 1. Comparison of Out-of-Sample Predictive Ability of the TVTP and VAR Models, January 1990-June 1990.

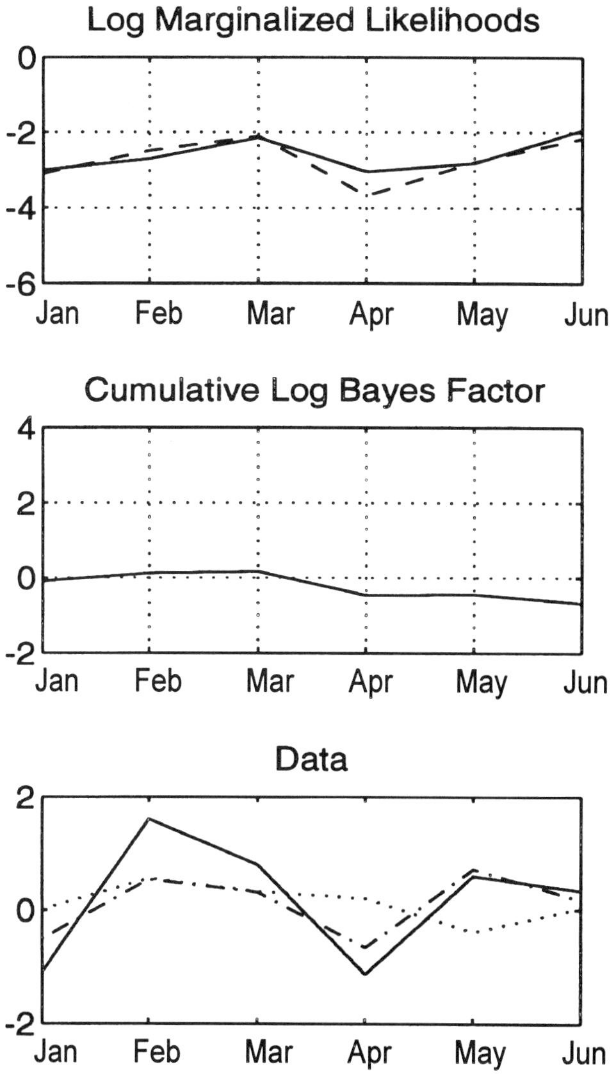

Notes: Top panel shows the log marginalized likelihoods for the TVTP model (dashed line) and VAR model (solid line). The middle panel shows the cumulative log Bayes factor; an increase indicates evidence in favor of the TVTP model. The lower panel shows the actual monthly observations for industrial production (solid line), income (dotted line), and sales (dashed line).

Figure 2. Comparison of Out-of-Sample Predictive Ability of the TVTP and SW Models, January 1990-June 1990.

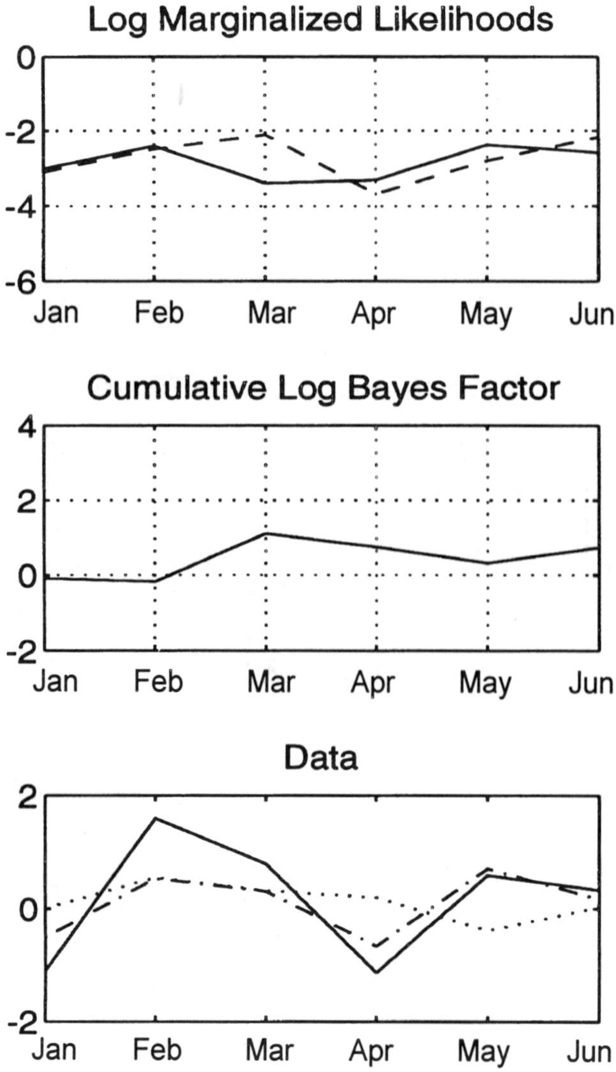

Notes: Top panel shows the log marginalized likelihoods for the TVTP model (dashed line) and SW model (solid line). The middle panel shows the cumulative log Bayes factor; an increase indicates evidence in favor of the TVTP model. The lower panel shows the actual monthly observations for industrial production (solid line), income (dotted line), and sales (dashed line).

Figure 3. Comparison of Out-of-Sample Predictive Ability of the TVTP and SW-TVTP Models, January 1990-June 1990.

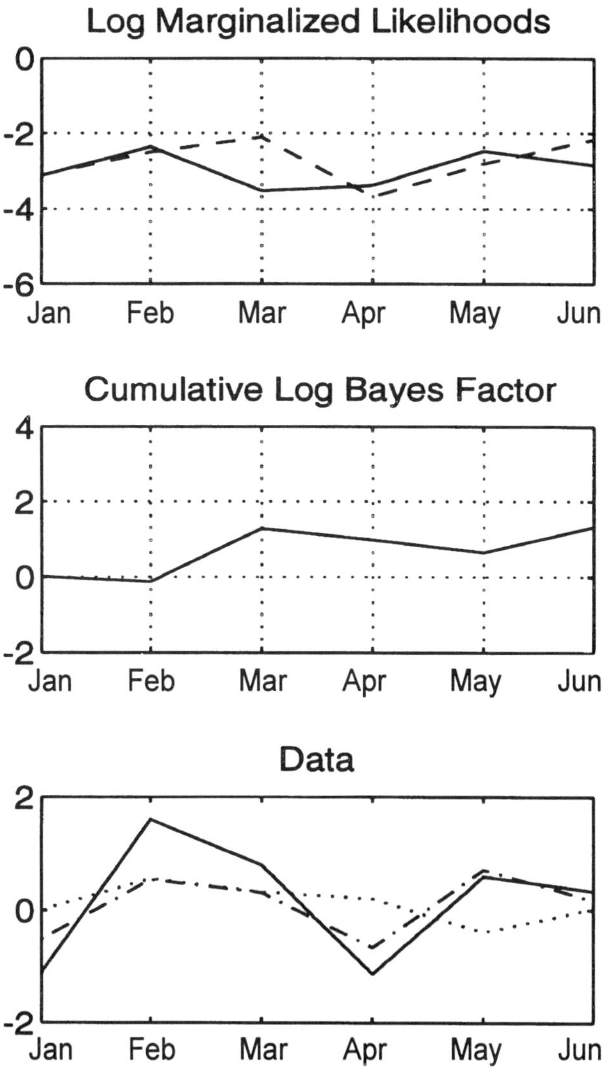

Notes: Top panel shows the log marginalized likelihoods for the TVTP model (dashed line) and SW-TVTP model (solid line). The middle panel shows the cumulative log Bayes factor; an increase indicates evidence in favor of the TVTP model. The lower panel shows the actual monthly observations for industrial production (solid line), income (dotted line), and sales (dashed line).

Figure 4. Comparison of Out-of-Sample Predictive Ability of the SW and SW-TVTP Models, January 1990-June 1990.

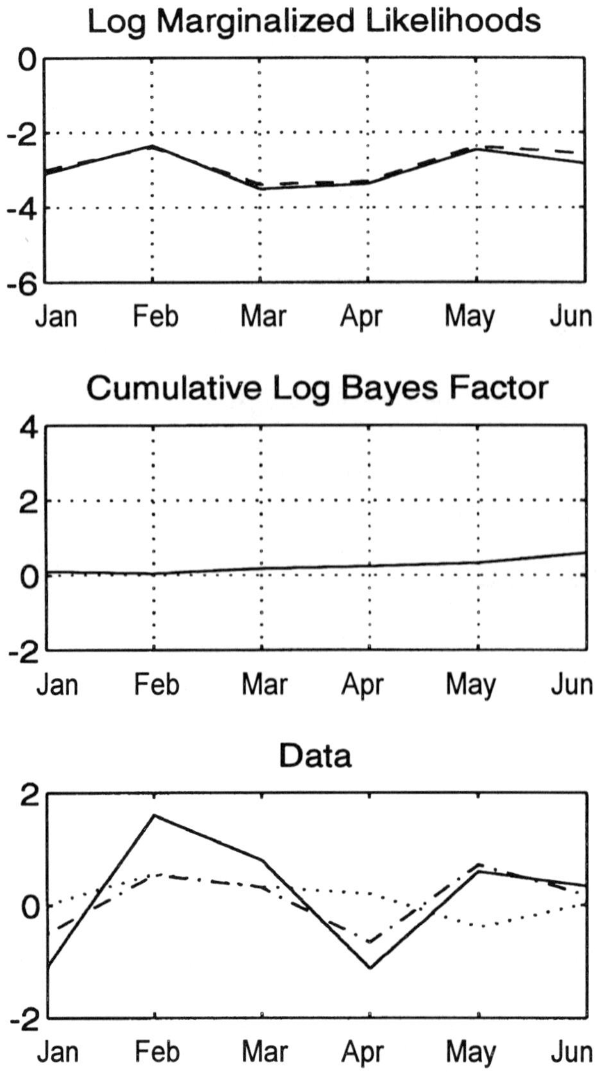

Notes: Top panel shows the log marginalized likelihoods for the SW model (dashed line) and SW-TVTP model (solid line). The middle panel shows the cumulative log Bayes factor; an increase indicates evidence in favor of the SW model. The lower panel shows the actual monthly observations for industrial production (solid line), income (dotted line), and sales (dashed line).

Figure 5. Comparison of Out-of-Sample Predictive Ability of the TVTP and Constant Mean Models, January 1990-June 1990.

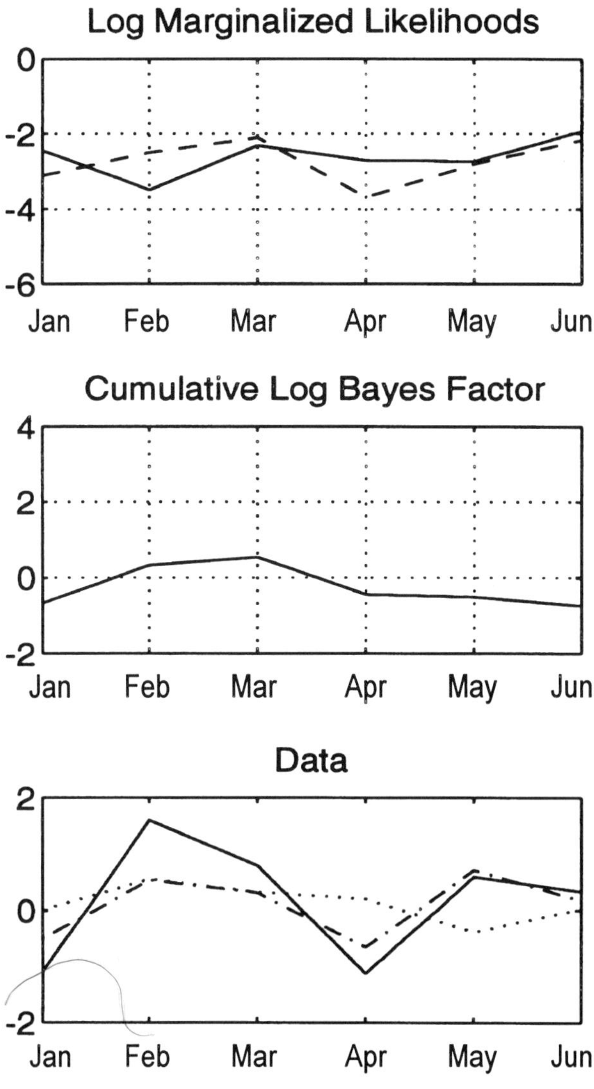

Notes: Top panel shows the log marginalized likelihoods for the TVTP model (dashed line) and Constant Mean model (solid line). The middle panel shows the cumulative log Bayes factor; an increase indicates evidence in favor of the TVTP model. The lower panel shows the actual monthly observations for industrial production (solid line), income (dotted line), and sales (dashed line).

Figure 6. Comparison of Out-of-Sample Predictive Ability of the TVTP and VAR Models, July 1990-December 1990.

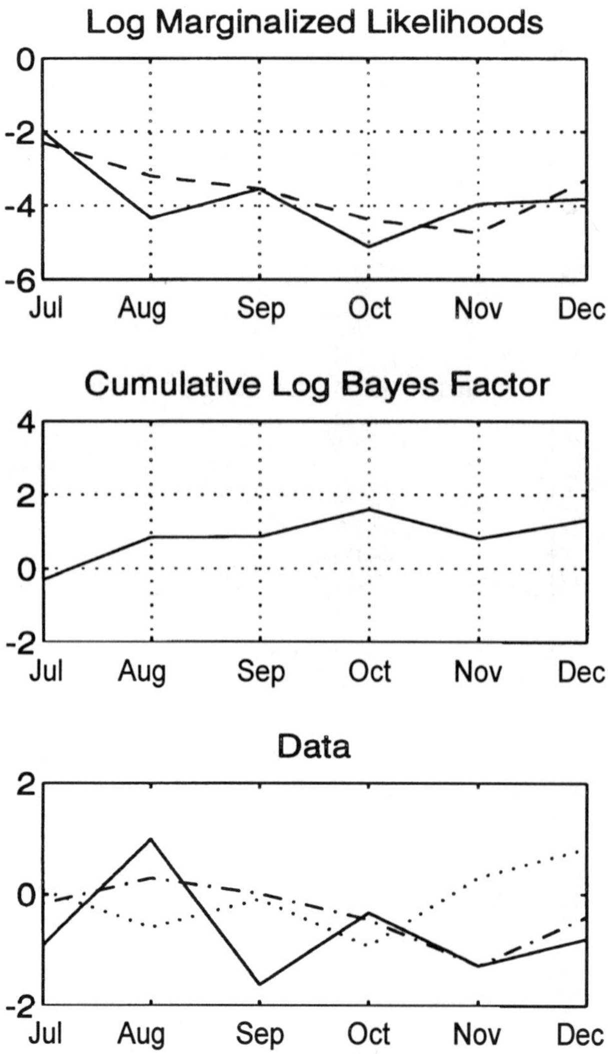

Notes: Top panel shows the log marginalized likelihoods for the TVTP model (dashed line) and VAR model (solid line). The middle panel shows the cumulative log Bayes factor; an increase indicates evidence in favor of the TVTP model. The lower panel shows the actual monthly observations for industrial production (solid line), income (dotted line), and sales (dashed line).

Figure 7. Comparison of Out-of-Sample Predictive Ability of the TVTP and SW Models, July 1990-December 1990.

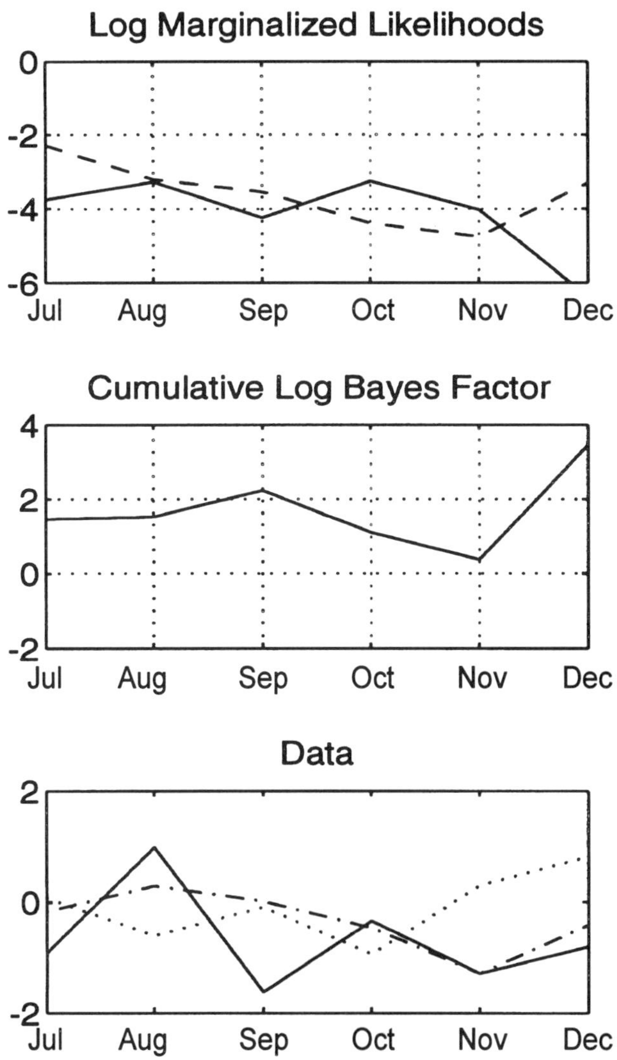

Notes: Top panel shows the log marginalized likelihoods for the TVTP model (dashed line) and SW model (solid line). The middle panel shows the cumulative log Bayes factor; an increase indicates evidence in favor of the TVTP model. The lower panel shows the actual monthly observations for industrial production (solid line), income (dotted line), and sales (dashed line).

Figure 8. Comparison of Out-of-Sample Predictive Ability of the TVTP and SW-TVTP Models, July 1990-December 1990.

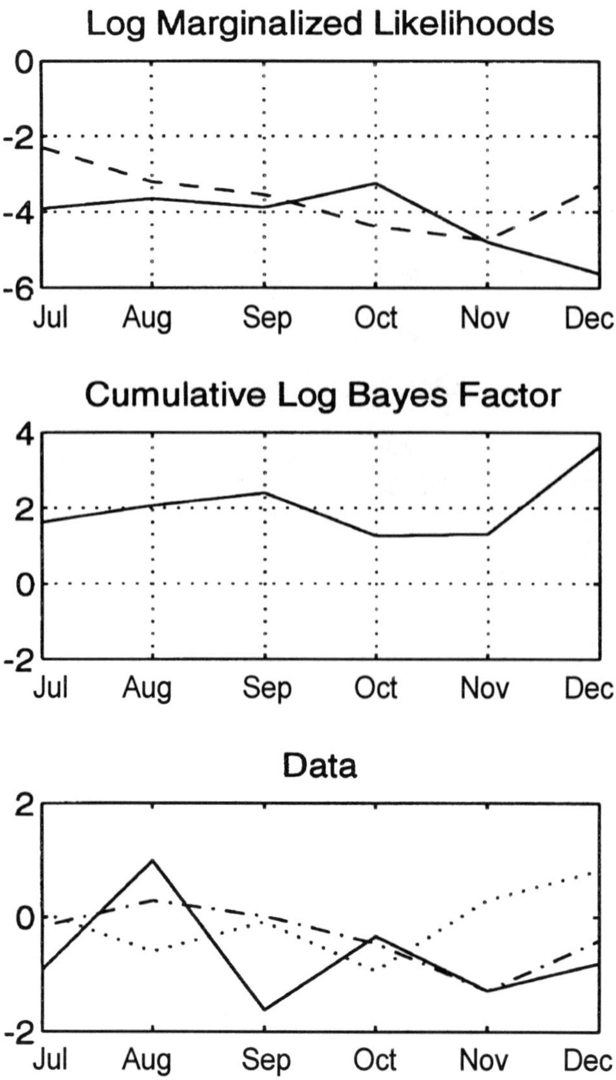

Notes: Top panel shows the log marginalized likelihoods for the TVTP model (dashed line) and SW-TVTP model (solid line). The middle panel shows the cumulative log Bayes factor; an increase indicates evidence in favor of the TVTP model. The lower panel shows the actual monthly observations for industrial production (solid line), income (dotted line), and sales (dashed line).

Figure 9. Comparison of Out-of-Sample Predictive Ability of the SW and SW-TVTP Models, July 1990-December 1990.

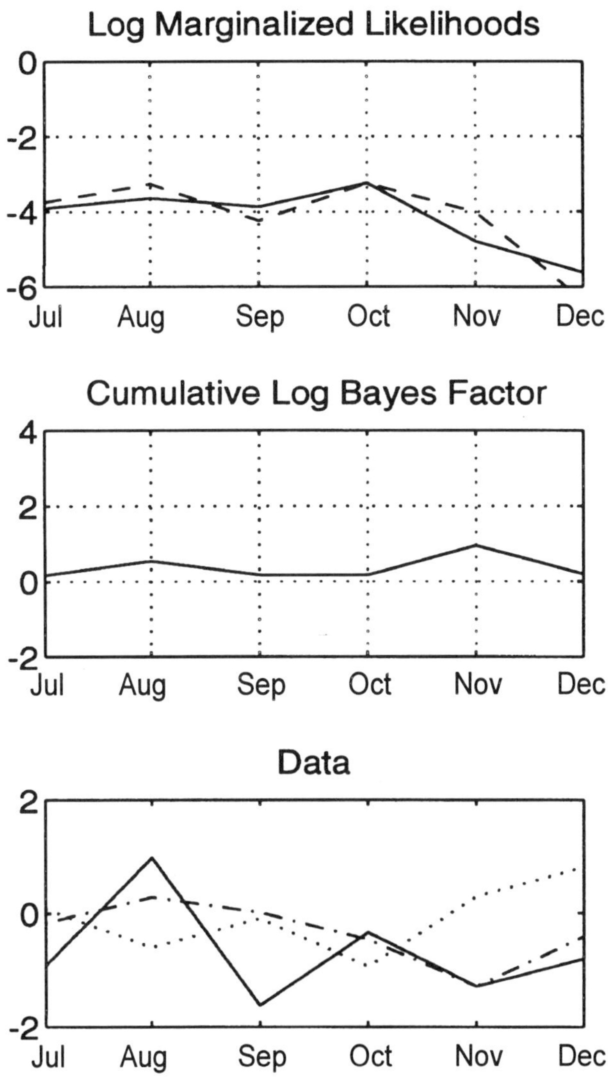

Notes: Top panel shows the log marginalized likelihoods for the SW model (dashed line) and SW-TVTP model (solid line). The middle panel shows the cumulative log Bayes factor; an increase indicates evidence in favor of the SW model. The lower panel shows the actual monthly observations for industrial production (solid line), income (dotted line), and sales (dashed line).

Figure 10. Comparison of Out-of-Sample Predictive Ability of the TVTP and Constant Mean Models, July 1990-December 1990.

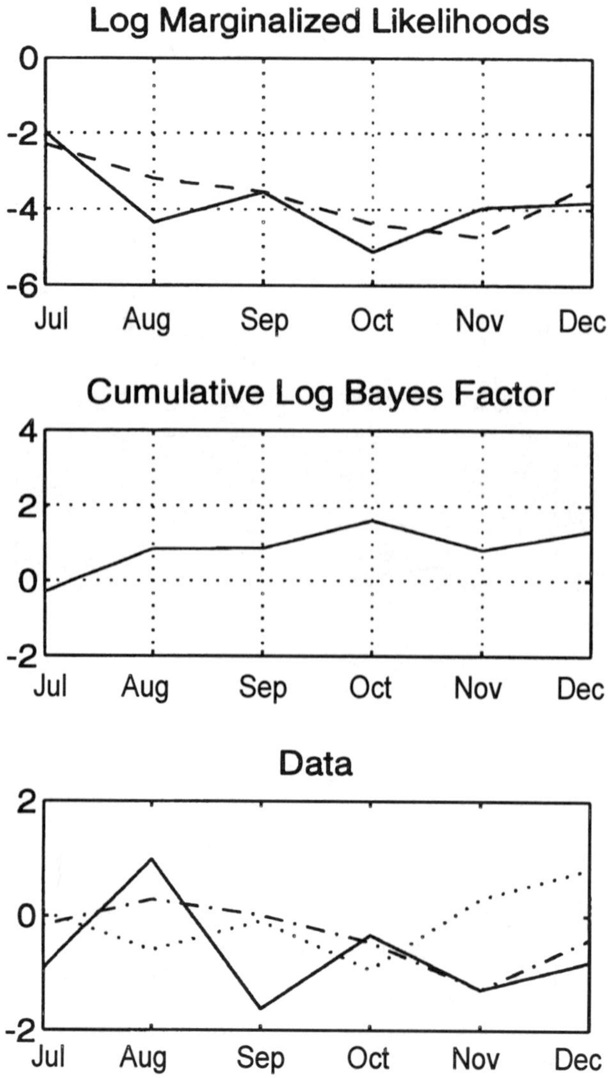

Notes: Top panel shows the log marginalized likelihoods for the TVTP model (dashed line) and Constant Mean model (solid line). The middle panel shows the cumulative log Bayes factor; an increase indicates evidence in favor of the TVTP model. The lower panel shows the actual monthly observations for industrial production (solid line), income (dotted line), and sales (dashed line).

during a turning point period. One possible explanation for its lackluster performance is that the statistical smoothing implicit in the unobserved component complicates the fitting of the Markovian switching. Moreover, both the SW and SW-TVTP models are beaten by the VAR. One might think that the VAR is simply a nested version of the linear and nonlinear SW models. However, identification restrictions on the unobserved component cause them to differ and may account for the results.

On the whole, the results suggest that the TVTP business cycle model performed relatively well in the second half of 1990—certainly better than in the first half of 1990. For those who are partial to nonlinear econometric modeling and to intrinsic business cycle modeling, the results are encouraging. Preliminary studies of earlier turning points show that the rankings of the models can change. More investigations are clearly called for.

Econometrically, the log marginalized predictive likelihoods and the cumulative log Bayes factors potentially provide many insights empirical business cycles. One nice feature is the tractability of non-nested model comparisons. Another nice feature is the simplicity of calculation. Even with computing speeds increasing and costs falling, direct numerical evaluation of the predictive likelihoods is quite burdensome. The methodology laid out by Geweke substantially reduces the costs of model comparison by using Markov chain Monte Carlo methods. One potential drawback of these methods is a bias that we found in comparing unobserved component models and Markov switching models. Models without unobserved variables may appear to outperform models with unobserved variables in a predictive Bayes factor sense even when the point forecasts are more accurate than those from models with unobserved variables. The bias arises from the averaging involved in calculating the simulated predictive likelihoods. This notion is essentially a Jensen's inequality issue. More investigation into the size of the bias is clearly called for.

8. Conclusions

This paper lays out an estimation and model comparison strategy to evaluate several standard empirical business cycle models. Estimation is successfully performed using Gibbs sampling methods and the estimated models are compared using predictive likelihood and Bayes factors. The empirical tractability of these methods to compare non-nested econometric models suggests gains from further research of business cycle behavior as well as many other interesting economic phenomena.

This paper also finds that nonlinear econometric modeling is potentially important for measuring and predicting business cycles. Intrinsic business cycle models are well suited to capture abrupt changes in economic activity, behavior that has traditionally been associated with business cycle turning points. However, the results must be tempered by the fact that linear models are quite valuable in explaining economic activity not associated with turning points. Ultimately, the methods used in this paper may offer insights into a synthesis of extrinsic and intrinsic business cycle models, and may provide a better understanding of business cycles.

Acknowledgements

The authors would like to thank Frank Diebold and Ken West for comments on an earlier draft of this paper, as well as seminar participants at the Canadian Economic Association Annual Meeting, Symposium on Nonlinear Dynamics and Econometrics, Federal Reserve System Macroeconomics Meeting, Federal Reserve Bank of Kansas City Workshop, and the Summer North American Econometric Society Meeting. Stephen F. Gordon also gratefully acknowledges financial support from Fonds FCAR (Quebec). The views expressed herein are those of the authors and do not necessarily reflect the views of the Federal Reserve Bank of Kansas City or of the Federal Reserve System

Notes

1. See Sims (1980).
2. See Stock and Watson (1989, 1991, and 1993) for more details. In their notation, $\beta_{SW} = D(1)\beta$ and $\Phi_{SW}(L) = D(L)-I$.
3. See Sections 3.4 and 4 of Geweke (1994) for details.
4. For the vectorized version, the matrix becomes $\mathbf{X} = I_k \otimes [1, x_{11}, \ldots, x_{ij}, \ldots, x_{kp}]$ where i indexes the set of variables and j indexes the number of lags in the VAR. The vector 1 is a column of ones and x_{ij} is the jth lag of the ith variable.
5. Stock and Watson also face this problem but restrict their model sufficiently to obtain stable estimation.
6. The predictive posterior odds ratio is a product of the prior odds ratio and the predictiveBayes factor. Equation (14) highlights the connection of the Bayes factor to the predictive (marginal) likelihoods for each model. Essentially, the Bayes factor is the ratio of the predictive likelihoods, thus indicating the Bayes factor's interpretation as a measure of the information in one model relative to another model.

References

Albert, J. and S. Chib (1993a). "Bayes Inference via Gibbs Sampling of Autoregressive Time Series Subject to Markov Mean and Variance Shifts, *Journal of Business and Economic Statistics*, 11, 1-15.

Albert, J. and S. Chib (1993b). "Bayesian Analysis of Binary and Polychotomous Response Data, *Journal of the American Statistical Association* 88, 669-679.

Blatt, J. (1978). "On the Econometric Approach to Business Cycle Analysis, *Oxford EconomicPapers,* 20, 292-300.

Blatt, J. (1980). "On the Frisch Model of Business Cycles, *Oxford Economic Papers,* 30, 468-479.

Boldin, M. (1992). "Using Switching Models to Study Business Cycle Asymmetries: Overview of Methodology and Application, Federal Reserve Bank of New York Research Paper no. 9211.

Bratley, P., B. Fox and L. Schrage (1987). *A Guide to Simulation.* New York: Springer-Verlag.

Brunner, A. (1992). "Conditional Symmetries in Real GNP: A Seminonparametric Approach, *Journal of Economics and Business Statistics*, 10, 65-72.

Casella, G. and E. George (1992). "Explaining the Gibbs Sampler," *The American Statistician,* 46, 167-174.

Chauvet, M. (1994). "An Econometric Characterization of Business Cycle Dynamics with Factor Structure and Regime Switching," unpublished manuscript, Department of Economics, University of Pennsylvania.

DeLong, B. and L. Summers (1986). "Are Business Cycles Symmetrical?," in Gordon, R. J. (ed.) *The American Business Cycle: Continuity and Change.* Studies in Business Cycles, no. 25. Chicago: The University of Chicago Press.

Diebold, F. and G. Rudebusch (1991a). "Turning Point Prediction with the Composite Leading Index: An Ex Ante Analysis," in Lahiri, K. and G. Moore, (eds.) *Leading Economic Indicators: New Approaches and Forecasting Records.* Cambridge, England and New York: Cambridge University Press.

Diebold, F. and G. Rudebusch (1991b). "Forecasting Output With the Composite Leading Index: A Real-time Analysis," *Journal of the American Statistical Association*, 86, 603-610.

Diebold, F. and G. Rudebusch (1996). "Measuring Business Cycles: A Modern Perspective," *Review of Economics and Statistics*, 78, 67-77.

Falk, B. (1986). "Further Evidence on the Asymmetric Behavior of Economic Time Series Over the Business Cycle," *Journal of Political Economy,* 94, 1096-1109.

Filardo, A. (1994). "Business Cycle Phases and Their Transitional Dynamics," *Journal of Business and Economic Statistics*, 12, 299-308.

Filardo, A. and S. Gordon (1998). "Business Cycle Durations," Journal of Econometrics, 85, 99-123.

Filardo, A. and S. Gordon (1994). "Co-movement of International Business Cycles," Federal Reserve Bank of Kansas City Working Paper.

Frisch, R. (1933). "Propagation and Impulse Problems in Dynamic Economies," in *Economic Essays in Honour of Gustav Cassel.* London: Allen and Unwin.

Gelfand, A. and A. Smith (1990). "Sampling-based Approaches to Calculating Marginal Densities," *Journal of the American Statistical Association*, 85, 398-409.

Geweke, J. (1994). "Bayesian Comparison of Econometric Models," Federal Reserve Bank of Minneapolis Working Paper.

Geweke, J., M. Keane and D. Runkle (1994). "Statistical Inference in the Multinomial Probit Model," *Journal of Econometrics*, 80, 125-165.

Granger, C., T. Teräsvirta and H. Anderson (1993). "Modeling NonLinearity Over the Business Cycle," in Stock, J. and M. Watson (eds.) *Business Cycles, Indicators, and Forecasting.* Chicago: The University of Chicago Press.

Hamilton, J. (1989). "A New Approach to the Economic Analysis of Nonstationary Time Series and the Business Cycle," *Econometrica*, 57, 357-384.

Hussey, R. (1992). "Nonparametric Evidence on Asymmetry in Business Cycles Using Aggregate Employment Time Series," *Journal of Econometrics*, 51, 217-231.

Hymans, S. (1973). "On the Use of Leading Indicators to Predict Cyclical Turning Points," *Brookings Papers On Economic Activity*, 2, 339-384.

Kling, J. (1987). "Predicting the Turning Points of Business and Economic Time Series," *Journal of Business*, 60, 201-238.

Lucas, R. (1977). "Understanding Business Cycles," in Brunner, K. and A. H. Meltzer (eds.) *Stabilization of the Domestic and International Economy*. Carnegie-Rochester Conference Series on Public Policy, vol. 5. Amsterdam: North Holland.

McCulloch, R. and R. Tsay (1994). "Statistical Analysis of Economic Time Series Via Markov Switching Models," *Journal of Time Series Analysis*, 15, 523-539.

McCulloch, R. and P. Rossi (1991). "An Exact Likelihood Analysis of the Multinomial Probit Model," *Journal of Econometrics*, 64, 207-240.

Mitchell, W. (1954). "Business Cycles: The Problem and Its Setting," *Studies in Business Cycles*, 1. New York: National Bureau of Economic Research.

Moore, G. and V. Zarnowitz (1982). "Sequential Signals of Recession and Recovery," *Journal of Business*, 55, 57-85.

Neftçi, S. (1980a). "Optimal Prediction of Cyclical Downturns," *Journal of Economic Dynamics and Control*, 4, 225-241.

Neftçi, S. (1980b). "Are Economic Time Series Asymmetric Over the Business Cycle?," *Journal of Political Economy*, 92, 307-328.

Sims, C. (1980). "Macroeconomics and Reality," *Econometrica*, 48, 1-49.

Stock, J. and M. Watson (1989). "New Indexes of Leading and Coincident Indicators," *NBER Macroeconomics Annual*, 351-94.

Stock, J. and M. Watson (1991). "A Probability Model of the Coincident Economic Indicators," in Lahiri, K. and G. H. Moore (eds.) *Leading Economic Indicators: New Approaches and Forecasting Records.* Cambridge, England and New York: Cambridge University Press.

Stock, J. and M. Watson (1989). "A Procedure for Predicting Recessions With Leading Indicators: Econometric issues and Recent Experience," in Stock, J. and M. Watson (eds.) *Business Cycles, Indicators, and Forecasting*. Chicago: The University of Chicago Press.

Tanner, M. and W. Wong (1987). "The Calculation of Posterior Distributions By Data Augmentations (With Discussion)," *Journal of the American Statistical Association*, 82, 528-550.

Watson, M (1994). "Business Cycle Durations and Postwar Stabilization of the U.S. Economy," *American Economic Review*, 84, 24-46.

Wecker, W. (1979). "Predicting the Turning Points of a Time Series," *Journal of Business*, 52, 35-50.

Zellner, A, C. Hong and G. Gulati (1990). "Turning Points in Economic Time Series, Loss Structures and Bayesian Forecasting," in Geisser, S., J.S. Hodges, S.J. Press, and A. Zellner (eds.) *Bayesian and Likelihood Methods in Statistics and Econometrics: Essays in Honor of George A. Barnard.* Amsterdam and New York: North Holland.

2 A MARKOV SWITCHING COOKBOOK

Bruce Mizrach
Rutgers University
James Watkins
American Express

1. Introduction

Economists continue to debate the importance of nonlinearity to their discipline. When it comes to forecasting levels, unit roots seems to be quite prevalent, and there has been a great deal of skepticism about nonlinear models. See the arguments pro and con in Ramsey (1996). The time series properties of higher moments have, however, led researchers to go beyond the standard linear, normally distributed world of the textbooks. The two most widely developed lines of research in this area are the ARCH volatility models of Engle (1982), and the asymmetric Markov-switching model of Hamilton (1989). Our focus in this paper concerns numerical procedures for the estimation of the MS type of models.

Hamilton extended Goldfeld and Quandt's (1973) Markov switching regression to the time series context. He analyzed the growth rate of U.S. real GNP. Hamilton's model not only accommodated the asymmetries first noted by Neftci (1984), but also succeeded in reproducing the business cycle turning points established by the NBER. The Markov-switching framework for output was later generalized to allow for time-varying, duration-dependent, and seasonally dependent transition probabilities (as in Filardo (1994), Durland and McCurdy (1994) and Ghysels (1994) respectively). Applications to interest rates by Hamilton (1988) and exchange rates by Engel and Hamilton (1990) illustrate the usefulness of the model outside of its initial application.

This paper is organized as follows. Section 2 briefly reviews Hamilton's Markov-switching framework and introduces a few general concepts. Estimation procedures are described in section 3, and the accompanying algorithms in sections 4 and 5. The relative merits of the two algorithms is discussed in section 6 and computational speed comparisons are presented in section 7. RATS code for both models is available from the authors at http://www-snde.rutgers.edu/ research.html

2. Time Series Models of Changes in Regime

A brief description of the Markov-switching (hereafter MS) framework is helpful to establish notation and vocabulary. The following description follows closely that of Hamilton (1993).

Consider for simplicity a first order autoregression where the mean value around which this series clusters may take on one of two values, $\mu^{(1)}$ and $\mu^{(2)}$:

$$y_t - \mu_t = \phi(y_{t-1} - \mu_{t-1}) + \varepsilon_t. \tag{1}$$

Suppose further that $\varepsilon_t \sim iidN(0, \sigma^2)$. A change in the value of μ alone is a change in regime (or state) in this simple case. It should be noted however, that all of the parameters of a model could be allowed to change with the state if thought appropriate. Hamilton's framework is rather agnostic regarding forces driving the change in regime: "...changes in regime are the result of processes largely unrelated to past realizations of the series and are not themselves directly observable." Hamilton (1993, p.234). It must be stressed that this does not mean that changes in regime are unrelated to the history of regimes; in fact, the state variable can have as long or longer a "memory" than the observation series y_t.

The state variable s_t is associated with the indices for the constant terms in equation 1; for instance, $s_t = 1$ is equivalent to saying $\mu_t = \mu^{(1)}$. Since the state variable is unobservable, we will need to form probabilistic inferences of its value, and in so doing form equivalent inferences regarding parameter values in 1. We assume that the state variable is governed by the Markov chain:

$$p(s_t = 1|s_{t-1} = 1) = p^{(11)}, \tag{2}$$

$$p(s_t = 2|s_{t-1} = 1) = p^{(12)}, \tag{3}$$

$$p(s_t = 1|s_{t-1} = 2) = p^{(21)}, \tag{4}$$

$$p(s_t = 2|s_{t-1} = 2) = p^{(22)}. \tag{5}$$

These transition probabilities are restricted so that $p^{(11)} + p^{(12)} = p^{(21)} + p^{(22)} = 1$. Hereafter, we will abbreviate notation when possible; the expression $p(s_t|s_{t-1})$ will refer to whichever of the above values is appropriate in the given context. The model given by 1, in conjunction with the assumptions regarding the transition probabilities, will be referred to as an MS(1) model. By increasing the autoregressive dimension, we can consider higher-order MS(r) models.

As will become apparent through the exposition of the estimation techniques, numerous probabilistic inferences can be computed at different points throughout the sample. For example, the inference $p(s_t = 1, \ldots, s_{t-r} = 2|Y_t)$ will refer to the probability that the unobserved state variable took on the values $2, \ldots, 1$ at times $t-r$ through t respectively, conditioned on data up to and including that from date t. Such a value may be abbreviated $p(s_t, \ldots, s_{t-r}|Y_t)$ when possible, but note that the chronological subset represented by Y_t will always be included to distinguish between filter inferences, smoothed inferences, and r-lag smoothed inferences (all to be defined later). Likewise, an example of an observation density conditional on states would be $p(y_t|s_t = 1, \ldots, s_{t--r} = 2, Y_{t-1})$ and as may be apparent from the notation, may depend on a finite history of past regimes. It will also be necessary to compute joint densities of states and observations, given by expressions such as $p(y_t, s_t = 1, \ldots, s_{t-r} = 2|Y_{t-1})$. Both the observation densities and joint densities are abbreviated in the same manner as the inferences regarding states. Finally, more familiar looking observation densities (i.e. $p(y_t|Y_{t-1})$) will become available as functions of the above densities.

Often the conditional density depends not only on the current regime, but also on past regimes. This requires us to make assumptions regarding the memory of the state variable. For simplicity, the state process is generally given the same time dimension as the observation process. In other words, it will be assumed that the state variable is AR(r). To still satisfy the property of a Markov system, we must restructure the two regimes as 2^{r+1} distinct regimes, to exhaust the combinations of states. For example, in an MS(2) model, the expression

$$(y_t - \mu^{(1)}) = \phi^{(11)}(y_{t-1} - \mu^{(2)}) + \phi^{(12)}(y_{t-2} - \mu^{(1)}) + \varepsilon_t^{(1)} \tag{6}$$

is associated with the event: $(s_t = 1, s_{t-1} = 2, s_{t-2} = 1)$.

3. The Filter

Recall the assumption that the state variable s_t is generally unobservable. In order to estimate the parameters of a MS model with this uncertainty, we must compute probabilities associated with each possible regime. Further, in the case of a MS model where the conditional density depends on both current and past regimes

$$p(y_t|s_t, Y_{t-1}) \neq p(y_t|s_t, s_{t-1}, s_{t-2}, \ldots, Y_{t-1}), \tag{7}$$

these inferences need to extend over several periods

$$p(s_t, s_{t-1}, \ldots, s_{t-r}|Y_t). \tag{8}$$

Such probabilities are estimated using Hamilton's recursive filter; We discuss the filter in the general case of an MS(r) model. This procedure will compute r and $r+1$ period inferences (2^r and 2^{r+1}distinct numbers for each t) and as a by-product, the conditional likelihood function. It is the conditional likelihood function that we seek for techniques such as Newton-Raphson (NR) or Davidon-Fletcher-Powell

(DFP), since the burdensome exact-likelihood method provides only a marginal improvement. The exposition is confined to the case where there are two states, and the initialization of the filter is reserved to the end of the discussion.

An arbitrary iteration of the filter begins by advancing an r period inference, available to us from the prior iteration,

$$p(s_{t+1}, s_t, \ldots, s_{t-r+1}|Y_t) = p(s_{t+1}|s_t) \cdot p(s_t, s_{t-1}, \ldots, s_{t-r+1}|Y_t). \tag{9}$$

We then use the appropriate density to find the joint probability inference of the current observation and the $r+1$ most recent states, conditional on last period's datum,

$$\begin{aligned} & p(y_{t+1}, s_{t+1}, s_t, \ldots, s_{t-r+1}|Y_t) \\ = \quad & p(y_{t+1}|s_{t+1}, s_t, \ldots, s_{t-r+1}, Y_t) \cdot p(s_{t+1}, s_t, \ldots, s_{t-r+1}|Y_t). \end{aligned} \tag{10}$$

Integrating over states, we find a density conditional only on prior data,

$$p(y_{t+1}|Y_t) = \sum_{s_{t+1}=1}^{2} \sum_{s_t=1}^{2} \sum_{s_{t-r+1}=1}^{2} (y_{t+1}, s_{t+1}, s_t, \ldots, s_{t-r+1}|Y_t). \tag{11}$$

We then have at our disposal an $r+1$ period inference conditional on current data

$$p(s_{t+1}, s_t, \ldots, s_{t-r+1}|Y_{t+1}) = \frac{p(y_{t+1}, s_{t+1}, s_t, \ldots, s_{t-r+1}|Y_t)}{p(y_{t+1}|Y_t)} \tag{12}$$

and by integration, an updated r period inference

$$\begin{aligned} p(s_{t+1}, s_t, \ldots, s_{t-r+2}|Y_{t+1}) \quad &= \\ p(s_{t+1}, s_t, \ldots, s_{t-r+1} \quad &= \quad 1|Y_{t+1}) + p(s_{t+1}, s_t, \ldots, s_{t-r+1} = 2|Y_{t+1}). \end{aligned} \tag{13}$$

The updated inference is then used as input for the next iteration. In later discussion, when we refer to "filter execution", this will mean that the entire sample is passed through the above process.

The filter is initialized with r-period unconditional probabilities,

$$p(s_r, s_{r-1}, \ldots, s_1) = p(s_t, s_{t-1}, \ldots, s_{t-r+1}). \tag{14}$$

To find these, we start by computing the ergodic probabilities, which are simply the unconditional estimates that the process will fall into each regime at an arbitrary date

$$\pi^{(j)} \equiv p(s_t = j) \text{ for } j = 1, 2. \tag{15}$$

These are found by solving the following set of equations:

$$p^{(1j)} \cdot \pi^{(1)} + p^{(2j)} \cdot \pi^{(2)} = \pi^{(j)} \text{ for } j = 1, 2 \tag{16}$$

$$\pi^{(1)} + \pi^{(2)} = 1. \tag{17}$$

Employing the appropriate transition probabilities, we can compute the necessary r-period unconditional probabilities, e.g.:

$$\begin{aligned} p(s_t \quad &= \quad 1, s_{t-1} = 2, s_{t-2} = 1) \\ &= \quad (1 - p(s_t = 2|s_{t-1} = 2)) \cdot (1 - p(s_{t-1} = 1|s_{t-2} = 1)) \cdot \pi^{(1)}. \end{aligned} \tag{18}$$

In the case of an MS(r) system, one needs to compute 2^r of these probabilities to initialize the filter.

After the entire sample has been passed through the filter, the computed observation densities can be used to form the conditional likelihood function

$$p(y_T, y_{T-1}, \ldots, y_{r+1}) = \prod_{t=r+1}^{T} p(y_t|Y_{t-1}). \tag{19}$$

3.1 The Full-Sample Smoother

The filter inferences provided above form a time series that can be volatile; they may in fact indicate changes in regime that have not occurred.. Smoothed inferences are more extensively conditioned by utilizing both past and future observations. This feature reduces the chance that we will misinterpret an outlier occurring in a particular regime for an actual change of state. Smoothed inferences can be distinguished from filter inferences by the time subscript on the relevant information set, Y_T instead of Y_t.

Suppose now that we wish to compute $r+1$ period smoothed inferences for our MS(r) model. After executing the filter to obtain $r+1$ period filter inferences, we expand those inferences to length $r+2$

$$\begin{aligned} & p(s_{t+1}, s_t, \ldots, s_{t-r}|Y_{t+1}) \\ = \quad & \frac{p(y_{t+1}|s_{t+1}, s_t, \ldots, s_{t-r+1}, Y_t) \cdot p(s_{t+1}|s_t) \cdot p(s_t, s_{t-1}, \ldots, s_{t-r}|Y_t)}{p(y_{t+1}|Y_t)} \end{aligned} \tag{20}$$

The $r+2$ period inferences can then be expanded in the same manner to $r+3$ period inferences, the $r+3$ period inferences into $r+4$ period inferences, and so on. This process is continued until we reach inferences of length $2r+2$ (2^{2r+2} distinct numbers for each t). At this point, we can integrate to find what might be referred to as non-adjacent probability inferences

$$\begin{aligned} & p(s_{t+r+1}, s_{t+r}, \ldots, s_{t+2}, s_t, \ldots, s_{t-r}|Y_{t+r+1}) \\ = \quad & \sum_{s_{t+1}=1}^{2} (s_{t+r+1}, s_{t+r}, \ldots, s_{t-r}|Y_{t+r+1}). \end{aligned} \tag{21}$$

These quantities can be advanced one period

$$\begin{aligned} & p(s_{t+r+2}, s_{t+r+1}, \ldots, s_{t+2}, s_t, \ldots, s_{t-r}|Y_{t+r+2}) \\ = \quad & \frac{p(y_{t+r+2}|s_{t+r+2}, s_{t+r+1}, \ldots, s_{t+2}, Y_{t+r+1})}{p(y_{t+r+2}|Y_{t+r+1})} \times \\ & p(s_{t+r+2}|s_{t+r+1}) \cdot p(s_{t+r+1}, s_{t+r}, \ldots, s_{t+2}, s_t, \ldots, s_{t-r}|Y_{t+r+1}), \end{aligned} \tag{22}$$

then integrated again to expand the gap in the inferences

$$\begin{aligned} & p(s_{t+r+2}, s_{t+r}, \ldots, s_{t+3}, s_t, \ldots, s_{t-r}|Y_{t+r+2}) \\ = \quad & \sum_{s_{t+2}=1}^{2} p(s_{t+r+2}, s_{t+r}, \ldots, s_{t+2}, s_t, \ldots, s_{t-r}|Y_{t+r+2}). \end{aligned} \tag{23}$$

Such iterations continue until the end of the sample is reached

$$p(s_T, s_{T-1}, \ldots, s_{T-r}, s_t, \ldots, s_{t-r}|Y_T),$$

allowing us, via integration, to find full-sample the smoothed inferences

$$p(s_t, s_{t-1}, \ldots, s_{t-r}|Y_T) \tag{24}$$

$$= \sum_{s_T=1}^{2} \sum_{s_{T-1}=1}^{2} \cdots \sum_{s_{T-r}=1}^{2} p(s_T, s_{T-1}, \ldots, s_{T-r}, s_t, s_{t-1}, \ldots, s_{t-r} | Y_T).$$

3.2 *Approximations to Full-Sample Smoothed Inferences*

Approximations to the above inference are available with significantly less computation. If we use partial conditioning on future observations, we may still arrive at a stable sequence of inferences. This possibility was recognized by Hamilton (1989). After executing the filter, one expands the inferences as would be done in the first step of the full-sample smoother. When inferences of length $2r + 1$ are reached, one integrates over $s_{t+1}, s_{t+2}, \ldots, s_{t+r}$ to find the r-lag smoothed inference

$$\begin{aligned} & p(s_t, s_{t-1}, \ldots, s_{t-r} | Y_{t+r}) \\ = & \sum_{s_{t+r}=1}^{2} \cdots \sum_{s_{t+1}=1}^{2} p(s_{t+r}, s_{t+r-1}, \ldots, s_{t+1}, s_t, s_{t-1}, \ldots, s_{t-r} | Y_{t+r}). \end{aligned} \tag{25}$$

These inferences get their name from the quantity of future data used for conditioning. The justification for this method is that we have considered data sufficiently far in the future that any further conditioning of the inference for date t does not utilize the observation from date t. Any further conditioning should have a negligible impact.

4. Hill Climbing

It is almost always the case that we are unsure as to when each regime was active in our sample. We therefore need to handle two types of uncertainty during estimation: uncertainty regarding parameter values and uncertainty regarding the path of the state variable. Recall that the filter computes, as a by-product, the conditional likelihood function

$$L(\theta) = \prod_{t=r+1}^{T} p(y_t | Y_{t-1}), \tag{26}$$

where

$$\begin{aligned} p(y_t | Y_{t-1}) = & \sum_{s_t=1}^{2} \cdots \sum_{s_{t-r}=1}^{2} p(y_t | s_t, s_{t-1}, \ldots, s_{t-r}, Y_{t-1}) \\ & \times p(s_t | s_{t-1}) \cdot p(s_{t-1}, s_{t-2}, \ldots, s_{t-r} | Y_{t-1}). \end{aligned} \tag{27}$$

The dual uncertainty of our estimation problem makes maximization of the above likelihood function is more complicated than it may first appear. If we knew the path of the state variable, we could simply maximize the above function with only small modifications to a canned software package. However, each chronological element of the likelihood function is a mixture distribution, composed of the densities representing each state. The weights for these densities (the probabilistic inferences associated with each regime or recent history of regimes) are functions of the parameters being estimated. As a result, the complete set of filter inferences will change with every perturbation of the parameter vector during gradient computation. For the sake of clarity, consider an arbitrary iteration of a DFP routine. We have at our disposal a tentative value for the set of parameters. Each element of the gradient requires that we execute the filter twice. Next, every stepsize to be considered requires a new execution of the filter, and only then can the updated parameter vector be chosen. We leave it to the reader to consider what is involved to compute the Hessian numerically.

Apart from this additional consideration, numerical maximum likelihood techniques, such as DFP are applied in the normal way.

5. The EM Algorithm

The EM algorithm (hereafter EMA), as outlined by Hamilton(1990), deals with the dual uncertainty problem in a different way. One begins with an initial guess for the vector of parameters, say $\boldsymbol{\theta}^{[0]}$. The filter and smoother, both parametrized by the extent of regime dependence, are executed to obtain inferences conditional on the entire sample of observations. The smoothed inferences are used as weights for coefficient updating, via minimization of the sum of weighted squared residuals. Improved estimates of the transition probabilities are simple functions of the smoothed probabilities. The set of updated values constitutes $\boldsymbol{\theta}^{[1]}$; we repeat the process until some convergence criterion is satisfied.

To outline the procedure in greater detail, consider a two-state MS(2) model:

$$(y_t - \mu^{(1)}) = \phi^{(11)}(y_{t-1} - \mu_{t-1}) + \phi^{(12)}(y_{t-2} - \mu_{t-2}) + \varepsilon_t^{(1)}, \tag{28}$$

$$(y_t - \mu^{(2)}) = \phi^{(21)}(y_{t-1} - \mu_{t-1}) + \phi^{(22)}(y_{t-2} - \mu_{t-2}) + \varepsilon_t^{(2)}. \tag{29}$$

$\varepsilon_{i,t} \sim N(0, (\sigma^{(i)})^2)$, $i = 1, 2$. Constructed this way, 10 parameters need to be estimated: $\mu^{(1)}$, $\mu^{(2)}$, $\phi^{(11)}$, $\phi^{(12)}$, $\phi^{(21)}$, $\phi^{(22)}$, $\sigma^{(1)}$, $\sigma^{(2)}$, $p^{(11)}$, $p^{(22)}$. A single iteration of the EM algorithm starts by executing the filter. Upon completion, the filter yields regime inferences for 2 periods

$$p(s_t, s_{t-1}|Y_t)$$

and for 3 periods

$$p(s_t^*|Y_t) = p(s_t, s_{t-1}, s_{t-2}|Y_t). \tag{30}$$

We use the $*$ to redefine the state in terms of the 8 permutations of the 3 lags, e.g. $s_t^* = 1$ implies $s_t = 1$, $s_{t-1} = 1$, $s_{t-2} = 1$. Also obtained are observational densities, both conditioned on states

$$p(y_t|s_t^*, Y_{t-1}) = p(y_t|s_t, s_{t-1}, s_{t-2}, Y_{t-1}) \tag{31}$$

and unconditional with regard to states

$$p(y_t|Y_{t-1}).$$

A typical observation density is written

$$\begin{aligned} p(y_t|s_t^* &= 3, Y_{t-1}) \\ &= \frac{1}{\sigma^{(1)}\sqrt{2\pi}} \exp\{[(y_t - \mu^{(1)}) - \phi^{(11)}(y_{t-1} - \mu^{(2)}) \\ &\quad -\phi^{(12)}(y_{t-2} - \mu^{(1)})]^2/2(\sigma^{(1)})^2\}. \end{aligned}$$

We next execute the smoother to find a probability series that is less volatile than that provided by the filter

$$p(s_t^*|Y_T) = p(s_t, s_{t-1}, s_{t-2}|Y_T), \tag{32}$$

while integration yields

$$p(s_t, s_{t-1}|Y_T) \text{ and } p(s_t|Y_T). \tag{33}$$

With smoothed inferences available, coefficients are updated numerically by minimizing the sum of weighted squared residuals

$$\left[\mu^{(1)}, \mu^{(2)}, \phi\right] = \text{arg min} \left\{\sum_{t=3}^{T} \sum_{k=1}^{8} (\nu_t^{(k)})^2 \cdot p(s_t^* = k|Y_T)\right\} \tag{34}$$

where $\phi = [\phi^{(11)}, \phi^{(12)}, \phi^{(21)}, \phi^{(22)}]$ and

$$\nu_t^{(k)} = y_t - E\left(y_t|s_t^* = k, Y_{t-1}\right). \tag{35}$$

An example of an error series element would be

$$\nu_t^{(2)} = (y_t - \mu^{(2)}) - \phi^{(21)}(y_{t-1} - \mu^{(1)}) - \phi^{(22)}(y_{t-2} - \mu^{(2)}).$$

Estimation of the regimes' variances requires similar weighting

$$\begin{aligned}(\sigma^{(j)})^2 &= \frac{p(s_t = j, s_{t-1}, s_{t-2}|Y_T)}{\sum_{t=3}^{T} p(s_t = j)} \\ &\times \sum_{t=3}^{T} \sum_{s_{t-1}=1}^{2} \sum_{s_{t-2}=1}^{2} [(y_t - \mu^{(j)}) - \phi^{(j1)}(y_{t-1} - \mu^{(s_{t-1})}) \\ &-\phi^{(j2)}(y_{t-2} - \mu^{(s_{t-2})})]^2\end{aligned}$$

for $j = 1, 2$. Updated transition probabilities also utilize smoothed inferences

$$p_{ij} = p(s_t = j|s_{t-1} = i) = \frac{\sum_{t=r+1}^{T} p(s_t = j, s_{t-1} = i|Y_T)}{\sum_{t=r+1}^{T} p(s_{t-1} = i|Y_T)} \text{ for } i, j = 1, 2. \tag{36}$$

We have completed a single iteration of the algorithm; we repeat until convergence.

6. Relative merits of Different Algorithms

The DFP routine described earlier is the same method used by Hamilton (1989) for his analysis of the business cycle. Hamilton (1990) proposed the above EMA as an alternative, to handle systems of greater complexity.

Problems may arise during gradient computation due to the shape of the likelihood surface associated with a MS model. Mixture distributions may have as many local maxima as regimes, and likelihood functions derived from these densities may be plagued by the same features. The EMA however, does not involve the examination of likelihood surfaces, and as such, may avoid both local maxima and singularities. Another positive attribute of the EMA noted in other applications is its ability to arrive in the neighborhood of the mode of the likelihood function in a few early steps, which can prove advantageous if one is performing a rough grid search to determine optimal starting values. Hamilton also argues that an EMA may not be as demanding numerically: "while one could calculate analytic derivatives from rote adaptation of the recursion ..., that approach would require burdensome additional computer programming and calculation time for each parameter." Hamilton (1990, p. 40).

If the EMA is indeed more robust than a DFP algorithm and computationally more efficient, one would clearly prefer the former. However, Hamilton's claim regarding the computational speed of the EMA is problematic. Consider the full-sample smoothing technique outlined earlier: this procedure requires computation of order $K^{2(r+1)}T^2$, where r is the relevant autoregressive dimension, K is the number of states and T is the sample size. In our experience, if r is 4 or larger, while $K = 2$ and T is a modest 100 to 200 points, a single iteration of the EMA can take several minutes on an 166 MHz Pentium, utilizing the 32 bit version of RATS. If several hundred iterations are necessary to achieve convergence, such investment in computer time becomes prohibitive. By contrast, a single iteration of a DFP routine requires several seconds.

Despite these concerns, the EMA may be preferable in many contexts. If no autoregressive dynamics are present, the full-sample smoother becomes much less of a problem, and the EMA's ability to avoid the difficulties of poorly shaped likelihood surfaces outweighs any additional CPU time. In their investigation of exchange rates, Engel and Hamilton (1990) successfully employed the EMA in the presence of numerous local maxima. The case for the EMA is also strengthened by the possibility of using the approximate smoother, as it is only marginally more demanding than the filter.

7. Computational Comparisons

To explore the computational issues more thoroughly, consider the Hamilton's (1989) analysis of real GNP growth. The model was fitted by DFP:

$$y_t - \mu_t = \sum_{p=1}^{4} \phi^{(ip)}(y_{t-p} - \mu_{t-p}) + \varepsilon_t,\ \varepsilon_t \sim iidN(0, \sigma^2). \tag{37}$$

The two constants $\mu^{(1)}$ and $\mu^{(2)}$ are associated with high and low rates of growth. We recreated Hamilton's results via three methods: DFP, the EMA with full-sample smoothing (EMA(1)), and the EMA using approximate smoothing (EMA(2)). Presented below are the computational demands of each algorithm for different convergence criteria.

Table 1 illustrates two important claims: first, that using the approximate EMA significantly reduces CPU time relative to the pure form of the EMA; second, when comparing gradient-search methods with EM methods, a tradeoff exists between the time that it takes to complete an iteration and the number of iterations needed to achieve convergence. While EM uses $2/3$ fewer iterations, each iteration takes from 5 to 20 times as long.

The usefulness of the approximate EM routine is of course dependent on the accuracy of the results. Comparing the approximate EM results with those provided by the DFP algorithm, we find that the mean absolute difference between the elements of the parameter vector decreases as the convergence criterion is tightened, and is no greater 0.0085 if we set $\delta = 10^{-5}$.

Table 1. Number of Iterations to Convergence

Criterion	**DFP**	**EMA(1)**	**EMA(2)**
$\delta = 10^{-3}$	61	22	27
$\delta = 10^{-4}$	166	50	51
$\delta = 10^{-5}$	254	69	75
sec./iter.	12	227	57

8. Conclusion

Markov-switching models have become widely used in accordance with the growing evidence of nonlinearity in economic time series. We have detailed the two leading algorithms for estimating MS models and have provided a discussion of the computational questions that arise during the course of an estimation problem.

Although the EM algorithm is usually more robust with respect to poor likelihood surfaces than gradient-based methods, the pure form of the former technique often places prohibitive demands on computer time. Fortunately, the availability of an approximation technique allows the EM method to stand as a practical alternative to commonly used hill-climbing routines.

References

Durland, J. Michael and Thomas H. McCurdy (1994). "Duration Dependent Transitions in a Markov Model of U.S. GNP Growth," *Journal of Business and Economic Statistics,* 12, 279-88.

Engel, Charles (1994). "Can the Markov Switching Model Forecast Exchange Rates?" *Journal of International Economics,* 36, 151-65.

Engel, Charles and James D. Hamilton (1990). "Long Swings in the Dollar: Are They in the Data and Do Markets Know It?" *American Economic Review,* 80, 689-713.

Engle, Robert F. (1982). "Autoregressive Conditional Heteroscedasticity with Estimates of the Variance of U.K. Inflation," *Econometrica,* 50, 987-1007.

Filardo, Andrew J. (1994). "Business Cycle Phases and their Transitional Dynamics," *Journal of Business and Economic Statistics,* 12, 299-308.

Ghysels, Eric (1994), "On the Periodic Structure of the Business Cycle," *Journal of Business and Economic Statistics,* 12, 289-98.

Goldfeld, Stephen M. and Richard E. Quandt (1973). "A Markov Model for Switching Regressions," *Journal of Econometrics,* 1, 3-16.

Goodwin, Thomas H. (1993). "Business Cycle Analysis with a Markov Switching Model," *Journal of Business and Economic Statistics,* 11, 331-39.

Hamilton, James D. (1989). "A New Approach to the Economic Analysis of Nonstationary Time Series and the Business Cycle," *Econometrica,* 57, 357-84.

Hamilton, James D. (1990). "Analysis of Time Series Subject to Regime Changes,"

Journal of Econometrics, 45, 39-70.
Hamilton, James D. (1993). "Estimation, Inference and Forecasting of Time Series Subject to Changes in Regime," in G.S. Maddala, C.R. Rao and H.D. Vinod, (eds.), *Handbook of Statistics* Vol. 11, Amsterdam: North Holland.
Neftci, Salih N. (1984). "Are Economic Time Series Asymmetric over the Business Cycle?" *Journal of Political Economy,* 92, 307-28.
Ramsey, James (1996). "If Nonlinear Models Can't Forecast, What Use are They?" *Studies in Nonlinear Dynamics and Econometrics,* 1, 65-86.

3 A REANALYSIS OF THE SPECTRAL PROPERTIES OF SOME ECONOMIC AND FINANCIAL TIME SERIES

James B. Ramsey
New York University
David J. Thomson
Bell Laboratories, Lucent Technologies

1. Introduction

In this paper we re-examine the spectral properties of some indices of production and of the Standard and Poor's 500 stock market index. We use some new procedures for estimating spectra that are more efficient and more powerful than the conventional fast Fourier transform (FFT) approach or those using the sample autocorrelation function. The data examined are the growth rates in the monthly production indices for durable goods and nondurable goods production, manufacturing, mining, and the monthly stock market index; the real indices begin in 1919 and end in 1988, the stock market index begins in 1926 and also ends in 1988. We also are able to examine the new series because the AR(1) coefficient is sufficiently below a unit root to enable us to obtain useful spectral results. Some new techniques are used to examine the extent of nonstationarity in these data.

The production indices have strong harmonics of the annual cycle at the frequencies of two, three, four, and five cycles per year. There is evidence for a cycle at 1.3 cycles per year, although not as strong as the periodic cycles of two to five cycles per year. There was only very weak evidence for cycles at lower and especially at business cycle frequencies. The background noise is non-white. The importance of this latter observation is that even after allowing for the main frequencies mentioned above, there is still a residual "predictable" component, even though this component would appear to be very complex. These data certainly cannot be treated merely as a linear filter applied to uncorrelated noise.

Our results confirm that at seasonal frequencies the market is efficient in that there are no harmonics in the stock market index and that the transfer function between the stock index and the real indices is flat except for nulls at the strong harmonics in the production data. There was some evidence of significant coherence at business cycle frequencies between the production indices and the stock market index; the coherences between the production indices themselves were very strong.[1] There was little evidence for any pronounced shifting in phase, especially over the post-war period. There was, however, very strong evidence of frequency shifts produced by the Second World War.

Our initial examination for the presence of nonstationarity in the series confirmed that for all the series considered there was significant nonstationarity, although somewhat surprisingly the nonstationarity for the stock market index seem to be less than for the production indices. Nonstationarity is particularly severe at business cycle frequencies.

The two basic approaches to the analysis of time series are the use of the "time domain" and "frequency domain."

Because these two approaches are mathematically related in that the autocorrelation function of the time domain approach is the Fourier transform of the spectrum of the frequency domain formulation and vice versa, it is often not clear why one approach should be preferred over the other. The choice between the two procedures depends on both technical and substantive issues. We discuss the technical issues first.

Any stationary time series without a purely deterministic component can be expressed as an expansion in terms of a causal moving average of an uncorrelated white noise process; this is Wold's general linear representation theorem, Priestley (1981); Granger and Newbold (1986):

$$X_t = \sum_0^{\infty} \alpha_i \epsilon_{t-i} \,. \tag{1}$$

Alternatively, a stationary series can be represented as a Fourier transformation of orthogonal random measures, this is the Cramér spectral representation theorem; see equation (2)

$$X_t = \int_{-\infty}^{\infty} e^{i\omega t} dZ(\omega) \,. \tag{2}$$

However, there are differences between the two approaches that are of importance to the researcher.

An initial reason for using the frequency domain approach is when the time series is presumed to be the sum of "harmonic components," that is, the sum of sinusoidal terms. If economic time series are generated, at least in part, by harmonic components, then the use of spectral analysis is an obvious tool. More relevantly, if economic or financial time series contain frequency bands of high power, the use of spectral analysis is a natural choice; this is the rational underlying the traditional N.B.E.R. definition of "business cycles." But the existence of strong harmonic components is not prevalent in economic data beyond time periods of one year and the presence of frequency bands with high power is still problematic so that spectral techniques would be hard to justify on these grounds alone.

Alternatively, the underlying presumption might be that the time series is generated by a linear transformation of random shocks, in which case the preferred approach is to use time domain procedures. Nevertheless, because ARMA(p, q) models can be easily represented in terms of spectral components, the spectral approach is often informative. For example, if the ARMA model is represented by:

$$\phi(B)X_t = \theta(B)Z_t, \quad \{Z_t\} \sim WN(0, \sigma^2) \tag{3}$$

then the spectral density function, Brockwell and Davis (1987) is given by:

$$f_X(\omega) = \frac{\sigma^2}{2\pi} \frac{|\theta(e^{-i\omega})|^2}{|\phi(e^{-i\omega})|^2}. \tag{4}$$

Correspondingly, one can re-express a continuous spectrum in terms of a causal AR process, or in terms of an invertible MA process.

An important reason for choosing spectral techniques is that the frequency domain approach does not assume a specific model, whereas the time domain approach does; that is, the frequency domain approach "lets the data tell the story." The cost is that there may be some difficulty in determining what the corresponding time domain model is, when the time domain formulation is believed to be the more applicable physical model. A more precise way of saying this is that the spectral approach provides a useful representation of the entire signal. The advantage of such a representation lies in the ease with which various operations can be handled in the transformed, frequency, domain as opposed to the original, time, domain. While the analogy should not be pressed, the transformation from the reals to the complex domain simplifies many problems and provides solutions to problems that may be much more difficult to solve in the restricted domain of the reals.

The benefit of the spectral approach is to be discovered in the properties of relationships between functions defined on frequencies relative to the corresponding relationships between functions defined on the time domain. For example, the fact that compression in the time domain is equivalent to expansion in the frequency domain, that a time shift yields a phase shift in the frequency domain and vice versa, and that convolution in the time domain is associated with products in the frequency domain provide opportunities for converting "hard problems" into simpler ones.

The resurgence of interest in dynamical analysis during the last few years argues for the use of spectral techniques as they provide a natural analytical link to the solution of differential equations describing dynamical systems. The ARIMA approach is tied to the linear formulation of difference equations, whereas the spectral approach can be used to explore both difference and differential equation formulations of dynamical systems and is not constrained to linearity in the difference or differential equations.

Finally, an important reason for choosing the frequency domain approach stems from the function analytic aspects of spectral analysis. The frequency domain expansion of a time series is an expansion in terms of eigenvectors in the discrete case and in terms of eigenfunctions in the continuous case. Every harmonic e^{ikx}, or $e^{i\omega x}$, is the eigenfunction of every derivative and of every finite difference. Further, the expansion of the series in terms of e^{ikx}, or $e^{i\omega x}$, provides a natural link with the theory of differential and difference equations, see for example, Kac (1959). The Fourier expansions have the advantage of being orthogonal, complete, and convenient.

The objective of this paper is to re-explore the use of spectral techniques given the recent innovations in the estimation of sample spectra that provide much greater efficiency than current, FFT based, techniques. The importance of these new procedures cannot be over emphasized, since the ever present problem of nonstationarity limits the effective sample size of any real time series, unless one can express the nonstationarity in terms of a parametric model. Thus, efficiency gains in estimation can be used to obtain useful estimates with relatively smaller sub-samples of the entire data set. This in turn enables the researcher to shorten the period of time over which approximate local stationarity is expected to hold.

Nonstationarity with respect to financial data has been clearly demonstrated in two recent papers, Loretan and Phillips (1994) and Pagan and Schwert (1990). As noted in both papers the assumption of stationarity is an important one and is one that should be settled before other, more model specific, approaches are used. Our efforts that are to be discussed below complement this work. Even in the absence of convincing evidence for nonstationarity, there are still a number of unresolved issues for the use of spectral analysis of economic and financial data. A major question is the extent to which one can usefully talk about business cycles at all; one might question the evidence for the existence of spectral power at business cycle frequencies.

In the financial literature, there are two issues that merit further analysis. The first issue is whether there are any power peaks in the stock market spectra. A second related issue concerns the extent and degree of coherence between the reals and the Standard and Poor's 500 index.

Section 2 introduces new techniques of spectral analysis that are available for the enhanced estimation of spectra. The following section, 3, briefly discusses the use and role of spectral techniques in the analysis of financial and economic data. The data series that are to be analyzed are described in Section 4 of the paper, which is followed by a detailed discussion of the specific empirical results that have been obtained. The paper continues with some comments on the evidence for nonstationarity and on the evidence for long period "cycles." The paper ends with a summary of the major empirical findings.

2. A Brief Review of a New Method to Estimate Spectra

The traditional method of calculating the spectrum since the development of the Fast Fourier Transform has been in two steps. The first step is to calculate a weighted Fourier transform of the raw time series, where $\omega = 2\pi f$ is the radian frequency, "f" is the natural frequency in cycles per year, $x(n)$ is the observed time series, N is the number of observations, and D_n is a data window as shown in equation (5). However, because $\hat{S}_D(\omega)$ is inconsistent in that its variance does not decrease with sample size, one has to use a smoothed spectral estimate by convoluting the estimates $\hat{S}_D(\omega)$ with another window, $G(\omega)$. A major problem with the smoothing function is that, unless the spectrum is known to be smooth, (that is, the spectrum is not mixed), acceptable "smoothers" have to be nonlinear. In addition, the current procedure does not use the phase information in the original data. Moreover, this procedure is statistically inefficient.

What is more important is that the current procedure does not emphasize that what we are trying to achieve is information about the statistical properties of $dZ(\omega)$, where $dZ(\omega)$ is the orthonormal basis for the Fourier representation of the time series $x(n)$. The essential problem in time series analysis is to

$$\hat{S}_D(\omega) = \left| \sum_{n=0}^{N-1} x(n) D_n e^{-i\omega n} \right|^2 \tag{5}$$

estimate the moments of $dZ(\omega)$ from the finite time series, $x(n)$, $n = 0, 1, 2, \ldots, N-1$. We begin to address this problem in the next sub-section.

Before continuing we should emphasize briefly the major gains from using our procedure. To mention that the procedure to be discussed is more efficient, does not really indicate the range of benefits. While the benefits of increased power for any given sample size are welcome, a much more important benefit is our ability to handle near unit root processes. A consensus is beginning to build up that many economic time series are in fact near unit roots, so that procedures that can estimate the spectra of such data without series leakage problems and without having first to prewhiten is of enormous consequence. Further, there is also a consensus that economic and financial data are not stationary and that nonstationarity goes beyond first and even second order effects. A major benefit of our approach is that we will be able to address the nonstationarity issue directly and will be able to indicate in precise terms the manner in which the data are nonstationary, thereby providing the possibility of discovering the economic mechanisms that generate the observed nonstationarity.

Notation and Definition of Multiple Window Estimation

In the first part of this paper we will assume that the available data are a representative sample from a stationary process, that is one whose statistics do not depend on the choice of time origin. Stationary processes are a subset of harmonizable processes, see Loève (1963) and have a Cramér, or spectral, representation

$$x(t) = \int_{-1/2}^{1/2} e^{i2\pi\nu t} dX(\nu) \,. \tag{6}$$

The stationarity assumption implies that the covariance $E\{x(t)\bar{x}(t')\}$ is a function only of the time difference, $R(t-t')$ and consequently the generalized spectral density function

$$\gamma(\nu,\nu')d\nu d\nu' = E\{dX(\nu)\overline{dX}(\nu')\} \,, \tag{7}$$

must be concentrated on the line $\nu = \nu'$, that is

$$\gamma(\nu,\nu') = S(\nu)\delta(\nu-\nu') \tag{8}$$

where $\delta(\cdot)$ is the Dirac delta function and $S(\nu)$ is the spectral density function. The Wiener-Khinchin theorem implies that:

$$R(\tau) = \int e^{i2\pi f\tau} S(f)df \,.$$

However, while the *probabilistic* properties of stationary processes have been thoroughly studied, see *e.g.* Doob (1953), the corresponding *statistical* problem of estimating the spectrum is not so well understood.

We turn to the estimation problem. Assume that we have observed a finite sample of N contiguous observations on a discrete-time process $x(t)$ for $t = 0, 1, \ldots, N-1$, assumed to be a representative sample from a stationary process, and that we desire to estimate the spectral density function from these observations. The literature on this problem is immense, so cannot be treated extensively here; good introductory texts, however, are Brillinger (1975); Priestley (1981); Percival and Walden (1993). Note, however, that we are attempting to estimate a *function* from a finite set of observations, so that the problem falls into the domain of inverse theory, and is *not* one of classical parameter estimation.

We may redefine the estimation problem as an "inverse problem," that is, the problem of "estimating" a function from a finite number of elements. There are two steps in the resolution of the estimation problem; definition of the fundamental integral equation to be solved, and the use of Slepian functions in the solution. As the first step, consider the discrete Fourier transform of the observed sequence $[x_t]$ as shown in equation (9).

Substitute the spectral representation of $x(t)$ given in equation (6) into equation (9). The result is the

$$x(f) = \sum_{t=0}^{N-1} x(t)e^{-i2\pi ft} \tag{9}$$

fundamental equation of spectral estimation:

$$x(f) = \int_{-1/2}^{1/2} \boldsymbol{K}_N(f-\nu)dX(\nu) \tag{10}$$

where $K_N(f)$ is a Dirichlet kernel. Equation (10) is the frequency domain expression of the projection from the infinite sequence generated by the Cramér representation shown in equation (1) onto the sample. The Dirichlet kernel is given by

$$\boldsymbol{K}_N(f) = \frac{\sin N\pi f}{\sin \pi f}\, e^{-i2\pi f \frac{N-1}{2}} . \tag{11}$$

A major difference between multiple window spectrum estimation and standard approaches is in how one *views* equation (10). Conventionally, it would be regarded as a convolution; in the multiple-window method that we pursue, it is considered as an integral equation. The second step is to solve this equation using an eigenfunction representation due to Slepian (1978).

Slepian Functions and Sequences

The key to obtaining "reasonable" solutions for (10) is a set of special functions known as *Discrete Prolate Spheroidal Wave Functions*, or, in honor of their inventor David Slepian, *Slepian Functions and Sequences.* Slepian sequences are defined as a real, unit-energy sequences on $[0, N-1]$ having the greatest energy in a frequency bandwidth $\boldsymbol{W}$. We use Slepian's [1978] notation, $V_n^{(k)}(N,W)$ for sample n of the kth sequence of length N observations concentrated in a bandwidth of width W, or for short, $\nu_n^{(k)}$. These sequences are solutions of the symmetric Toeplitz matrix eigenvalue equation; the solutions are given in equation (12) for $0 \le n \le N-1$ and are defined by this equation for n outside this interval. The Slepian functions $V_k(N,W;f)$

$$\lambda_k \nu_n^{(k)} = \sum_{m=0}^{N-1} \frac{\sin 2\pi W(n-m)}{\pi(n-m)}\, \nu_m^{(k)} \tag{12}$$

Thomson (1990) are the Fourier transforms of the corresponding sequences and are defined by equation (13)

$$V_k(f) = \sum_{n=0}^{N-1} \nu_n^{(k)}(N,W)e^{-i2\pi n f} . \tag{13}$$

As before we have dropped the explicit dependence on sample size and bandwidth. These functions satisfy the homogeneous integral equation (14) with the same kernel that appeared in (10).

$$\lambda_k V_k(f) = \int_{-W}^{W} \boldsymbol{K}_N(f-\nu)V_k(\nu)d\nu \tag{14}$$

The definition of the Slepian functions used here, $V_k(f)$ differs from the $U_k(f)$'s defined in Slepian (1978) only in the definition of the Fourier transform.

The eigenvalues, λ_k, more precisely $\lambda_k(N, W)$ in equations (12) and (14) give the fraction of the energy contained in the bandwidth $(-W, W)$ and consequently are bounded between 0 and 1. Of these, the first

$$K \approx \lfloor 2NW \rfloor \tag{15}$$

where $\lfloor \cdot \rfloor$ indicates "largest integer less than," are "large," *i.e.*, nearly 1, and define the dimensionality of the time-frequency subspace. The complementary quantity, $1 - \lambda_k$ is the "leakage" from the kth window. Because the Slepian sequences were formulated to maximize energy concentration, the collective leakage properties of these windows is the best possible. The sequences are orthonormal on $[0, N-1]$ and the functions are orthonormal on $[-\frac{1}{2}, \frac{1}{2})$ and are *also* orthogonal on $(-W, W)$.

Spectrum Estimation as an Inverse Problem

Recall the fundamental equation (10). We attempt an eigen-solution of the integral equation on the interval $(f-W, f+W)$ using standard methods for the solution of a Fredholm equation of the first kind. By the assumed stationarity, the part of dX in the inner, or local, domain, $(f-W, f+W)$ is uncorrelated with dX on the rest of the frequency domain. The standard form for the expansion coefficients is given in equation (16). From the integral equation and from the double orthogonality

$$x_k(f) = \frac{1}{\lambda} \int_{-W}^{W} V_k(\nu) x(f-\nu) d\nu \tag{16}$$

properties equation (16) becomes, in the time domain, equation (17)

$$x_k(f) = \sum_{n=0}^{N-1} e^{-i2\pi f n} \nu_n^{(k)}(N, W) x(n)\,. \tag{17}$$

The expansion coefficients are obtained by windowing the data with a Slepian sequence and Fourier transforming. Obviously, an FFT may be used for efficient computation of the expansion coefficients, or eigencoefficients.

Given the eigencoefficients, the simplest multiple-window spectrum estimate is

$$\bar{S}(f) = \frac{1}{K} \sum_{k=0}^{K-1} |x_k(f)|^2\,. \tag{18}$$

Note that each of the K terms in this equation is a conventional direct spectrum estimate, *i.e.*

$$\hat{S}_k(f) = \left| \sum_{n=0}^{N-1} e^{-i2\pi f n} \nu_N^{(k)} x(n) \right|^2$$

and for $NW = 2$ or 4, the lowest order estimate, $\hat{S}_0(f)$ closely resembles what would be obtained with a standard Hamming or Parzen window respectively. Here, however, there are K windows and, if the spectrum is flat in $(f - W, f + W)$ the coefficients are uncorrelated because the windows are orthogonal. Thus each of the $S_k(f)$'s contributes 2 degrees-of-freedom, so that estimate (18) has 2^K degrees-of-freedom.

In practice, equation (18) is not used because the leakage properties of the kth window becomes successively poorer as k approaches $2NW$. For stability, it is desirable to use as many windows as possible, but one does not want estimates in regions where the spectrum is small to be biased by leakage from frequencies where the spectrum is large.

If one uses a local least-squares solution for the integral equation instead of simply truncating the series after the first K coefficients, one obtains a data adaptive weighting that gives superior protection against leakage and bias. Details of this adaptive weighting procedure are given in Thomson (1982); Thomson (1990). Spectra estimated using these adaptive weighting schemes are not simple quadratic forms in the data, but the solution of an equation that is a rational fraction of quadratic forms. As such, they have some surprising properties; in particular, estimates of autocovariances obtained by Fourier transforming the adaptively weighted spectrum can be non-zero for lags greater than the data length, and so have some of the properties of the Papoulis extrapolation estimates, Papoulis (1975).

To see the dependence on the bandwidth W of the estimate, recall that there are $K \approx \lfloor 2NW \rfloor$ windows with eigenvalues near 1. If W is too small, one has poor statistical stability, but if W is too large, the estimate has poor frequency resolution. Typically W is chosen between $1.5/N$ and $20/N$ with a time-bandwidth product of 4 or 5 being a common starting point. Thus $W = 4/N$ or $5/N$ with corresponding values for K of 6 or 8 gives estimates with 12 or 16 degrees-of-freedom. The case $NW = 4$ is interesting for comparison. The bandwidth is the same as that obtained when a Parzen or Papoulis window is used as a data taper. With these classical windows, the resulting spectral estimate has a Chi-square distribution with two degrees of freedom. In contrast, the eight windows used in the multiple-window estimate typically give 12 to 16 degrees of freedom, so that the variance is much lower than that for the single window estimates. Quadratic inverse theory, Thomson (1990), gives a test for deciding if W is too large.

Somewhat more recently, Bronez (1992) made a statistical analysis of the performance differences between the multi-taper analysis discussed above and weighted overlapped segment averaging. The performance measures of leakage, variance, and resolution were considered. Multi-taper analysis always performed better than the weighted overlapped segment averaging of spectral components. For example at the same level of resolution and leakage, the variance of the multi-taper method was one half to three quarters of the more usual nonparametric procedure.

Variance of Spectrum Estimates

A major disadvantage of conventional, smoothed single-window, estimates is that confidence intervals for these estimates have been made assuming the data to be stationary and Gaussian; these assumptions are often widely optimistic. By contrast a major trend in modern statistical work has been the development of nonparametric confidence regions derived from resampling ideas, most notably the jackknife and bootstrap.

The simplest way to apply these ideas to multiple-window estimates is to resample the eigencoefficients, that is, the $x_k(f)$, in an estimate of $\ln S(F)$ or of coherence. Details are given in Thomson and Chave (1991) with a refinement in the scale given in Thomson (1994).

Harmonic F Tests for Periodic Signals

One of the major triumphs in the theory of stationary stochastic processes is the Wold decomposition that states that such a process may be uniquely split into deterministic and non-deterministic parts. By contrast, a major problem with most spectrum estimation procedures is that they ignore differences between the first and second moments. First moments describe, loosely, the non-zero average and periodic components that would persist between different realizations, while the second moments describe the non-deterministic part of the process. Most statistical procedures include some form of centering that, as in the robustness study of Andrews (1972), or in the multivariate Stein estimates, Stein (1956) can be quite elaborate. In time series work, however, it is rare for "centering" to consist of much more than the removal of a simple average or, perhaps, a polynomial trend.

Conceptually, however, "centering" a time series is much more complicated and should be done as a function of frequency. The usual procedure of subtracting the arithmetic average in the time domain is only marginally effective at zero-frequency. An obvious example is provided by the durable goods index that is examined later in this paper. In addition to the standard average, there are significant mean value terms at multiples of one cycle per year. These result in sharp spikes, or lines, in the spectrum and so are referred to as lines or line components.

Here the multiple window method is advantageous because it separates line and continuum components and permits an analysis-of-variance test for significance of line components. If we assume a single periodic signal at a frequency f_0 in the frequency band $(f-W, f+W)$, the expected value of the eigencoefficients at a frequency f is

$$E\{x_k(f)\} = \mu V_k(f - f_0)\,. \tag{19}$$

The complex amplitude, μ can be found by least-squares on the observed coefficients, that is by minimizing

$$e^2 = \sum_{k=0}^{K-1} \left| x_k(f) - \hat{\mu}(f) V_k(f - f_0) \right|^2 \tag{20}$$

with the result:

To test the significance of a mean one commonly uses a "t" statistic, but as μ is complex, a multivariate T statistic must be used. This is equivalent to an F test for the regression of the V_k's on the x_k's or just the ratio of the energy explained by the assumption of a line component at the given

$$\hat{\mu}(f) = \frac{\sum_{k=0}^{K-1} \overline{V_k(f-f_0)}\, x_k(f)}{\sum_{k=0}^{K-1} |V_k(f-f_0)|^2} \tag{21}$$

frequency to the residual energy:

$$F(f) = \frac{(K-1)|\hat{\mu}(f)|^2 \sum_{k=0}^{K-1} |V_k(f-f_0)|^2}{\sum_{k=0}^{K-1} |x_k(f) - \hat{\mu}(f) V_k(f-f_0)|^2} . \tag{22}$$

Because the eigencoefficients are complex, there are initially $2K$ degrees of freedom. Two are used in the estimate of μ, so $F(f)$ has an F distribution with 2 and $2K - 2df$. Moreover, if the frequency where $F(f)$ is maximum is used as an estimate of line frequency, it's variance is within 5–10% of the Cramér-Rao bound. One should be cautious, however, because when F is plotted as a function of frequency there are usually many apparently highly significant peaks. One must remember that, in a random sample of size N, there are approximately N independent frequency bins and so one should expect to see an F test above the $100(1 - 1/N)$ percent level just from sampling considerations. Thus, from a sample of 1000 observations from a purely non-deterministic process, one would expect some frequency to have an F test exceeding the 99.9% point so that one should not try to interpret every peak. Alternatively, when frequencies are specified *a priori* to an accuracy of approximately the Cramér-Rao bound for frequency estimation, the nominal significance levels apply.

When periodic components are detected with high significance, it is often desirable to "reshape" the spectrum. The major reason for this is that, if one does not reshape as discussed below, the leakage from the strong periodic signals biases the estimation of the power of the remaining frequencies.

To reshape the spectrum at a periodic component at a frequency of f_0 with an estimated amplitude of $\hat{p}(f_0)$, replace the eigencoefficients by

$$X_k^{(r)} = X_k(f) - \hat{\mu}(f_0) V_k(f - f_0)$$

for frequencies in the neighborhood of f_0. Next, replace the power so removed in the spectrum at the single frequency f_0.

Outlier Detection

A pervasive problem in the analysis of time-series is distortions of the estimated spectrum caused by "outliers" in the observations. A given series may be intrinsically non-Gaussian, so that higher-order spectra are needed to characterize it accurately. More commonly, as is the case with the data considered in this paper, a majority of the observations seem to be reasonably Gaussian, but a few isolated observations and intervals do not. Detection of outliers in time series using a "robust filter" approach has been treated in Thomson (1977); Kleiner et al. (1979); Martin and Thomson (1982).

This method begins by computing an estimate of the spectrum as discussed above, Fourier transforming this estimate to a corresponding estimate of the autocovariance sequence, and finding an autoregressive model that matches the first few autocorrelations. Next, the one-step prediction errors are computed from the AR model and the original data and outlying, i.e. non-predictable, points replaced with their predictions. This "cleaned" series is then used to make a better estimate of the spectrum, and so on. While this method usually works well, the predictor can get "lost" when outliers appear in patches, and a somewhat better method is to replace the AR model and predictor with an empirical Karhunen-Loève expansion. Thus, we compute the Karhunen-Loève eigensystem for a sequence of $M \ll N$ points. Denote the eigenvector $\psi_n(t)$ of the autocovariance estimate by:

$$\theta_n \psi_n(t) = \sum_{m=0}^{M-1} R(t-u)\psi_n(u)$$

standardized by:

$$\sum_{u=0}^{M-1} \psi_n(u)\psi_k(u) = \delta_{n,k}\,.$$

Next, we replace the predictive stage in the robust filter by expanding the time series in a sliding block of length M starting at a base time t_0; the expansion coefficients are

$$c_n(t_0) = \sum_{u=0}^{M-1} \psi_n(u)x(t_0+u)\,.$$

The second-order statistics of these coefficients are given by:

$$\boldsymbol{E}\{c_n(t_0)c_j(t_0)\} = \theta_n \delta_{n,j}$$

so that under the Gaussian assumption the data-dependent part of the log-likelihood function is

$$L(t_0) = \sum_{n=0}^{M-1} \frac{c_n^2(t_0)}{\theta_n}$$

which should have a χ^2_M distribution. The sensitivity of the likelihood at the base time t_0 to a particular observation, say at point i, is just

$$\begin{aligned}\frac{\partial L(t_0)}{\partial x(t)} &= \sum_{n=0}^{M-1} \frac{c_n(t_0)}{\theta_n} \frac{\partial c_n(t_0)}{\partial x(t)} \\ &= \sum_{n=0}^{M-1} \frac{c_n(t_0)}{\theta_n} \psi_n(i - t_0)\end{aligned}$$

and the dependence $x(i)\frac{\partial L(t_0)}{\partial x(i)}$. We average over base times to obtain

$$\bar{D}(i) = x(i) \cdot \frac{1}{M} \sum_{t_0=i+1-M}^{i} \frac{\partial L(t_0)}{\partial x(i)} .$$

This process is nonrecursive and so cannot get "lost" as the predictor does, is as effective at identifying discrepant observations, and is somewhat more sensitive. For outliers, M or more points from the ends of the series, the asymptotic sensitivity of this procedure tends to the interpolation variance of the process, as opposed to the prediction variance of the robust filter, and is consequently more sensitive. In the plots shown here we have taken M to be 12, that is, a one year span.

3. The Economic Issues and the Role of Spectral Analysis

Notwithstanding the long history of the success of spectral techniques in many disciplines and notwithstanding the pioneering work of Granger, see for example, Granger and Hatanaka (1964); Granger and Morgenstern (1963); Granger and Rees (1968), until recently there has been little interest shown by economists in the technique. This may be due to the traditional reluctance of economists to use tools that are not model specific. The discussion in the previous section may provide some motivation for economists to reconsider this position.

The key issues in the finance literature for time series of stock prices have been the relevance of the random walk hypothesis, the more recent concept of mean reversion, the relationship of stock prices to the business cycle, and the potential for forecastability. Granger and Morgenstern (1963) demonstrated that seasonal and business cycle components were of little importance, the random walk hypothesis was reasonable for short to intermediate run time scales, and that there was very weak evidence for forecastable components. Durlauf (1991) returned to the analysis of the "martingale" hypothesis for stock prices and found some evidence against a random walk, as did Lo and MacKinlay (1988). Daniel and Torous (1990) used conventional spectral techniques to evaluate the relationship between stock price returns and the business cycle; they concluded that there was evidence of spectral coherence at business cycle frequencies, two to eight year periods, but not elsewhere. Stronger results were obtained for "excess returns"; that is, for stock returns that are in excess of a bench mark rate of return.

One fundamental aspect of the situation is that if there are predictable variations in the forecast valuations of future dividends, the operation of an efficient market will arbitrage away the predictable elements. If then, the profitability of the real sector involves a degree of periodicity, the action of an efficient market will remove any such evidence of periodicity in the stock prices. In short, the market acts like a filter on the profitability input signal. This insight leads us to suspect that the transfer function relating the reals of the system, for example, the index of nondurable goods production, to the stock price index will produce nulls in the transfer function for any periodicities that exist in the real production data.

The bulk of the literature on these topics has been based mainly on the use of autocorrelations and ARIMA type models, with, or without, an allowance for conditional heteroskedasticity. The literature is vast, so that we cannot begin to summarize the whole. A few representative and classical contributions will illustrate the broad outlines of the debate. Fama and French (1988) in a classical agenerated by the random walk with drift. Mankiw et al. (1991) reevaluated the analysis allowing for small sample biases in the estimates to conclude that the rejections of the random walk hypothesis were only marginally rejected. Kim and Nelson (1991) claimed that the evidence for mean reversion in the previous articles was the pre war component of the data and that the post war period in contrast was mildly mean averting. Jegadeesh (1990) analyzed individual securities post war to conclude that there is significant negative first order serial correlation, even after allowing for the "January" effect, and that there was significant positive correlation at higher lags.

Another series of papers explored the relationship between variations in excess stock returns and the business cycle, Fama and French (1989); Fama (1990) came to the conclusion that there is a weak relationship at business cycle frequencies. In an interesting paper that generalizes the scope of the analysis from stocks and bonds to foreign exchange, real estate, and precious metals, Cutler et al. (1991) found that excess returns are positively correlated at high frequencies and negatively correlated at low frequencies.

While all of this literature is interesting and informative, there is still considerable doubt about the "stylized facts," especially in so far as the empirical results are at variance with "efficient market" hypotheses. There is sufficient disagreement about the basic facts that alternative ways of analyzing the data would prove to be useful to help resolve the issue. In large part that is what we shall try to do using the improved spectral analytical tools discussed above.

4. The Economic and Financial Data Used in the Analysis

We examined two types of variables. The first, traditional, variable of interest is the "growth rate" of whatever economic or financial datum is to be considered. The growth rate is defined by:

$$X(t) = \frac{y_t - y_{t-1}}{y_{t-1}} \tag{23}$$

where y_t is the tth value of the raw data series and $X(t)$ is the required growth rate. The main reasons for looking at growth rates are that scale effects are eliminated and that, as is well known, the statistical distributions of growth rates are particularly simple; exhibiting at most some low level moving average structure, with a small positive mean value, or "drift term." In all cases the mean was subtracted so that all series have mean zero and are approximately mean stationary. However, the series are not necessarily higher order stationary and often are obviously not second order stationary.

We also examined the logarithms of the raw indices. We can do this using our procedures, because the estimated AR(1) coefficients, while uniformly high, are still sufficiently less than one to enable us to obtain useful spectral results.

The variables that were examined using the new procedures included:

- monthly Standard & Poor's 500 stock market index, 1926–1987
- monthly index of durable goods production, 1919–1987
- monthly index of nondurable goods production, 1919–1987
- monthly index of manufacturing output, 1919–1987
- monthly index of mining output, 1919–1987

The production index for "total production" was initially considered as well as the series listed above. However, we discovered that the degree of coherence between this series and that for manufacturing was extremely high and that the phase difference was approximately zero at all frequencies. Consequently, we concluded that very little additional information would be obtained from including this series. All further analysis of the "total production" series was dropped.

5. The Empirical Results

Figures 1 and 2 illustrate for nondurable goods and for the S&P 500 indices the time paths of the data, both in terms of the raw data, that is, in terms of the indices themselves, and in terms of the growth rates. The corresponding figures for manufacturing, mining, and durable goods are similar to that for nondurable goods. Figure 1 shows the time series for consumer nondurable goods production. Two aspects of the data are immediately obvious; the decline in the amplitude of oscillation of the growth rates during World War II and the enhanced periodicity of the growth rates post World War II. The plots for the other real time series, while similar, have the following differences that should be noted. For the manufacturing index the variation of the growth rate declined during the war, but the level increased substantially during that period. For the mining index, the variation of the growth index did not decline during the war, but did do so rapidly after the war. Further, the mining series is unique in appearing to have peaked out in the levels by 1980.

The S&P 500 index in contrast to the indices for the reals shows an increase in variability in growth rates during the late post war period and the rate of increase in the levels has increased substantially since the beginning of the '80's, see Figure 2.

Initial Spectral Analysis and Outlier Detection

Figures 3 to 9 summarize the initial results of the spectral analysis.

As was to be expected, the growth rate index for the S&P 500 stock price shows very little evidence of any cycles, although relative to spectra for noise the S&P 500 spectrum has more of a shape that is characterized by narrow bands of power separated by steep declines. The spectrum has too much variation to be regarded as noise. The spectrum for "noise" would be more irregular on a fine scale, but without the sequence of steep declines in power that are apparent in Figure 3. This figure also portrays the jackknifed estimates of the standard deviation in terms of an approximate 95% confidence region. Further, the width of the confidence region between the jackknifed bounds is wider than one would expect from a noise process. In the F-test for periodic components there are some frequencies that are statistically significant, although the frequencies do not seem to be recognizable. One might well conclude that these frequencies are to be expected by chance, but the clustering may indicate some more systematic underlying mechanism.

In Section 2, we discussed the relative ability of our procedure to handle time series with convergent, but slowly decaying autocorrelation functions. Figure 4 demonstrates this contention with respect to the logarithms of the Standard and Poor data. Figure 4 shows the same type of plot as in Figure 3, but in this instance for the logarithms of the Standard and Poor data; that is, these plots are for the undifferenced data. In Figure 4a, we see that the spectrum drops off sharply from very high values at zero frequency, about 10^4, but that the rate of decline slows dramatically after about 1.5 cycles per year. Clearly these data are not representative of 1/f noise. Also shown in Figure 4a is an overlay of the results from estimating a smoothed periodogram using the regular FFT procedures; this graph helps the reader to appreciate the inferential gains from the use of our procedure. Figure 4b indicates the results of the F test for periodic components using the undifferenced data. Interestingly, the difference in results is to indicate relatively greater weight for lower frequencies to be significant, than for the higher frequencies. However, there is no particular pattern to the rejections to the null hypothesis so that it is not clear whether these results are due solely to chance.

Figure 5 for the log data presents four graphs that in turn indicate the autocorrelation decay, the corresponding partial autocorrelation function, and the degree of fit and the relative prediction error resulting from an AR fit of order 26.

As is to be expected of these stock market data in log form, the autocorrelation dies fairly slowly, but the partial is nearly zero after the first lag. A 26 order AR fit provides a close reproduction of the original series at least as

Figure 1. Durable goods index and the associated growth rate plotted as time series.

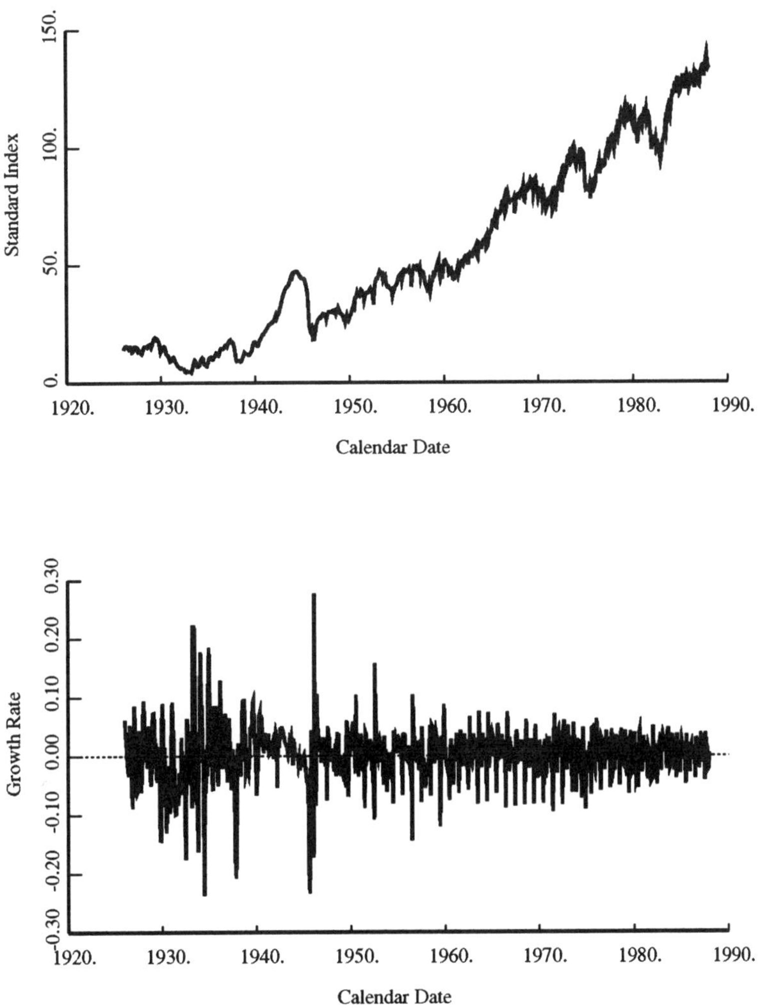

measured by a comparison of the spectra of the raw series and a simulation of 26 lags. The plot of relative prediction error indicates no predictive power at any lag as was to be expected.

To one extent or another all the real indices reflect the presence of "seasonal cycles" with frequencies of one to five cycles per year. The spectral plot for the nondurable goods growth rate index is shown in Figure 6; where the spectrum is presented after reshaping. The "seasonal frequencies" of two, three, four

Figure 2. Standard and Poor's 500 stock market index and associated growth rate plotted as time series.

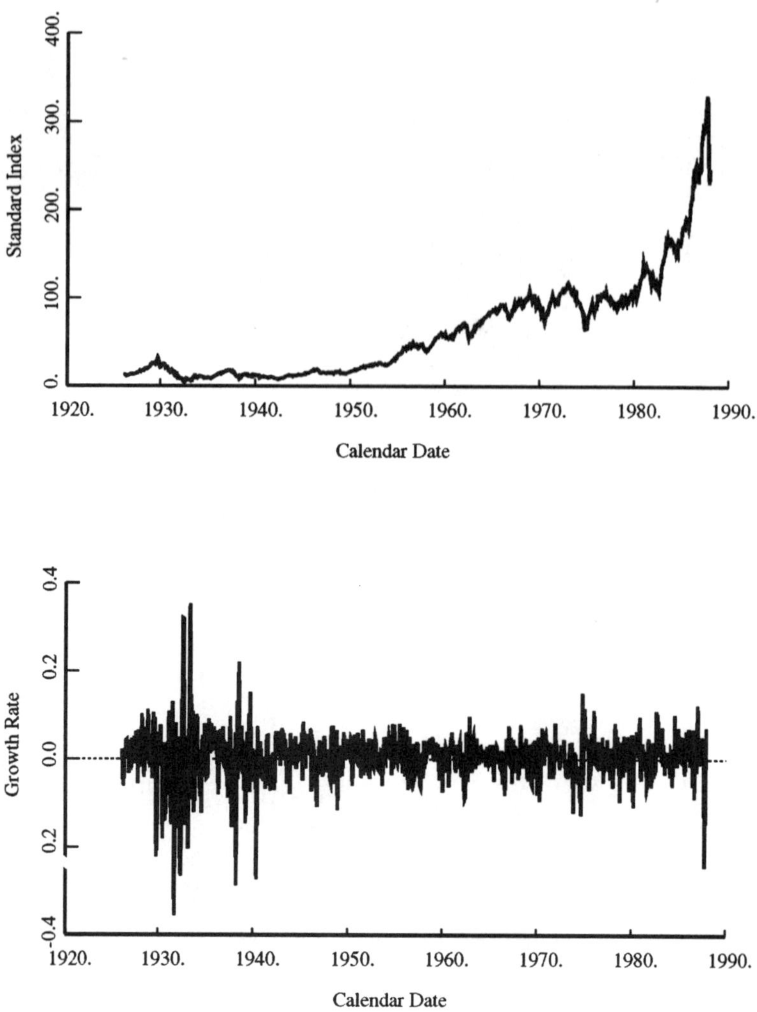

Figure 3. The upper frame shows the spectrum (solid curve) for the Standard & Poors growth curve series. The lower and upper dashed curves are the 5 and 95% confidence limits determined by jackknifing over windows. The lower panel shows the F-test for periodicities, peaks where the test is significant above the 95% point are labelled.

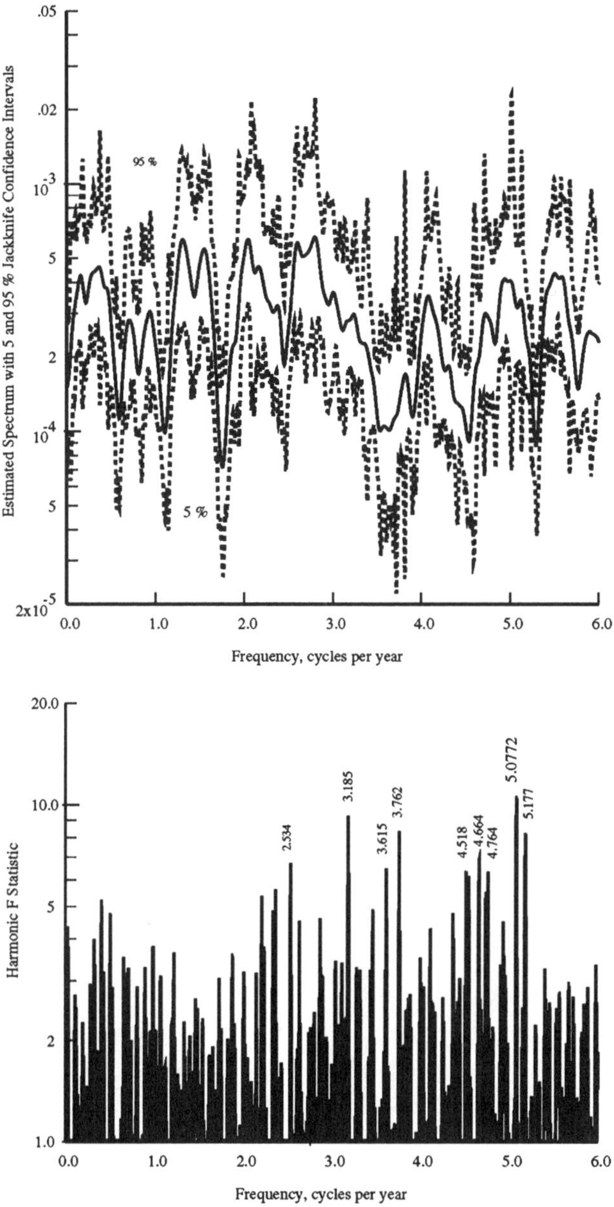

Figure 4. Spectrum and F-test for the logarithm of the S&P standard index. The dashed curve in the top curve is a standard periodogram. The offset at higher frequencies is bias.

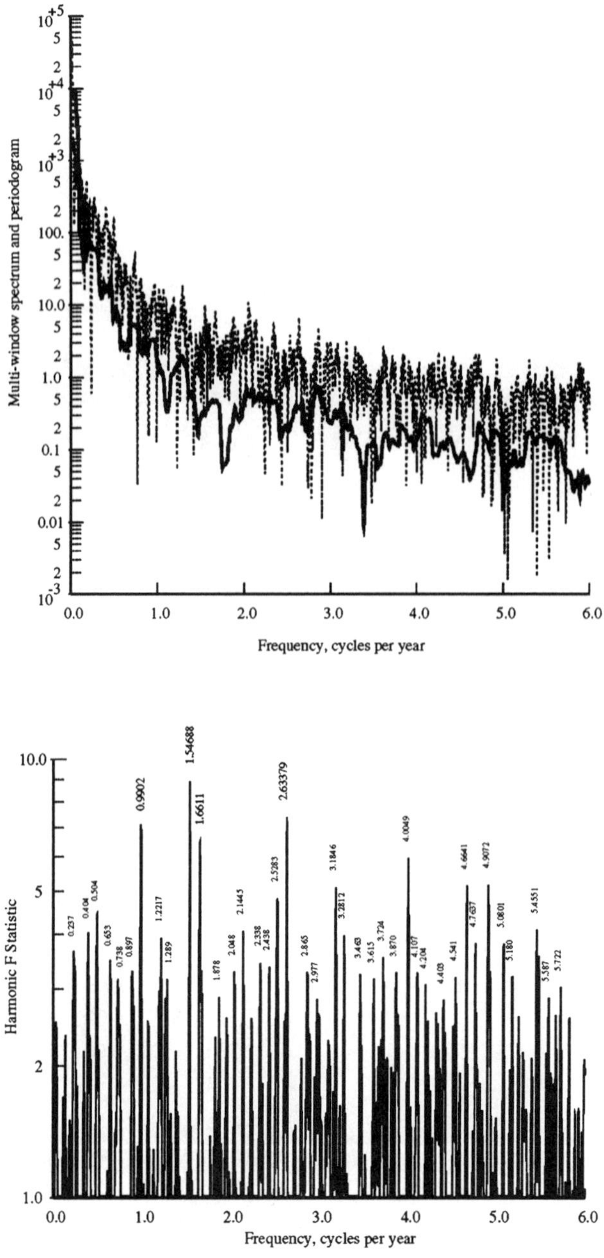

Figure 5. Multitaper autocorrelation, and related quantities for the logarithm of the S&P index.

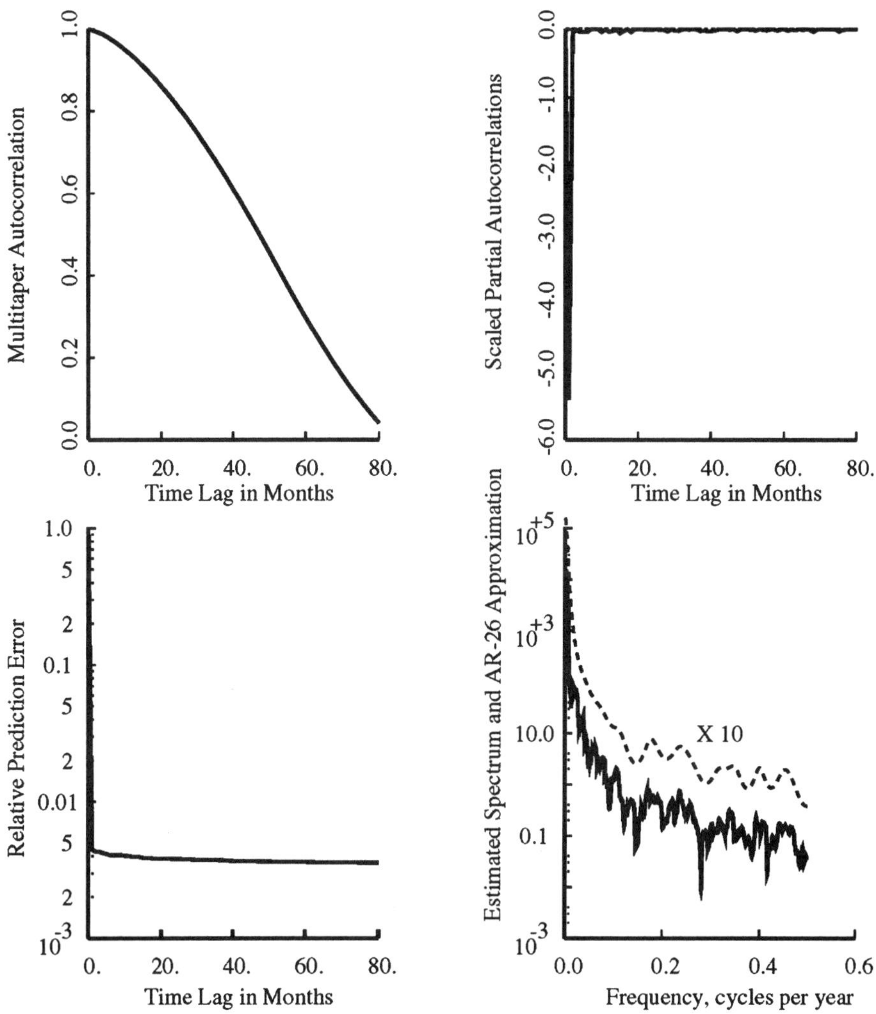

and five cycles per year are very apparent and the F-test results confirm the visual impression. The extent of the "shoulders" in the neighborhood of each periodic frequency is also evident. "Shoulders" represent significant levels of power at frequencies in the neighborhood of the line spectrum. There is weak evidence of power at business cycle frequencies and the background variation is not white.

Both from the general shape of the spectrum and from the harmonic F-test statistics shown in Figure 6, the evidence is clear that the background variation is non-white.

The "shoulders" at the base of the periodic spectral lines that appears in both sets of plots, Figures 6 and 7, indicate that the frequencies are not purely periodic; this is in agreement with the analysis of Ramsey (1992) in which the variation of nondurable and durable goods indices are modeled as forced oscillators; the forcing terms are the line frequencies that are below a year in period and the oscillator reaction to the forcing modifies the pure sinusoidal appearance of the forcing terms.

In contrast, Figure 7 shows the corresponding results for the undifferenced nondurable goods data. In Figure 7a, we have once again overlaid our spectrum with that produced by the usual smoothed periodogram; the benefits from our procedure should be very clear, even visually. In comparing Figures 6 and 7, one notices that the seasonal frequencies in the undifferenced form show clear evidence of narrow bands of frequencies. The background is still not white and the shoulders are more clearly delineated. In comparing the harmonic F tests using the differenced and the nondifferenced data, we observe more structure in the nondifferenced data.

The fundamental frequency and its harmonics are still present, but there is now a second level of frequencies. The second level includes a relatively low frequency of 0.428 and one that is approximately 1.3, but more interesting is the presence of the sequence of 2.159, 3.188, 4.187. At the moment we do not know the source of this sequence of frequencies.

In Figure 8, we see the autocorrelation and partial autocorrelation plots for nondurable goods. We note that in addition to the expected first order lag, there is also a substantial partial autocorrelation at a lag of twelve months that induces a corresponding significant drop in the relative prediction error. Finally, the spectral analysis indicates that at an order for the AR process of 26, we observe that there is very little difference between the data and the fitted AR process.

Because the graphs for manufacturing and durable goods are very similar to those for nondurable goods, we have not presented the graphs to support the discussion in this paragraph. With respect to the index of manufacturing, the analysis of the differenced data gave weak evidence of cycles of length 0.14 and 0.275 cycles per year, that is, cycles of periods of about seven and three and a half years; in addition of course to the well pronounced cycles produced by the harmonics of one twelfth of a year.

If we examine the graphs for the analysis of the nondifferenced index of manufacturing we observe, in addition to results that are very similar to those discussed above, considerably more significant frequencies below one cycle per year. Further, we also observe the sequence of frequencies, 2.1885, 3.1875, and 4.1865. This second set of evidence lends credence to the view that this set of frequencies represents some unsuspected mechanism in the short term dynamics of production.

Figure 6. Reshaped spectral plot and associated F statistics for non-durable goods index.

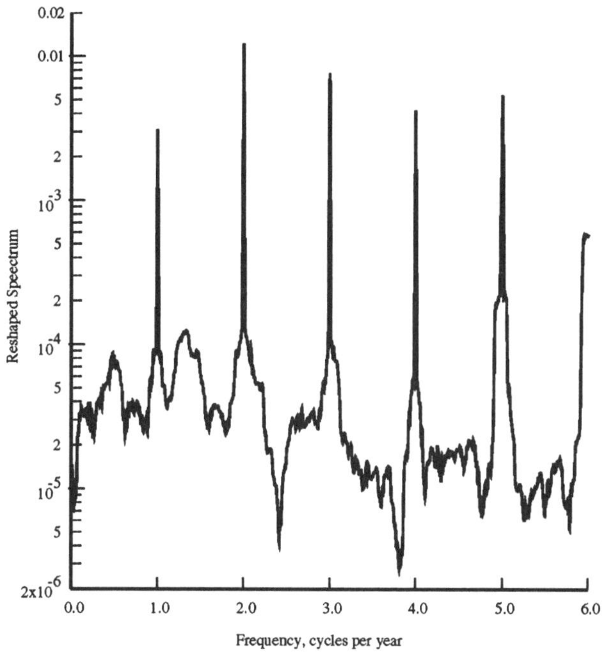

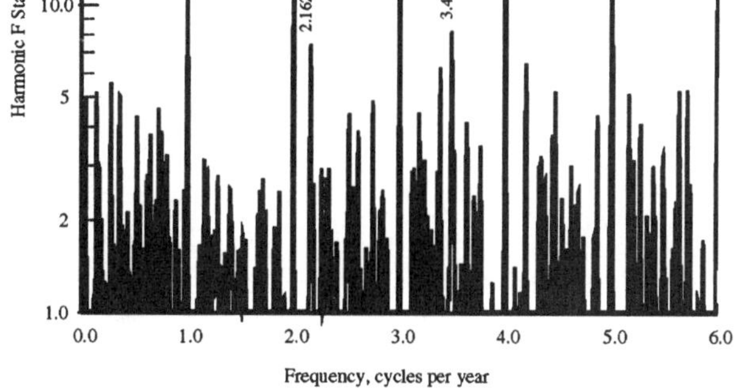

Figure 7. Multitaper spectrum and periodogram (upper panel) and F-test (lower panel) for the non-durable goods series.

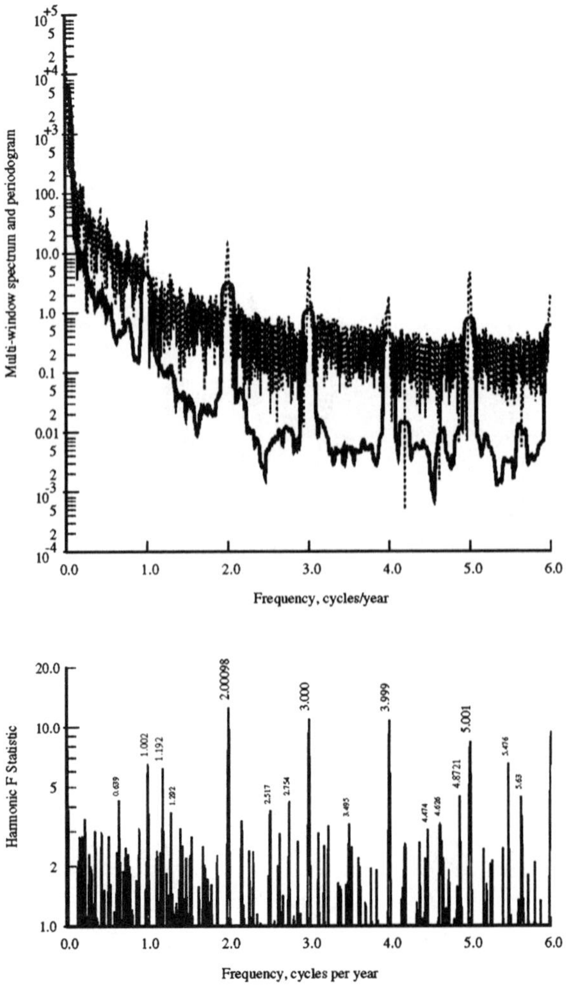

The analysis of outliers that is illustrated in Figure 9, indicates a substantial degree of conformity between the various real indices. There are few surprises in these results. 1946 and 1947 have strong outliers in both durable and manufacturing indices; the 1930's are outliers for both the reals and for the Standard and Poor's 500 index. Another interesting result is that for non-durable goods there is a moderately strong outlier that matches the period of the minor stock market shock in the mid-seventies.

The outliers for the mining index are somewhat different. 1922 is an extremely large outlier, that is no doubt due to the extensive and bitter labor

Figure 8. Multitaper autocorrelation and related quantities for the non-durable goods index. Note the large partial coherence at a 13-month delay.

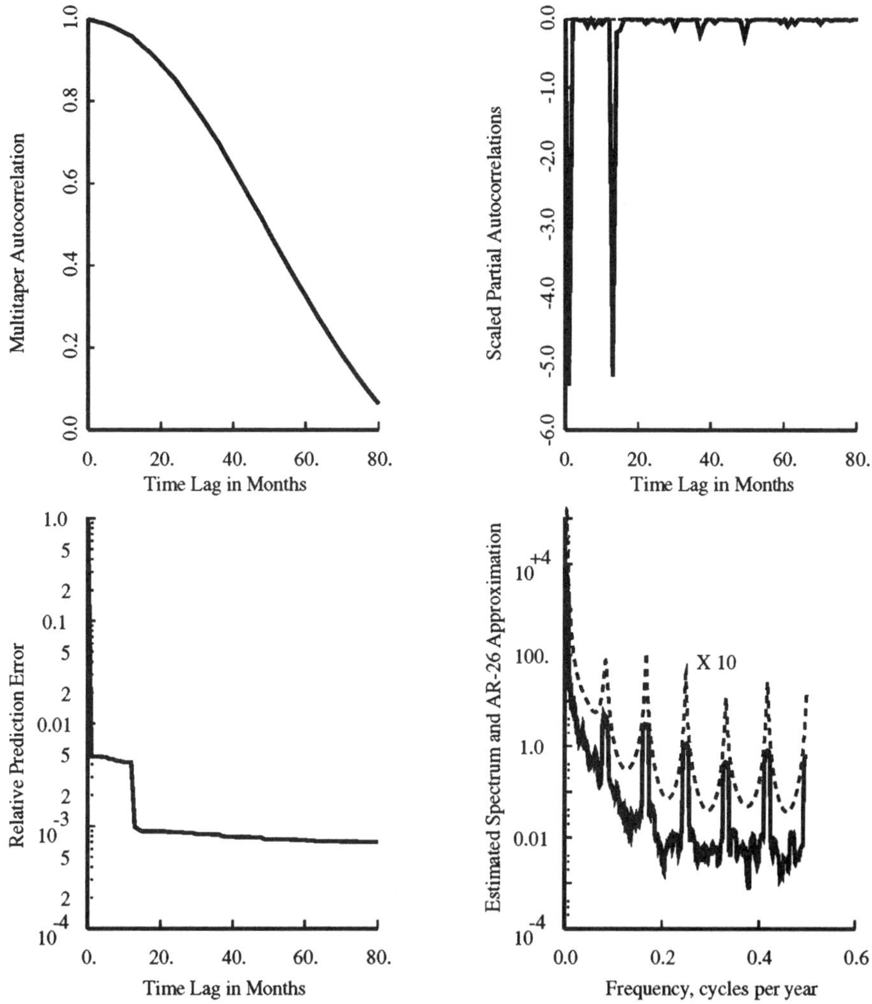

disputes during that year. The remaining outliers are relatively minor, mainly in 1941, 1946, and 1950.

Figure 9. Time series plot of the outlier test for two time series made using a 12-month moving Karhunen-Loève expansion. Upper panel: S&P 500, Lower panel: Durable goods index.

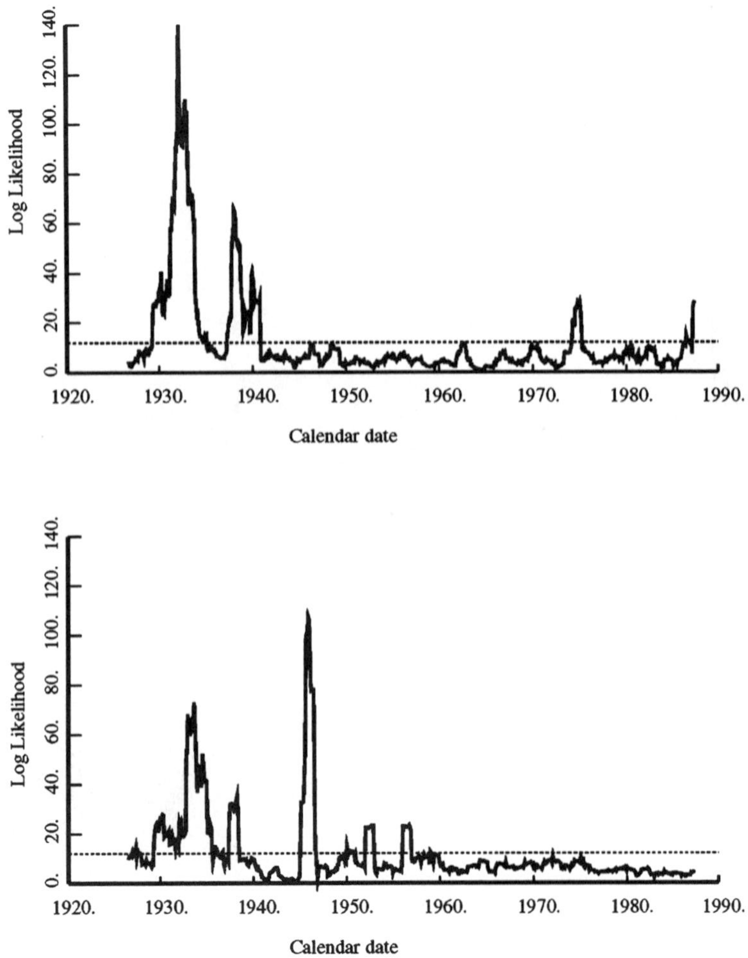

6. Analysis of Coherence and Phase Between Pairs of Series

Results for the Growth Rate Series

Figure 10 summarizes the results of the analysis of the coherence and relative phase between pairs of series for the differenced data. This figure shows the coherency calculations for the post war data, that is, from January 1948 to January 1988. These results are compared in the discussion that follows to the results that were obtained using both the pre and post war periods. In all the plots in Figure 10, the dark lines are the coherences and the dotted lines are one standard deviation below the coherency estimate; the standard deviation was estimated by the jackknife procedure.

Figure 10. Four coherence plots. Post-war growth curves are used here. The outer left-hand scale shows magnitude-squared coherence; the inner left an inverse tanh transform to standard normal units, and the right-hand scale the cumulative distribution for independent series. The solid curves are the estimate, and the dashed curve is one jackknife standard deviation lower.

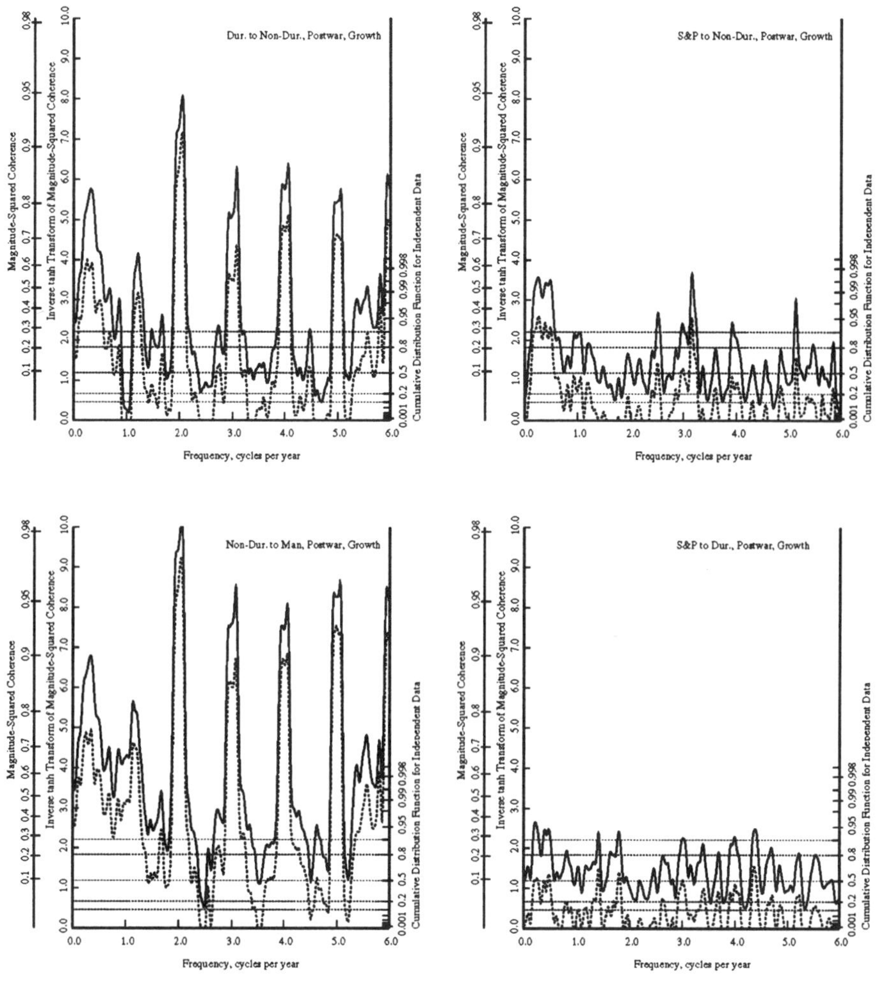

Non-Durable Goods and Manufacturing; (Figure 10, lower left). The overall level of coherence is very high. The frequencies of 0.33, 1.3, 2.0, and 5.5 stand out. Extending the sample period back to 1926 strengthens these results. The phase oscillates about zero, indicating that there is very little overall time difference between the two series. The spikes in the phase plot are correlated with the lowest coherencies where the resolution of the estimates for the phase are very low, so that they are of little significance.

Manufacturing and Durable Goods; (not shown). The overall degree of coherency is very strong, much stronger than for nondurables. The null hypothesis of zero coherence is rejected at better than a probability of 0.001 for all frequencies. The key resonating frequencies between the two series are at 0.33, 1.3, and 5.5 cycles per annum. Including the pre war period modifies these results slightly in that a resonance at 0.1 is much stronger than the 0.33 for the post war period and the seasonal peaks at 3 and 4 cycles per annum are clearer for the whole period. For the post war period, phase remains essentially constant.

Manufacturing and Mining; (not shown). While the results are not as clear as in the previous two cases and the overall level of coherency is not as strong, the same major frequencies appear; 0.33, 1.3, 5.5 (5.4 in this case). Given the lower level of the degree of coherence, it is not surprising that the phase should be more erratic, the large deviations from zero being correlated with the large drops in coherency. Including the pre-war period does not substantially modify these conclusions about coherency. However, while the two series are in phase up to two cycles per annum, the two series rotate out of phase over the two to three cycle frequency and again between 4.5 and up. In short, there seems to have been some change in the higher frequency dynamics of mining between the pre and post war periods. The onus of change has been placed on mining since there is no evidence that the change involved manufacturing when comparing phase shifts between manufacturing and mining and between mining and the other series.

Non-Durable and Durable Goods; (Figure 10, upper left). The overall level of coherency is very high, although not as high as between the manufacturing and durable goods indices. The overwhelmingly largest coherence is at 0.33 cycles per annum, although 1.3, 2.0, and 5.5 are also prominent. Phase is essentially constant over all frequencies.

S&P 500 and Durable Goods; (Figure 10, lower right). As might be expected given the analysis so far, the major frequency that is coherent is 0.33, the business cycle frequency. As is true for all the coherency plots involving the S&P 500 index, the overall background level of coherency is low, whereas the background coherence between the reals is high. The phase diagram for the phase relationship between S&P 500 and durable goods indices oscillates about zero; but the amplitudes are large due to the low level of strength in the overall coherency. The pre and post war coherency and phase plots are similar.

Non-Durable Goods and S&P 500; (Figure 10, upper right). The degree of coherence is overall greater for the non-durable goods index relative to the durable goods index. The prominent frequencies are 0.33, 3.15, and 5.1, although there is far more power at the low, business cycle, frequency level than elsewhere. Adding the pre war data yields more statistically prominent frequencies, but these are the same frequencies with a few other seasonal frequencies that were observed for the pre and post war data for the S&P 500 and Durable Goods indices. For all three sets of real indices the entire data sets, that is, the union of the pre and post war series, yield the frequencies of 0.1, 0.35, 0.6, 0.9, and 1.3 cycles per year. For the post war period only, these frequencies collapse onto a broad base level of power at 0.33. The phase variation for the combined pre and post war data shows that the frequencies below one cycle per annum are in phase.

S&P 500 and Manufacturing; (not shown). For all of the remaining results that involve coherencies between the S&P 500 index and the reals, the overall level of coherence drops dramatically. However, there are some moderately strong coherencies at certain frequencies; but most often at low, business cycle, frequencies. The union of the pre and post war periods produced stronger coherencies overall. The variation in phase for S&P 500 and Manufacturing is very similar to that for the Durable Goods index.

S&P 500 and Mining; (not shown). The coherency between S&P 500 and the Mining indices demonstrates the least amount of overall coherence. Only two frequencies exceed the 0.05 critical region for the test of the null hypothesis of zero coherence and those only just. The phase plot is not informative as the level of coherence is so small.

Results for the Nondifferenced Logarithms

In this section we summarize the differences in results for coherency measures that were obtained by using the undifferenced, but logarithmic transformed, data. The results are summarized in Figure 11, a panel of four graphs that match Figure 10 in its layout to facilitate comparisons. For the coherences between nondurable goods and manufacturing and for that between durable and nondurable, see the panels on the left, the most striking features are that the coherences are much higher and the resolution at the harmonics of two, three, four, and five cycles per year are remarkable. There is still a significant coherence between durable and nondurable goods at a cycle of 1.3 cycles per year. Evidence for coherence between the Standard and Poor series and durable and nondurable goods is much stronger for the nondifferenced data at all frequencies. But this is particularly so at the lower frequencies; perhaps an unremarkable result under the circumstances. At the annual harmonics it is only at three and four cycles per year that there is strong coherence between the Standard and Poor index and those for durable and nondurable goods.

An Initial Examination of Nonstationarity

Nonstationarity is a pervasive problem, especially in economic and financial data. So far in this paper we have pretended that the data form a sample from a stationary process. In this section we review some preliminary analysis of these difficulties with respect to the growth rates. A more intensive and detailed analysis will be presented in a subsequent paper. We can begin by noting a few obvious facts.

First, plots of the growth curves show obvious nonstationarity. Compare, for example, the era from about 1936 to 1946 for the durable goods growth rate, see Figure 1, with that from 1950 to 1975.

Second, the outlier tests, see Figure 9 are, except for the 1973 oil shock, much more volatile during the depression and World War II than they have been since about 1960.

Third, the jackknife variance estimates for the various spectral density estimates are larger than one would expect for a stationary Gaussian process. Strict normality of the growth rates themselves is not particularly important in this regard. This is because we are dealing with large samples and because the eigencoefficients represent narrow band filters applied to the data so that their distribution will be close to Gaussian, Mallows (1967).

We approached the problem of nonstationarity in several different ways. As a first step, the spectrum was estimated by epochs for each series and the spectral results compared over epochs. The second procedure was to examine the variation in frequencies over time by complex demodulation. Particular attention was paid to "business cycle" frequencies, that is, we examined demodulates at a frequency of 0.35 cycles per year with a low and high pass band of 0.2 to 0.5 cycles per year. We supplemented these results with some formal tests for nonstationarity against the null of stationarity.

The spectra for both Non-Durable and Durable Goods indices were reestimated for the time blocks 1919–1939, 1939–1959, 1959–1979. In both cases the degree and sharpness of the estimates of the within year frequency components were enhanced each time in moving from the earlier to the later periods; the Non-Durable Goods index improving faster than the Durable Goods index. An annual cycle became more pronounced for the Non-Durable Goods index, but not for the Durable Goods index.

Manufacturing and Mining indices were reexamined for epochs of 1926–1946, 1946–1966, 1966–1986. The spectral variations for the Manufacturing index were similar to those obtained for the Durable Goods index. The spectral variations for Mining are similar to that for Manufacturing, except that the low frequencies, especially a one year cycle, were substantially reduced in power.

The three epochs for the S&P 500 index were 1926–1946, 1946–1966, and 1966–1986. There is considerable variation over the three epochs. The first epoch has far less structure than the other two. While the overall level of power

Figure 11. Coherences computed between logarithms of the basic series. The scales are as in Figure 10.

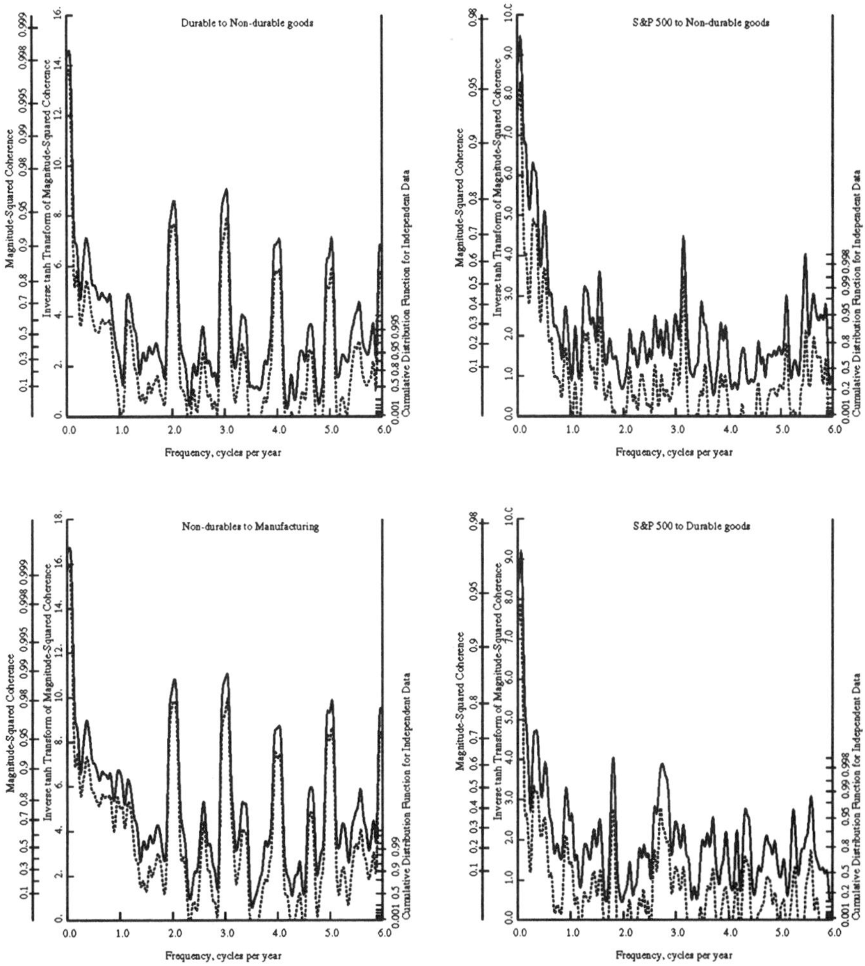

has declined since the 'thirties, the low frequency components have become relatively more pronounced. Cycles of about three, four, and five cycles per year were pronounced during the early post war recovery, that precision of cycle estimate was reduced thereafter.

The result of examining complex demodulates can be illustrated by comparing at two benchmark frequencies the demodulation of the S&P 500 and nondurable indices. Figure 12 compares the demodulates for a frequency band of 1.1 to 1.75 cycles per year.

Figure 12. Complex demodulation of the spectrum and its variance for two time series at a frequency with high coherency; 1.1 to 1.75 cycles per year.

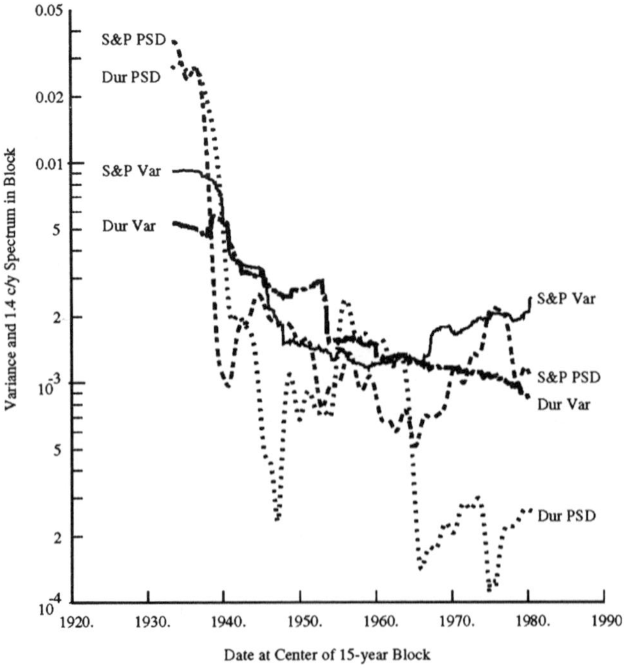

Figure 12 shows the variances and spectrum estimates made on moving 15-year blocks for each of the Standard and Poor's stock index and the nondurable goods index. The spectrum estimates were made using four windows on each block. In Figure 12 the rapid decline in the amplitudes of both spectra coming out of the war is clearly evident. The variances of both series are remarkably constant. While the amplitude of the spectrum of the S&P 500 index remains quite steady for the remainder of the period, the amplitude for the spectrum of the nondurable index declines dramatically to the mid-sixties, after which point in time the amplitude rises again quite rapidly. These results are even more pronounced for the business cycle frequency comparison of 0.2 to 0.5 cycles per year. The spectrum and variance estimates for the S&P 500 track each other reasonably, but the range of variation for both over time is too large for the process to be considered stationary. In contrast, the variance of the nondurables growth rate remains reasonably constant, within a factor of 2, whereas the spectrum in both frequency bands changes by more than an order of magnitude; once again the results are inconsistent with stationarity.

Instantaneous Frequency and Nonstationarity

In Figure 13 we present two examples of calculating the "instantaneous frequency" variation about a benchmark frequency using complex demodulation. If we were observing a stationary process, we would observe in the absence of random fluctuations, a straight line. The extent to which there is oscillation about the benchmark frequency during the historical period indicates the degree of nonstationarity of the data. Using the logarithm transformed data, we present just two illustrative examples for manufacturing for the frequencies denoted by two and three cycles per year. The results for the other time series were very similar.

The most striking feature is the variation in frequencies induced by the reorganization of production precipitated by World War II. There is also some, but much weaker, evidence for a similar oscillation in frequency induced by the Korean war. While it is almost a "truism" to state that economic and financial data are nonstationary, it is an advance to be able to relate specific forms of nonstationarity, in this case frequency drift, to specific real world events. Previous, but unpublished work, by the authors using the same data, but different techniques, revealed that there seems to be a characteristic war spectrum; that is, the frequency spectrum during war periods is different to that prevailing in peace time. The results illustrated in Figure 13 give an example of this process. An interesting question raised by these results is whether a close examination of European data would reveal even more striking results.

Nonstationary Quadratic-Inverse Estimates

In order to motivate quadratic inverse theory, recall that the basic problem in spectral estimation is that of estimating a function, the spectrum, from a finite set of observations. Spectra need not be continuous nor amenable to description by a finite set of parameters. Consequently, the problem as stated is ill posed. Linear inverse theory as sketched in equations (16)–(18) in terms of the solution of equation (10), is similar to the Backus-Gilbert theory. The spectrum is a quadratic function that describes the second moments of the data. The purpose of quadratic inverse theory is to estimate these moments efficiently.

The first satisfactory quadratic-inverse estimates for stationary processes appeared in Thomson (1990). The stationary test statistic $Q_N(f)$ that is defined in equation (24) has $L-1$ degrees of freedom. Nonstationary quadratic inverse theory is briefly summarized in an Appendix.

$$Q_N(f) = \sum_{l=0}^{L-1} \alpha_l \left[\frac{\hat{a}_l(f)}{\bar{S}(f)} - tr\, A^{(l)} \right]^2 \tag{24}$$

Figure 14 (upper panel) is a plot of this test against frequency for the durable goods growth curve. The test statistic that is defined in equation (24) should be, and in simulations appears to be, asymptotically distributed

Figure 13. Instantaneous frequencies of the manufacturing series demodulated at 2 and 4 cycles per year. Dashed curve – raw, solid smoothed.

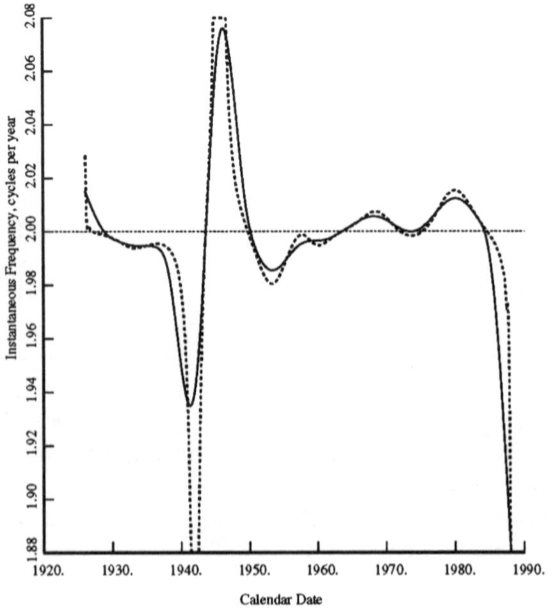

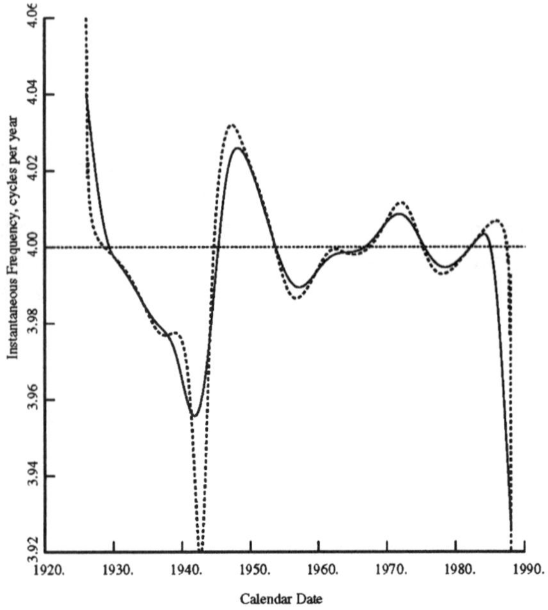

as χ_5^2. For the highest levels of the frequency band the test statistic values are not unreasonable under the hypothesis of stationarity. The frequency regions in narrow bands about multiples of one cycle per year are indicative of the "superstationary" behavior of a periodic component; that is, there is less variation than one would expect in a stationary process. The lower frequency region, however, particularly the part from .5 to nearly 3 cycles per year, but excluding 2 cycles per year, are highly nonstationary and the test statistic is above the 95% level over most of this range.

Figure 14 (lower panel) shows one component of this test, the ratio of the nonstationary quadratic-inverse coefficient $a_1(f)$ to the standard multiple-window spectrum estimate $\bar{S}(f)$. This ratio is, loosely speaking, the *time derivative* of the log spectral density. For stationary processes such a concept is not meaningful but, once the process is admitted to be nonstationary, this ratio shows the relative change in power at a given frequency with time. In this situation the second and fifth harmonics of the annual cycle are relatively stable, the third and fourth are increasing, while the background continuum process has been mostly decreasing. The marked series of spikes near 0.25, 0.5 and 0.75 cycles per year are interesting; although we do not as yet know the source of these results.

The same test $Q_N(f)$ was applied to the S&P 500 index for the entire 1926–1988 period. There is evidence for nonstationarity at low frequencies, in particular at periods of about five years. There is also evidence of nonstationarity at low frequencies, in particular at periods of about five years. There is also evidence of nonstationarity for frequencies of from one to three cycles per year. The test values imply that variations between periods of one month to a quarter have been reasonably stable.

7. Conclusions

The real indices have high coherence between themselves and exhibit strong seasonal components at the harmonic frequencies of one to five cycles per year. There is also evidence, but not as strong, for a cycle of 1.3 cycles per year. However, these frequencies are not entirely harmonic and the background noise is not white. There is weak evidence of business cycle frequencies at both three and a half years as well as at seven years for some sequences.

The Standard and Poor's 500 growth rate index does not contain any seasonal frequencies. This is in accord with the efficient markets hypothesis. Nonetheless, there is evidence of coherence between the stock market index and the indices of production at business cycle frequencies.

Using the nondifferenced data modifies these coherence results somewhat. First, the degree of general coherence is much greater and the seasonal coherences are much more sharply delineated. Further, there is now evidence of coherence between the nondifferenced Standard and Poor index and the

Figure 14. (Upper Panel) A chi-squared (5 d.f.) test statistic for nonstationarity in the durable goods index, plotted as a function of frequency. (Lower Panel) A quadratic-inverse estimate of the time derivative of the log power spectral density.

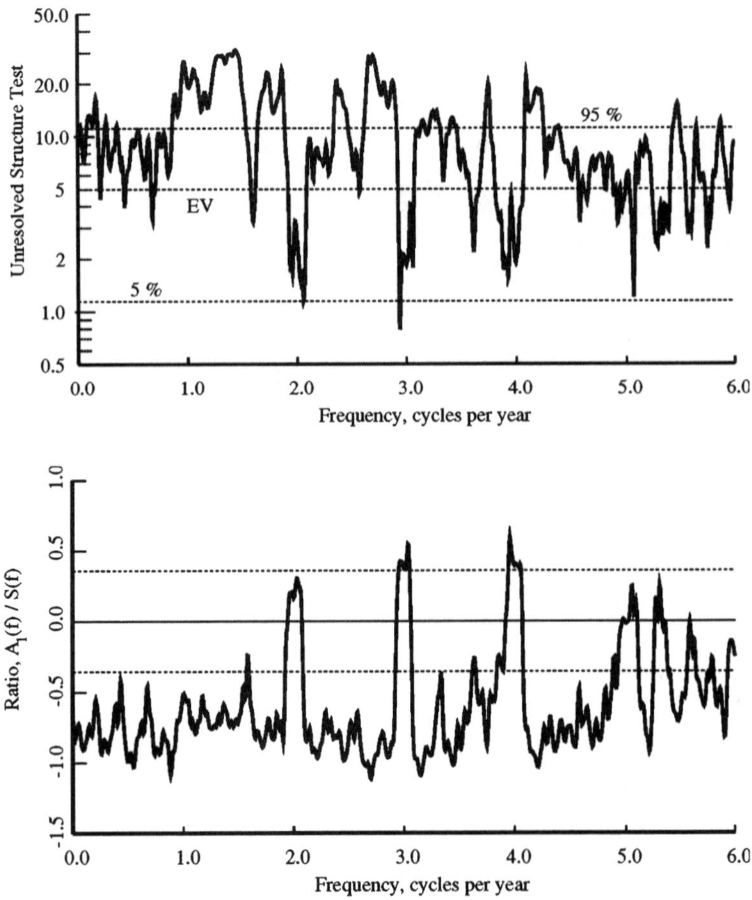

indices for durable and nondurable goods at three and four cycles per year and very strong coherence at very low frequencies.

The evidence for nonstationarity is clear. Besides strong visual evidence and the results from comparing spectra over sub-epochs, we calculated the complex demodulation of the spectra at business cycle frequencies. All the spectra suffered substantial declines in power over the post war period. In addition, the power for the nondurable goods production continued to decline until the mid 'sixties. The post war variances of the spectra remained remarkably constant after the initial decline from the war years. We also examined a Chi-squared test for nonstationarity and found that the evidence for nonstationarity for the durable goods index was strongest at the frequencies between one and three cycles per year. Examining the time derivatives of the spectra, we observe that the rate of increase is positive for the harmonic frequencies and is negative for the very low frequencies; the background noise while declining in power, has been doing so at a decreasing rate.

We also observed using complex demodulation that around certain important benchmark frequencies, there was a very strong oscillation in each benchmark frequency that was induced by World War II and some evidence for a similar effect induced by the Korean war.

The overall lessons are as follows. Both the real production and the S&P 500 index data are nonstationary, particularly at business cycle frequencies. The stock market is definitely efficient at seasonal frequencies in growth rate terms, but has noticeable levels of coherence with the reals at business cycle frequencies. Coherence effects are enhanced using nondifferenced data. The reals are all highly coherent, both at business cycle and at seasonal frequencies. The background noise for all the series is non-white. Finally, our results indicate the importance of being able to use the nondifferenced data in our analysis that revealed more structure in the data and enhanced the clarity and resolution of the results obtained from the differenced data.

Acknowledgments

The financial and institutional support of the C.V. Starr Center for Applied Economics is gratefully acknowledged.

Appendix: The Quadratic-Inverse Test for Lack of Stationarity

Quadratic-inverse theory for stationary processes is described in Thomson (1990). The Cramér spectral representation shown in equation (2), is also valid for nonstationary processes, Loève (1963). To understand the heuristics of the nonstationary theory, Thomson (1993), consider a useful, but little used special case, which is that of a white nonstationary process. This is the time-domain mirror image of the relationship shown in equation (8).

$$E[x(t)x(t')] = P(t)\delta(t - t')$$

where $P(t)$ indicates the power in the signal as a function of time. The eigencoefficients shown in equation (17) that are computed for such a process are:

$$C_{jk}(f) = E[y_j(f)y_k^*(f)] \tag{A.1}$$

$$C_{jk}(f) = \sum_{n=0}^{N-1} \nu_n^{(j)} \nu_n^{(k)} P(n) \tag{A.2}$$

These eigencoefficients are correlated and can be represented by equation (A.2). Estimates of this correlation may be used to estimate $P(t)$. Omitting details, if one generalizes $P(t)$ to the expression $P(f,t)$ and expands it as:

$$P(f,t) = \sum_{m=0} a_m(f) A_m(t) \tag{A.3}$$

then equation (A.1) becomes:

$$C_{jk}(f) = \sum_{m=0} a_m(f) A_{jk}^{(m)} \tag{A.4}$$

where the $A_{jk}^{(m)}$ are a set of matrices obtained by substituting (A.3) into (A.2). This expansion is similar to that used in components of variance. If, mirroring the quadratic-inverse theory, one chooses to define the real basis functions $A_m(t)$ by: (see Thomson (1993))

$$a_m A_m(n) = \sum_{r=0}^{N-1} \left[\frac{\sin 2\pi W(n-r)}{\pi(n-r)} \right]^2 A_m(r) \tag{A.5}$$

subject to the standardization that:

$$\sum_{r=0}^{N-1} A_n(r) A_m(r) = \delta_{n,m} .$$

The corresponding basis matrices used in (A.4) are:

$$A_{jk}^{(m)} = \sqrt{\lambda_j \lambda_k} \sum_{n=0}^{N-1} \nu_n^{(j)} \nu_n^{(k)} A_m(n) \tag{A.6}$$

are real, symmetric and trace-orthogonal; that is:

$$tr\,[A^{(m)} A^{(r)}] = a_m \delta_{mr} . \tag{A.7}$$

If the process is stationary, $P(f,t)$ as a function of t is constant by definition and has a value that is approximately $2W\,S(f)$. One has:

$$E[a_m(f)] = e_m \bar{P}(f)$$

where the bar indicates taking the first moment and:

$$e_m = \sum_{t=0}^{N-1} A_m(t)$$

After some algebra, one can obtain:

$$\operatorname{cov}\left[a_m(f),\ a_n(f)\right] = \frac{1}{a_m}\bar{P}^2(f)\delta_{m,n}$$

Because the eigenvalues a_m are approximately $2NW - m/2$, the a_0 coefficient has approximately the same variance as the mean of $S(f)$ and, for reasonable time-bandwidth products, the next few a_m also have low variance. From these moments, one can show that the individual terms in the equation:

$$Q_T(f) = \sum_{t=0}^{L-1} a_m \left[\frac{a_m(f)}{\bar{P}(f)} - e_m\right]^2$$

are each distributed as χ_1^2 and because P is the estimate of its mean, the sum is approximately distributed as χ_{L-1}^2.

Notes

1. Using the nondifferenced data revealed even stronger and more sharply delineated coherencies.

References

Andrews, D. (1972). *Robust Estimates of Location.* Princeton, NJ: Princeton University Press.

Brillinger, D. (1975). *Time Series, Data Analysis, and Theory.* New York: Holt, Rinehart & Winston.

Brockwell, P.J. and R.A. Davis (1987). *Time Series: Theory and Methods.* New York: Springer-Verlag.

Bronez, T.P. (1992). "On the Performance Advantage of Multiple Spectral Analysis," *IEEE Transactions of Signal Processing*, 40, 2941–2946.

Cutler, D.M., J.M. Poterba, and L. Summers (1991). "Speculative Dynamics," *Review of Economic Studies*, 58, 529–546.

Daniel, K. and W. Torous (1990). "Common Stock Returns and the Business Cycle," working paper, Department of Economics, UCLA.

Doob, J. (1953). *Stochastic Processes.* New York: J. Wiley & Sons.

Durlauf, S.N. (1991). "Spectral Based Testing of the Martingale Hypothesis," *Journal of Econometrics*, 50, 335–376.

Fama, E. and K.R. French (1988). "Business Cycles and the Behavior of Metals Prices," *The Journal of Finance*, XLIII, 1075–1093.

Fama, E.F. (1990). "Stock Returns, Expected Returns, and Real Activity," *The Journal of Finance*, 45, 1089–1108.

Fama, E.F. and K.R. French (1989). "Business Conditions and Expected Returns on Stocks and Bonds," *Journal of Financial Economics*, 25, 23–49.

Granger, C. and M. Hatanaka (1964). *Spectral Analysis of Economic Time Series.* Princeton, NJ: Princeton University Press.

Granger, C. and O. Morgenstern (1963). "Spectral Analysis of New York Stock Market Prices," *Kyklos*, 16, 1–27.

Granger, C.J. and H.J. Rees (1968). "Spectral Analysis of the Term Structure of Interest Rates," *Review of Economic Studies*, 35, 67–76.

Granger, C.W. and P. Newbold (1986). *Forecasting Economic Time Series*. New York: Academic.

Jegadeesh, N. (1990). "Evidence of Predictable Behavior of Security Returns," *The Journal of Finance*, 45, 881–898.

Kac, M. (1959). "Statistical independence in probability, analysis and number theory," in *Carus Mathematical Monographs*, vol. 12. Mathematical Assoc. of America.

Kim, M.J. and C.R. Nelson (1991). "Mean Reversion in Stock Prices? A Reappraisal of the Empirical Evidence," *Review of Economic Studies*, 58, 515–528.

Kleiner, B., R. Martin, and D. Thomson (1979). "Robust Estimation of Power Spectra with Discussion," *J. Royal Statist. Soc.*, 41, 313–351, B.

Lo, A. and A.C. MacKinlay (1988). "Stock Market Prices Do Not Follow Random Walks: Evidence from a Simple Specification Test," *The Review of Financial Studies*, 1, 41–66.

Loève, M. (1963). *Probability Theory*. New York: D. Van Nostrand.

Loretan, M. and P.C. Phillips (1994). "Testing the Covariance Stationary of Heavy-tailed Time Series," *Journal of Empirical Finance*, 1, 211–248.

Mallows, C.L. (1967). "Linear Processes Are Nearly Gaussian," *Journal of Applied Probability*, 4, 313–329.

Mankiw, N.G., D. Romer, and M.D. Shapiro (1991). "Stock Market Forecastability and Volatility: A Statistical Appraisal," *Review of Economic Studies*, 58, 455–577.

Martin, R. and D. Thomson (1982). "Robust-Resistant Spectrum Estimation," *Proc. IEEE*, 70, 1097–1115.

Pagan, A. and G. Schwert (1990). "Testing for Covariance Stationarity in Stock Market Data," *Economics Letters*, 33, 165–170.

Papoulis, A. (1975). "A New Algorithm in Spectral Analysis and Band Limited Extrapolation," *IEEE Transactions on Circuits and Systems*, 22, 735–742.

Percival, D. and A. Walden (1993). *Spectral Analysis for Physical Applications: Multitaper and Conventional Univariate Techniques*. Cambridge, MA: Cambridge University Press.

Priestley, M. (1981). *Spectral Analysis and Time Series*. New York: Academic Press.

Ramsey, J.B. (1992). "Seasonal Economic Data as Approximate Harmonic Oscillators," working paper No. 92-16, C.V. Starr Center for Applied Economics, New York University.

Slepian, D. (1978). "Prolate Spheroidal Wave Functions, Fourier Analysis, and Uncertainty – V: The Discrete Case," *Bell System Technical Journal*, 57, 1371–1429.

Stein, C. (1956). "Inadmissibility of the Usual Estimator for the Mean of a Multivariate Normal Distribution," in Neyman, J. (ed), *Proc. Third Berkeley Symp. Math. Stat. Prob.*, 197–206.

Thomson, D. (1994). "An Overview of Multi-Window Quadratic-Inverse Spectrum Estimation," *IEEE International Conference on Acoustics, Speech and Signal Processing*, 6, 185–194.

Thomson, D.J. (1977). "Spectrum Estimation Techniques for Characterization and Development of WT4 Waveguide," *Bell System Tech. J.*, 56, Part 1 1769–1815; Part 2 1983–2005.

Thomson, D.J. (1982). "Spectrum Estimation and Harmonic Analysis," *Proceedings of the IEEE*, 70, 1055–1096.

Thomson, D.J. (1990). "Quadratic-Inverse Spectrum Estimates; Applications to Paleoclimatology," *Phil. Trans. of the Royal Society London A*, 332, 539–597.

Thomson, D.J. (1993). "Nonstationary Fluctuations in Stationary Time Series," *Proc. SPIE*, 2027, 236–244.

Thomson, D.J. and A. D. Chave (1991). "Jackknifed Error Estimates for Spectra, Coherences, and Transfer Functions," Chap. 2 in Haykin, S. (ed), *Advances in Spectrum Estimation.* New York: Prentice Hall.

4 NONLINEAR ECONOMETRIC MODELLING: A SELECTIVE REVIEW

Norman R. Swanson
Pennsylvania State University
Philip Hans Franses
Erasmus University, Rotterdam

1. Introduction

In recent years nonlinear econometric modelling has received increasing attention in econometrics. This is not to say, however, that nonlinear phenomena have not been considered an important aspect of economics for many decades. Rather, as statistical and econometric methodology has developed, our ability to formulate, and more importantly, estimate nonlinear models have increased. However, as with all sciences and social sciences, nonlinear theories and models have been around virtually from the beginning. Although many examples can be brought to bear in support of this notion, one need really simply consider business cycles, which have long been known to exhibit various seasonal and cyclical nonlinearities (for example, see Burns and Mitchell (1946)).

In the discipline of economics, the vast array of economic theories which are available to economists often contain nonlinear elements (e.g., first order conditions, functional forms, etc.). In the past, however, many of the testable implications of economic theories, for example were formulated and derived with one of the main objectives being the specification of linear econometric models. This is perhaps not surprising for at least two reasons. First, up until quite recently, the statistical tools available for the examination of economic variables were better able to handle estimation and inference within a linear context. Second, and perhaps just as importantly, the computational ability of early computers (and even earlier slide rules) was limited enough to make it infeasible to estimate either large or complex econometric models. However, as computers have become more efficient, so have the tools, algorithms, tests, and modelling strategies used by increasing numbers of practicing economists also become more complex. Indeed, one feature of the econometrics profession in recent years seems to be that we are always able to develop estimation and inferential procedures (and associated nonlinear econometric models) which are complex enough to fully tax the capability of even the most powerful computers. Of course, this is not a feature of economics alone, as physics, biology, chemistry, and other "harder" sciences have clearly done likewise. Furthermore, it should be noted that we are not merely developing new and more complex theories in order to take advantage of computational ability. Rather, we are taking advantage of the opportunity of "better" empirically modelling a system which we have always known to be immensely complex. In this sense, the growth in nonlinear modelling in applied economics, for example, seems quite natural. Put another way, the development of nonlinear econometrics and applied economics is a feature of modern econometrics which is not at all undesirable, as it can easily be argued that we are simply attempting to better model an inherently nonlinear world; and as computational capability increases, so naturally does the complexity of our models of economic behavior.

One question is often asked is: "Are we modelling economic behavior any better by using more complex constructs?" This question is difficult to answer decisively at this early juncture in the "nonlinear revolution." However, many of the references given in this chapter attempt to provide at least partial answers to this question by comparing the performance of alternative models and theories. This is often done, for example, by examining the predictive ability of linear and related nonlinear models, or by comparing the ability of linear models to "match" the data as well as more complex nonlinear models.

To fix some ideas concerning what is commonly meant by "nonlinearity", it is perhaps useful to consider the following simple model:

$$y_t = g(x_{t-i},\ i \geq 1) + h(\epsilon_{t-i},\ i \geq 0),\ i = 1,...,p. \tag{1}$$

If $g(\cdot)$ and $h(\cdot)$ are affine functions, then y_t follows a standard ARMA or autoregressive integrated moving average (ARIMA) model, depending on whether the process is second order stationary (assuming Gaussianity, for example) or not. If, on the other hand, $x_t = y_t$, and $h(\cdot) = \epsilon_t$, then y_t follows a simple nonlinear autoregression of order p (NLAR(p)), where p refers to the highest order of lags of y_t appearing in $g(\cdot)$. This process is often referred to as "nonlinear in mean", for obvious reasons. Also, the process y_t is usually called "short-memory" if $\Phi_t(x{:}y_0)$ approaches $\Phi_h(x)$ as t approaches h, for h large, so that the starting value no longer affects the marginal distribution after the process has been running for a long time, where $\Phi_t(x{:}y_0)$ is the conditional distribution function of y_t given y_0 ($Prob(y_t \leq x \mid y_0) = \Phi_t(x{:}y_0)$), and $\Phi_t(x)$ is the corresponding marginal distribution of y_t. This process is called "stable" if y_t and y_s have the same marginal distribution for t, s large.

On the other hand, we may have $x_t = y_t$, and $g(\cdot)$ an affine function, say of order p, then one might write

$$y_t = a_0 + a_1 y_{t-1} + \cdots + a_p y_{t-p} + \epsilon_t. \quad (2)$$

If in addition, ϵ_t is distributed as $N(0,h_t)$, and

$$h_t = \alpha_0 + \alpha_1 h_{t-1} + \cdots + \alpha_1 h_{t-m}, \quad (3)$$

the resulting model is an autoregressive conditional heteroskedasticity (ARCH) process, as examined by Engle (1982). Indeed, $g(\cdot)$ and $h(\cdot)$ can be further generalized to essentially allow for an infinitely wide class of nonlinear processes. Examples include threshold autoregressive models (Tong(1983, 1990)), bilinear models (Granger and Andersen (1978)), smooth transition autoregression (STAR) and related models (Luukkonen, Saikkonen, and Teräsvirta (1988a, b) and Granger and Teräsvirta (1993)), Markov switching models (Hamilton (1989) and the references in Hamilton (1994)), models related to chaos theory (Brock, Hsieh, and LeBaron (1991) and the references therein), random parameter models, and time varying parameter models, to name but a very few. A number of nonlinear models are also examined in the papers which comprise the special issue of the *Journal of Applied Econometrics* on Nonlinear Dynamics and Econometrics (1991), including for example Teräsvirta and Anderson (1991), Mizrach (1991), and Rothman (1991), where issues as diverse as constructing nonlinearity tests, modelling nonlinearities in U.S. macroeconomic variables, and forecasting using nearest neighbor techniques are examined. Indeed, it is not unreasonable to say that one can find an example of nonlinear modelling in economics by simply picking up a recent issue of virtually any economics journal. In the above papers and books, various recent nonlinearity tests are also examined and developed, including time reversibility tests, bispectrum tests, neural network nonlinearity tests (see also Lee, White, and Granger (1993)), STAR tests, BDS tests (Brock, Dechert, and Scheinkman (1996), and various others. As misspecification can be in the form of neglected nonlinearity, many other tests can in some cases be interpreted as tests of nonlinearity, including ARCH tests, White heteroskedasticity tests, and Ramsey (1969) RESET tests, to name but a few. Also, nonlinearity can appear in the form of asymmetrically adjusting error-correction models, and various other forms of "nonlinear cointegration" (see Balke and Fomby (1997), Granger and Swanson (1996), and Corradi, Swanson, and White (1997), for example). A very small subset of other recent papers which address nonlinearity in economic variables include: Diebold and Nason (1990), Granger (1993), Neftci (1984), Potter (1995), and Weiss (1996). Also, a particularly good starting point is Granger and Teräsvirta (1993).

In this chapter, we do not attempt to disseminate the vast accumulated knowledge in the area of nonlinear econometrics. For this, the reader is referred to the above works. Rather, we discuss two varieties of nonlinear models which may be of some interest to econometricians. In Section 2, we consider what are referred to as "stochastic unit root" (STUR) models. In these models, coefficients of AR processes are assumed to vary (according to some well defined stochastic process) with average or mean values of unity, and are not fixed to be unity (or to sum to unity in the case of an AR(p) model). These models are not "near unit root" models, however, as the unit root in a simple AR(1) process, for example, is not assumed to approach unity as the sample size gets large. Rather, STUR models focus on a reasonable claim that very few economic series are precisely characterized as containing "pure" or constant unit roots, and attempt to examine whether some economic series can be better forecast using more general nonlinear models. Also, STUR models may provide a good starting point for multivariate nonlinear VAR models. In Section 3, we discuss the class of nonlinear models call artificial neural networks (ANNs). These models are nonparametric in the sense that the number of parameters one fits to the data increases with the sample size. ANN models already have a rich history in the biological sciences, and were first constructed in an effort to model the inner workings of the human brain. Estimation and modelling using ANNs is discussed, and some earlier findings are presented. In Section 4 we review some practical issues in empirical modelling, such as the effects of outliers, and model selection.

2. Stochastic Unit Root Processes

2.1 Theoretical Considerations

Consider a series x_t generated by

$$x_t = a_t x_{t-1} + \epsilon_t , \tag{4}$$

where ϵ_t is zero mean, *i.i.d.* with variance σ^2_ϵ, and where

$$a_t + \exp(\alpha_t) \tag{5}$$

with α_t a Gaussian stationary series having mean m, variance σ^2_α and power spectrum $g_\alpha(\omega)$. This is an example of a doubly stochastic process as considered by Tjøstheim (1986), and has been specifically analyzed by Brandt (1986), and Pourahmadi (1986, 1988). Model (4) can also be viewed as an extension of the periodic integration model (Franses (1996)) where α_t takes different values in different seasons. However, the exponential form (5) is a convenient case for consideration of the long-run properties of the series. A particular example that will be used to illustrate the results is the case where α_t is given by an AR(1) process

$$\alpha_t = \mu + \rho\alpha_{t-1} + \eta_t \tag{6}$$

where $|\rho|<1$, and η_t is *i.i.d.* normally distributed $N(0,\sigma^2_\eta)$ and is independent of the series ϵ_t. It will be assumed throughout this section that α_t is generated exogenously from x_t, so that

$$g(\alpha_{t-1}|\alpha_{t-j}, x_{t-j}, j \geq 0) = g(\alpha_{t+1}|\alpha_{t-j}, j \geq 0)$$

where $g(\alpha|I)$ is a conditional distribution. For the example (6), $m = \mu/(1-\rho)$ and $\sigma^2_\alpha = \sigma^2_\eta/(1-\rho^2)$. (4) can be solved as

$$x_t = \epsilon_t + a_t\epsilon_{t-1} + a_t a_{t-1}\epsilon_{t-2} + ... + a_t a_{t-1} \cdots a_{t-k+2}\epsilon_{t-k+1} + a_t a_{t-1} \cdots a_{t-k+1} x_{t-k}$$

or,

$$x_t = \epsilon_t + \pi_{t,1}\epsilon_{t-1} + \pi_{t,2}\epsilon_{t-2} + \ldots + \pi_{t,k-1}\epsilon_{t-k+1} + \pi_{t,k}x_{t-k} \tag{7}$$

for any integer k, $0 \le k \le t$, where,

$$\pi_{t,j} = \exp(S_{\alpha t}(j)),$$

and,

$$S_{\alpha t}(j) = \sum_{i=0}^{j-1} \alpha_{t-i}$$

with the notation $S_{\alpha t}(0) = 1$. It should be noted that for $j > 0$

$$E_j \equiv E[S_{\alpha t}(j)] = jm, \tag{8}$$

and,

$$V_j = var[S_{\alpha t}(j)] = \sigma_\alpha^2[j + \sum_{r=1}^{j-1}(j-r)\rho_\alpha(r)] \tag{9}$$

where $\rho_\alpha(r) = corr(\alpha_t, \alpha_{t-r})$. For j large (9) can be approximated as

$$V_j \approx jf_\alpha(0) \tag{10}$$

as shown in Koopmans (1974, p. 171), where $f_a(w) = \frac{1}{2\pi}g_\alpha(w)$.

A standard result that is useful in this analysis is that if a random variable X is distributed $N(m,\sigma^2)$ then

$$E[\exp(kX)] = \exp(km + ½k^2\sigma^2). \tag{11}$$

It follows that for k large,

$$E[\pi_{t,k}] = \exp(E_k + ½V_k) \approx \exp(k\theta), \tag{12}$$

where,

$$\theta = m + ½f_\alpha(0). \tag{13}$$

Assuming for convenience that x_0 is deterministic, and taking expectations of (7) with $k = t$ gives

$$E[x_t] = x_0\exp(t\theta)$$

and, as the ϵ_t are *i.i.d.*,

$$var(x_t) = \sigma_\epsilon^2 \sum_{j=0}^{t-1} E[\pi_{t,j}^2] = \sigma_\epsilon^2 \sum_{j=0}^{t-1} \exp(2jm + 2V_j) \approx \sigma_\epsilon^2 \sum_{j=0}^{t-1} \exp(2j\phi), \tag{14}$$

where, $\phi = m + f_\alpha(0)^1$. Thus,

$$\left.\begin{aligned} var(x_t) &= \sigma_\epsilon^2 \left[\frac{1-\exp(t2\phi)}{1-\exp(2\phi)}\right] && if\ \phi \neq 0 \\ &= t\,\sigma_\epsilon^2 && if\ \phi = 0 \end{aligned}\right\}. \tag{15}$$

Turning to consideration of (linear) regression coefficients $\beta_{t,k}$ as in:

$$x_t = \beta_{t,k} x_{t-k} + \text{error}$$

it can be noted from (7) that

$$E_L(x_t | x_{t-j}, j \geq k) = x_{t-k} E_L[\pi_{t,k} | x_{t-j}, j>k] = x_{t-k} E[\pi_{t,k}] \tag{16}$$

where $E_L(\cdot)$ is an expectation using only the components of the conditioning sets in a linear form and $E(\cdot)$ is an unconditional expectation. Thus, from (12),

$$\beta_{t,k} = E[\pi_{t,k}] = \exp(k\theta), \text{ for large k.} \tag{17}$$

Using the identity $corr(x_t, x_{t-k}) = \beta_{t,k}\sqrt{\dfrac{var(x_{t-k})}{var(x_t)}}$ thus gives the following approximate expressions for this autocorrelation when t and k are both large:

	$\theta < 0$	$\theta = 0$	$\theta > 0$
$\phi > 0$	$\exp(-kf)$	$\exp(-kf)$	$\exp(-kf)$
$\phi = 0$	$\exp(k\theta)(1-k/t)^{1/2}$	1	-
$\phi < 0$	$\exp(k\theta)$	-	-

where $\theta = m + \frac{1}{2}f$, and $f \equiv f_\alpha(0) = \dfrac{1}{2\pi} g_\alpha(0)$ where $g_\alpha(0)$ is the spectrum of α_t at zero frequency. Empty cells are cases that cannot occur, since by definition, $\phi \geq \theta$.

The autocorrelations decline exponentially (approximately) in all cases except when $\theta = \phi = 0$, so that $f_\alpha(0) = 0$ and $m = 0$. This special case occurs, for example, if $\alpha_t = 0$, all t, since it then follows that $E[e^{\alpha_t}] = 1$, and x_t is a perfect unit root process.

Two alternative characterizations of the STUR process given by (4) and (5) are (i) STURA: α_t is such that $E[e^{\alpha_t}] = 1$ so that

$$m + \tfrac{1}{2}\sigma_\alpha^2 = 0 \tag{18}$$

(ii) STURB: α_t is such that $E[\pi_{t,k}] = 1$, k large, where $\pi_{t,k}$ is given in (7), so that

$$\theta = 0 \quad \text{or} \quad m + \tfrac{1}{2} f_\alpha(0) = 0. \tag{19}$$

The persistence which arises when the restrictions given by (18) or (19) are met is similar to the persistence exhibited by perfect unit root processes. For instance, as in the case of a random walk, shocks from the distant past continue to have an impact on current values of the process.

More precisely, the properties of STURA depend on the relative size of σ_α^2 and $f_\alpha(0)$. Suppose that $f_\alpha(0) > \sigma_\alpha^2$, so that, essentially, α_t is smoother than white noise. It follows that θ and ϕ are both positive, so that from (15), the variance of x_t is increasing explosively, although probably mildly so, and $E[\pi_{t,k}]$ is increasing with k. Thus, distant shocks have greater impact on the current value of the series than do recent shocks. This property also holds for linear I(d), $d > 1$, processes, for example, and might be called "super-persistence".

A STURB process will have $E[e^{\alpha_t}] < 1$ if $f_\alpha(0) > \sigma_\alpha^2$, so that a regression of

$$x_t = \beta_{t,k} x_{t-k} + \text{error}$$

will yield $\beta_{t,k}$ values which are nearer to unity for large values of k than for small values of k. The persistence property is now particularly interesting. Recall that

$$x_t = \sum_{j=0}^{k-1} \pi_{t,j} \epsilon_{t-j} + \pi_{t,k} x_0$$

and

$$\pi_{t,j} = \exp[S_{\alpha t}(j)],\ j \geq 1 \text{ and } \pi_{t,0} = 1$$

where, $S_{\alpha t}(j)$ has mean jm, variance $jf_\alpha(0)$, for j large. If α_t is normally distributed, then $\pi_{t,j}$ will have a lognormal distribution. With the constraint that $E[\pi_{t,j}] = 1$, which implies that m is negative, standard results then give

$$variance\ (\pi_{t,j}) = \exp(jf_\alpha(0)) - 1$$

$$mode\ (\pi_{t,j}) = \exp(-3/2\, jf_\alpha(0))$$

$$median(\pi_{t,j}) = \exp(-1/2\, jf_\alpha(0)).$$

Also, a 95% confidence interval (about the median) is

$$\exp[-\tfrac{1}{2} jf_\alpha(0) \pm 2(jf_\alpha(0))^{\frac{1}{2}}].$$

Thus, as j increases, $\pi_{t,j}$ has a constant mean, an increasing variance, a mode and median which tend towards zero, and a $100(1-\delta)\%$ confidence interval that tends towards very small values. This implies that the form of persistence which is displayed is very fragile.

From a theoretical perspective, stochastic unit roots may be seen to arise quite naturally in economics. As an example, it is well known that many basic formulations of the permanent income hypothesis involve the term $(1+\delta)/(1+R_t)$ (see Hall (1978), for example). Using the standard approach, δ is the marginal rate of time preference, and R_t is the real rate of interest. A frequently made assumption is that this term is identically unity. However, when that assumption is relaxed, and R_t is allowed to vary stochastically, then the term may well vary around unity, and probably has a mean very close to unity, as expected in the STUR case.

To summarize, the above results indicate that the properties of STUR processes are often markedly different from comparable properties of perfect unit root processes. For example, some properties mimic those of the ARCH model in (3). Another characteristic of stochastic unit roots is that they are quite difficult to distinguish from perfect unit roots. This is not surprising given that evidence presented below indicates that variances of stochastic unit roots are often quite small. In this sense the usual power failures associated with unit root tests should apply. Nevertheless, before discussing an appropriate testing strategy for the presence of STUR, we provide an example of a STUR process which is readily mistaken for a pure unit root process based on a standard Dickey-Fuller test.

2.2 Empirical Considerations

In this section, we discuss estimation of and test for the presence of stochastic unit root processes. First, one method for estimating the parameters of STUR processes is summarized. For further details, and discussion of alternative estimation strategies, the reader is referred to Granger and Swanson (1997). The method which we discuss below is closely related to the approximate maximum likelihood method discussed in Guyton, Zhang, and Foutz (1986). In contrast to Guyton, *et al.*, though, conditions are not found for the existence of a *stationary* generalized autoregressive process. Rather, the parameters of a nonstationary STUR process are estimated. Assume that

$$x_t = a_t x_{t-1} + \epsilon, \quad a_t = \exp(\alpha_t) \text{ and } E(a_t) = 1 \text{ or } \theta = 0, \qquad (20)$$

where

$$\alpha_t = \mu + \sum_{i=1}^{p} \rho_i \alpha_{t-i} + \eta_t,$$

and α_t is a stationary stochastic process, thus generalizing (6) to be an AR(p) process. (The STUR model may also be extended in the current context by adding lags of x_t, for example.) Of course, for the sake of estimation, we will not assume that $E(a_t) = 1$ or that $\theta = 0$, as we do not wish to impose that the process must be STURA or STURB according to (18) or (19). Another possible variation of the above STUR process is the addition of a time varying intercept, say c_t, to the measurement equation. In this way, the upward trending behavior of many economic series could, potentially, be better modelled. As we will see below, such generalizations pose no problem for the approximate maximum likelihood (AML) estimation technique.

Assume that the ϵ_t and the η_t are normally distributed with zero means, and variances σ_ϵ^2 and σ_η^2, respectively and are independent. Thus, a_t is distributed as a lognormal random variable. The unknown parameters are gathered into a vector, $\Omega = (\mu, \rho, \sigma_\epsilon^2, \sigma_\eta^2)$, where $\rho = (\rho_1, \rho_2, \ldots, \rho_p)$. In order to construct a likelihood function, use the well known property that $f(x, a \mid \Omega) = f(x \mid a, \Omega) f(a \mid \Omega)$, where $f(\cdot)$ is a joint destiny. Then, noting that likelihood is proportional to probability, assuming a value of unity for the constant of proportionality, carrying the conditioning argument one step further, and introducing a randomization parameter, K, yields

$$L(x \mid \Omega) = \int L(x, a \mid \Omega) da = \int L(x \mid a, \Omega) f(a \mid \Omega) da = E_{a \mid \Omega} L(x \mid a, \Omega)$$

and,

$$\hat{E}_{a \mid \Omega} L(x \mid a, \Omega) = \hat{L}(x \mid \Omega) = \frac{1}{K} \sum_{k=1}^{K} \left[\frac{1}{\sqrt{2\pi}} \right]^T \sigma_\epsilon^{-T} e^{-\frac{1}{2\sigma_\epsilon^2} \sum_{t=1}^{T} \hat{\epsilon}_{t,k}^2} \tag{21}$$

$$\hat{\epsilon}_{t,k} = x_t - \hat{a}_{k,t} x_{t-1} \tag{22}$$

where by $E_{a \mid \Omega}$ we mean the expectation over a conditional on the parameter Ω, and assuming that the density of x_1 is the same as the conditional density of x_t, given x_{t-1}, and that $x_0 = 0$. This is a standard result, except that the a_t are unobserved so that we construct an estimate, say $\hat{L}(\cdot)$, by randomly sampling across different distributions of a_t.

In order to calculate values of the likelihood function for various points in the parameter space, various sequences, $a_k = \{a_{k,t}\}$, are generated for each $\{\mu, \rho_i, i = 1, ..., \rho, \sigma_\eta^2\}$ in the parameter space. Thus, the approximate maximum likelihood function given as (21) is a random variable, and is an estimator of $E_{a \mid \Omega} L(x \mid a, \Omega)$. Maximization of (21) proceeds by first generating K sequences $\{\eta_{t,k}\}$, $t = 1, ..., T$. Then, $\hat{L}(\cdot \mid \cdot)$ and its derivatives are calculated for each point in the parameter space $\hat{\Omega} = (\hat{\mu}, \hat{\rho}_i, i = 1, ..., p, \hat{\sigma}_\epsilon^2, \hat{\sigma}_\eta^2)$. This iterative procedure is carried out until a $\hat{\Omega}$ is found which optimizes $\hat{L}(\cdot \mid \cdot)$, according to some convergence criterion. Following accepted techniques from the simulation-based estimation literature, the *same* K sequences, $\{\eta_{t,k}\}$, $t = 1, ..., T$, are used throughout the optimization. Approximate maximum likelihood methods are also considered in the disequilibrium literature. For example, Laroque and Salanie (1994) show that simulated pseudo ML estimators have good finite sample properties, even for values of K as low as 15. It should be noted that σ_ϵ^2 cannot be concentrated out of the approximate likelihood function (since the sum of logs is not equal to the log of sums).

The finite sample properties of the AML method were examined using a series of Monte Carlo experiments. STUR processes were generated according to (20) with $p = 1$, $\rho_1 = 0.6$, $E(a_t) = 1$, $\eta_t \sim i.i.d.$ N(0,0.0001) and N(0,0.000001), and $\epsilon_t \sim i.i.d.$ N(0,1.0). One thousand trials were carried out, for sample sizes of 100, 250, and 500 observations, and for randomization parameter, K, values of 10, 25, 100, and 250. Parameters were estimated using the GAUSS computer program and the Broyden, Fletcher, Goldfarb, Shanno (BFGS) algorithm. The results of this simulation exercise are found in Granger and Swanson (1997), and their findings can be summarized as follows. First, the AML method appears to produce fairly reasonable average point estimates of the STUR parameters, in particular for the larger sample sizes of 250 and 500, and for the larger K values of 100 and 250. For example, when $K = 10$ and 25 the average point estimates of σ_η^2 are not as close to their true values as when higher randomization parameters are used. Interestingly, selecting $K = 250$ does not generally result in substantial improvement over the $K = 100$ case. Second, σ_ϵ^2 is estimated very well, even for small K and T. Finally, it should be noted that the parameters are often estimated fairly imprecisely, particularly when smaller values of σ_η^2 are used. This suggests that AML estimates should be used with caution, in practical applications.

Turning now to the issue of testing, it is first worth pointing out that ADF tests have low power against some STUR alternatives, since the test is well known to exhibit limited power against a variety of "near" I(1) processes. In the STUR case it would be ideal if one were to construct a test which has STUR as the null hypothesis. This is difficult, though, given the rather complex nature of STUR processes, and is not considered here. The approach adopted here has difference stationarity as the null hypothesis, and a random coefficient AR (p + 1) model as the alternative. Leybourne, McCabe, and Tremayne (1994) develop the LM type test which we will discuss. Interestingly, the test is also applicable to STUR processes, and is robust to various departures under the null such as ARCH and GARCH. Of note is that the test starts by assuming that a series is non-stationary. In this way, the test has the null of a fixed unit root (or a linear I(1) process) and the alternative of some more general non-stationary data generating mechanism.

The model treated by Leybourne, *et al.* has

$$x_t = \sum_{i=1}^{q+1} \delta_{i,t} x_{t-i} + \epsilon_t \tag{23}$$

$$\delta_{1,t} = (a_t + \pi_1), \delta_{i,t} = (\pi_i - a_t\pi_{i-1}), i = 2, \ldots, q, \delta_{q+1,t} = -a_t\pi_q,$$

where $a_t \sim i.i.d.\ (1, \omega^2)$ and $\epsilon_t \sim i.i.d(1, \sigma_\epsilon^2)$, and q is a non-negative integer. Trends may also be incorporated into (23), and some dependence is allowed between the a_t and ϵ_t. Also, note that under the null hypothesis that $\omega^2 = 0$, x_t is seen to be an AR(q + 1) process with a unit root, since all of the coefficients sum to unity. The version of (23) which we consider has $q = 1$ and $\pi_1 = 1$:

$$x_t = a_t x_{t-1} + \epsilon_t, \tag{24}$$

and is examined by Leybourne, McCabe and Tremayne (1994) and McCabe and Tremayne (1995). In this framework, consider: H_0: $\omega^2 = 0$ against H_1: $\omega^2 > 0$. Before writing the statistic, three features of the test should be mentioned. First, Leybourne, *et al.* show that the test statistic below remains unchanged when the natural log of a_t in (24) is allowed to follow a stationary AR(1) process, as in (4), (5) and (6). Second, randomized coefficient models such as (23) and (24) can be written as models of conditional variation. However, the lagged level of the process itself drives the conditional heteroskedasticity, while in many G(ARCH) formulations it is the innovations that drive the conditional variation. Leybourne, *et al.* have shown that assuming the ϵ_t are stationary is sufficient to ensure that the test has no power against G(ARCH). They have also shown that the empirical size of the test in the G(ARCH) case is less than the nominal value so that the test is conservative. Unfortunately, this result does not hold when the form of heteroskedasticity is IGARCH. These and other related issues will be left to future work. Third, it should also be noted that the test is augmented in the same way as ADF tests to allow for various forms of serial correlation in the differences of x_t.

Assuming normality, using the standard conditioning argument, and substituting $\omega_2 = 0$ in the first order condition, yields

$$\partial L(\cdot)/\partial\omega^2\big|_{\omega^2=0} = \sigma_\epsilon^{-4}\sum_{t=q+2}^{T}\Big(\sum_{j=1}^{t-1}\epsilon_j\Big)^2(\epsilon_t^2-\sigma_\epsilon^2)+o_p(1)$$

The statistic is

$$\hat{Z}_T = T^{\frac{-3}{2}}\hat{\sigma}_\epsilon^{-2}\hat{\kappa}^{-1}\sum_{t=q+3}^{T}\Big(\sum_{j=q+2}^{t-1}\hat{\epsilon}_j\Big)^2(\hat{\epsilon}_t^2-\hat{\sigma}_\epsilon^2),$$

where, $\hat{\epsilon}_t$ is the residual from a regression of Δx_t on a constant, a trend (and lags of Δx_t), with the number of lags, q, chosen accordingly (i.e., so that the ϵ_t are not autocorrelated), $\hat{\sigma}_\epsilon^2 = \frac{1}{T}\Sigma\hat{\epsilon}_t^2$ and $\hat{\kappa}^2 = \frac{1}{T}\Sigma(\hat{\epsilon}_t^2-\hat{\sigma}_\epsilon^2)^2$. Leybourne, *et al.* show that the size and power of the statistic is not significantly affected by overfitting. Thus, they suggest that to err on the side of too many lags is better than underspecifying the number of lags. $\hat{Z}_T$ converges weakly to a sum of functionals of generalized Brownian bridges and Brownian motions, given by

$$\int_0^1 G_1(r)^2dG_2(r) - \int_0^1 G_1(s)^2dsG_2(1),$$

where

$$G_1(r) = W_1(r) - rW_1(1) + 6r(1-r)[W_1(1)/2 - \int_0^1 W_1(s)ds],\ G_2(r) = \psi W_1 + (1-\psi^2)^{½}W_2(r),$$

the limit processes W_1 and W_2 are independent Brownian motions, G_1 is a generalized Brownian bridge, and the nuisance parameter ψ is the covariance between ϵ_t and ϵ_t^2. The critical values for this test statistic have been simulated under the assumption that ϵ_t is symmetrically distributed so that $\psi = 0$ (see Leybourne, *et al.* (1994)). However, the distribution could be tabulated for different values of ψ, and an estimated value used to calibrate the critical value. Interestingly, if $\psi = 0$, $\hat{Z}_t$ can be modified so that it converges to a random variable with a standard normal distribution. In particular, note that the asymptotic distribution given above can be written as

$$\int_0^1 G_1(r)^2dG_2^*(r) = \int_0^1 J(r)dG_2(r),$$

where

$$G_2^*(r) = G_2(r) - rG(1)$$

and

$$J(r) = G_1(r)^2 - \int_0^1 G_1(s)^2ds.$$

Thus, $\hat{Z}_T$ is characterized by a mixture of normals distribution, N(0, $\int_0^1 J(r)^2dr$). This suggests that an alternative test statistic is:

$$\hat{Z}_T^* = [\sum_{t=p+3}^{T} \hat{S}_{t-1}^2(\hat{\eta}_t^2 - \hat{\sigma}_v^2)]/[\hat{\kappa}(\sum_{t=p+3}^{T} \hat{S}_{t-1}^4 - (\sum_{t=p+3}^{T} \hat{S}_{t-1}^2)^2)]^{½}$$

$$= [\sum_{t=p+3}^{T} \hat{J}_{t-1}^* \hat{v}_t^2]/[\hat{\kappa}(\sum_{t=p+3}^{T} \hat{J}_{t-1}^{*2})^{½}],$$

where $\hat{S}_t = \sum_{j=p+2}^{t} \hat{v}_j$, $\hat{J}_t = \hat{S}_t^2$, and $\hat{J}_t^* = \hat{J}_t - T^{-1}\sum_{j=1}^{T}\hat{J}_j$. Then it follows that $\hat{Z}_T^*$ converges weakly to

$$[\int_0^1 J(r)dG_2(r)]/[(\int_0^1 J(r)^2 dr)^{½}]$$

which is the N(0,1) distribution in the special case where $\psi = 0$. This result suggests that the above modified STUR test is very easy to apply, and does not require the simulation of critical values for a non-standard distribution, when v_t is known to have a symmetric distribution.

Based on Monte Carlo simulations of the STUR model specified by (4), (5) and (6), where the alternative hypothesis considered is (23), but with $a_t = \exp(\alpha_t)$ and

$$\alpha_t = \mu + \rho\alpha_{t-1} + \eta_t$$

where $|\rho| < 1$, $\eta_t \sim i.i.d.$ $N(0, \sigma_\eta^2)$ and is independent of the series ϵ_t, and $E(a_t) = 1$, the following findings were reported in Granger and Swanson (1997). Since the test has difference stationarity as the null, it may have power against STUR, as STUR processes are clearly not difference stationary, in theory. As might be expected $\hat{Z}_T$ has very good power against the STUR alternative for larger values of σ_η^2 and for larger sample sizes (i.e., $\sigma_\eta^2 = 0.1 - 0.001$, $T = 500$, and 1000). As σ_η^2 falls, the power of the test tails off significantly. Also, overall, the evidence from a number of other ADF test simulations suggests that the difference-stationarity test has more power against STUR alternatives than standard ADF tests. For example, for a sample size of 250 observations, the power of the $\hat{Z}_T$ test is fairly low (0.152 for nominal size of 5%), but is still around three times as high as the power of the ADF test. In summary, while the above test is only a starting point, it nevertheless serves as a good indicator of whether (macro)economic time series are well modelled as containing stochastic unit roots.

Using the above test as a guide, Granger and Swanson (1997) find that approximately one half of a randomly selected group of macroeconomic variables reject the null of difference stationarity in favor of the alternative (STUR). Also, a forecasting experiment based on 1- and 4-step ahead forecasts suggests that the mean-square forecast error (MSFE) "best" model is STUR just as often as it is a Kalman filter based model, a simple linear autoregressive model (with lag structure chosen so as to *whiten* the residuals of the estimated model), or a random walk model. This finding applies particularly to various interest rate and price series. On the other hand, linear models are preferred when forecasting unemployment.

3. Artificial Neural Network Models

3.1 Methodology

Cognitive scientists have proposed a class of flexible nonlinear models inspired by certain features of the way that the brain processes information. (A good introduction to the cognitive science literature is Rumelhart and McClelland (1986).) Because of their biological inspiration, these models are referred to as "artificial neural networks" (ANNs). Because of their flexibility and simplicity, and because of demonstrated successes in a variety of empirical applications where linear models fail to perform well (see White (1989) and Kuan and White (1994) for some specifics), ANNs have become the focus of considerable attention as a possible vehicle for forecasting financial variables as well as for pattern recognition. Among recent applications are those of White (1988), Dutta and Shekhar (1988), Moody and Utans (1991), Dorsey, Johnson and van Boening (1991), Dropsy (1992), Franses and Draisma (1997), Franses and Van Homelen (1997), Kuan and Liu (1992), and Ripley (1994). See also the recent book by Bishop (1995) and Trippi and Turbau (1993). For those interested in a detailed discussion of ANNs and their econometric applicability, we refer to Kuan and White (1994). For this chapter, it suffices to treat these models as a potentially interesting black box, delivering a specific class of nonlinear regression models. In particular, the ANN nonlinear regression models considered here have the form:

$$f(x,0) = \tilde{x}'\alpha + \sum_{j=1}^{q} G(\tilde{x}'\gamma_j)\beta_j \tag{25}$$

where $\tilde{x}$ is a (column) vector of explanatory variables, $\tilde{x} = (1, x')'$ augments x by the inclusion of a constant term, $\theta = (\alpha', \beta', \gamma')'$, $\beta = (\beta_1, \dots, \beta_q)'$, $\gamma = (\gamma'_1, \dots, \gamma'_q)'$, q is a given integer and G is a given nonlinear function, in our case, the logistic cumulative distribution function (c.d.f.) $G(z) = 1/(1 + \exp(-z))$.

A network interpretation of (25) is as follows. "Input units" send signals $x_0(=1), x_1, \dots, x_r$ over "connections" that amplify or attenuate the signals by a factor ("weight") γ_{ji}, $i = 0, \dots, r, j = 1, \dots, q$. The signals arriving at "intermediate" or "hidden" units are first summed (resulting in $\tilde{x}'\gamma_j$) and then converted to a "hidden unit activation" $G(\tilde{x}'\gamma_j)$ by the operation of the "hidden unit activation function" G. The next layer operates similarly, with hidden activations sent over connections to the "output unit." As before, signals are attenuated or amplified by weights β_j and summed. In addition, signals are sent directly from input to output over connections with weights α. A nonlinear activation transformation at the output is also possible, but we avoid it here for simplicity. Of note is that two or more hidden layers, as well as alternative activation functions are also frequently used (for example, see Chen and Swanson (1997)).

In network terminology, $f(x, 0)$ is the "network output activation" of a "hidden layer feedforward network" with "inputs" x and "network weights" θ. The parameters γ_j are called "input to hidden unit weights," while the parameters β_j are called "hidden to output unit weights." The parameters α are called "input to output unit weights."

Hornik, Stinchcombe and White (1989, 1990) (among others, see also Cybenko 1989, Carroll and Dickinson 1989 and Funahashi 1989) have shown that functions of the form (25) are capable of approximating arbitrary functions of x arbitrarily well given q

sufficiently large and a suitable choice of θ. This "universal approximation" property is one reason for the successful application of ANNs. In fact, White (1990) establishes that ANN models can be used to perform nonparametric regression, consistently estimating any unknown square integrable conditional expectation function.

In Swanson and White (1995), model (25) is applied to the problem of forecasting $R_{t+\tau} - R_{t+1}$ or $R_{t+\tau} - R_{t+\tau-1}$, lags of these variables, as well as forward rates observed at time *t*, where $R_{t+\tau}$, the 1-month spot interest rate observed at time $t+\tau-1$, and $F_{\tau,t}$, the forward rate for month $t+\tau$ observed at time *t*, and all data are end of month U.S. Treasury bill rates (see e.g., Mishkin (1988)). The basic idea behind this approach is that the inclusion of the nonlinear terms $G(x'\gamma_j)$ should enhance forecasting ability relative to standard linear models, if overfitting is properly avoided and the data truly have nonlinear features. In their analysis, a model selection approach is taken to fitting the artificial neural network models, thus shedding light not only on the usefulness of the current forward rate for predicting the future spot rate, but also on the usefulness of the Schwarz Information Criterion (SIC) for selecting "best" forecasting models ("best" in their paper refers to the best model in the context of confusion matrices, MSFEs, and various profit measures - see Swanson and White (1995, 1997) for further details of these types of "model selection" criteria). Their findings can be summarized as follows: First, for a relatively long out-of -sample period, the "best" model contains forward rates in four out of five forecast horizons, based on a forecast mean squared error performance measure, and five out of five horizons, when chosen using a confusion performance measure. The best models from a trading profitability standpoint contain the forward rate for three of the five horizons. Forward rates are thus useful in predicting future changes in spot rates, to this extent. Second, windows of observations (used in the estimation of the models) of less than maximal size occasionally appear as forecast mean squared error-optimal, generally appear as confusion-optimal and often appear as profit-best, suggesting instability in the relationships of interest. Third, the in-sample SIC does not appear to be a reliable guide to out-of-sample performance, so it fails to offer a convenient shortcut to true out-of-sample performance measures for selecting models, and for configuring nonlinear artificial neural network models. Finally, artificial neural network (ANN) models appear to be promising for use in this forecasting context, particularly as they often lead to the "least-confused" models, and further refinement and application of ANN methods is warranted.

In a closely related analysis, Swanson and White (1997) use a model selection approach to compare real-time forecasts (see Swanson (1996)) from 9 macroeconomic variables using various adaptive and nonadaptive models, linear and potentially nonlinear (artificial neural network) models, and professional constructed forecasts (from the Survey of Professional Forecasters (SPF), see Croushore (1993)). As above, they rely on various model selection criteria, for which summaries and discussions can be found in Diebold and Mariano (1995), Engle and Brown (1986), Fair and Shiller (1990), Henriksson and Merton (1981), Keane and Runkle (1990), Leitch and Tanner (1991), Pesaran and Timmerman (1994a, b), Stekler (1991, 1994), and Zarnowitz and Braun (1992). Their conclusions can be summarized as follows: First, even when their econometric models are constrained to include information available only on a real-time basis, econometric predictions still outperform SPF predictions for many of the variables, based on mean squared forecast error and mean absolute deviation measures. In particular, adaptive multivariate linear (and to a lesser extent nonlinear artificial neural network) models tend to outperform SPF, no change, and nonadaptive univariate and multivariate linear models. It should be noted, though, that the SPF forecasts appear to perform more or less as well as a number of econometric models when comparing predictions of the direction of change in a variable, and when minimizing the mean absolute percentage error model selection criterion. Nevertheless, based on predictions of the direction of change, the multivariate adaptive linear and nonlinear artificial neural network models (when grouped together) dominate all other (linear) models combined, providing the least confused forecasts for the majority of variables examined, for forecast horizons of both 1 quarter and 1 year. Overall, their results, which include Diebold-Mariano loss differential tests, χ^2 tests of independence, and sign tests, indicate that model selection should proceed on a case by case basis, with adaptive, nonadaptive, and SPF forecasts alternately dominating depending on which variable is being examined. Second, windows of observations less than the maximal size rarely appear in prediction-"best" models, suggesting relative stability in the relationships of interest. Third, they again find that the in-sample SIC does not appear to offer a convenient shortcut to true out-of-sample performance measures for selecting models, or for configuring adaptive neural network models, when forecasting macroeconomic variables. Fourth, the use of unrevised data in real-time forecasting appears to offer a valid guide for comparing real-time professionally available forecasts with econometric predictions. This is contrary to the common practice of building econometric models using the latest fully revised data, which is a mixture of unrevised, partially revised, and fully revised data, and has the feature that future data may have been used (perhaps inadvertently) to revise earlier data (such as when revised seasonal factors and revised benchmark figures are used). Finally, multivariate adaptive models appear to be promising for use in this context, although it should perhaps be noted that little evidence that explicitly supports the exclusive use of artificial neural network models is found. Nevertheless, it is suggested that further refinement and application of adaptive linear and nonlinear methods for modelling macroeconomic variables appears warranted, particularly in the context of truly *ex ante* or real-time forecasts.

In addition to being useful for modelling conditional mean, ANN models can also be used to help in the examination of conditional variance. For example, Franses and Van Homelen (1997) apply ANNs to forecasting exchange rates. Such financial data often have GARCH (generalized autoregressive heteroskedasticity) features, and it is well known that GARCH (if not corrected for) can lead to the suggestion (based on standard nonlinearity tests) that the data should be modelled nonlinearly in mean. Franses and Van Homelen (1997) show using Monte Carlo analysis, that when data exhibit nonlinearity only in the form of undetected GARCH, then ANNs do not lead to improved forecasts, relative to simpler linear models. However, when data are nonlinear in mean, ANNs will exploit this, yielding improved forecasts. Franses and Van Homelen (1997) then consider using ANNs to model daily exchange rate data, and find that ANNs do not lead to improved forecasts relative to linear models, when the models are compared using tests of independence in the context of 2x2 contingency tables which measure the "directional forecast accuracy" (see Pesaran and Timmerman (1992), and above). Finally, as ANNs appear not to be fooled by GARCH, Franses and Van Homelen (1997) recommend the evaluation of out-of-sample forecasting using ANNs as a diagnostic for checking for nonlinearity.

3.2 Estimation Considerations

In estimating ANN models of the form discussed above, it is inappropriate to simply fit the network parameters with $q = 4$ hidden units by least squares, as the resulting network typically will have more parameters than observations, achieving a perfect or nearly perfect fit in sample, with disastrous performance out-of-sample. To enforce a parsimonious fit, the ANN models should be estimated by a process of forward stepwise (nonlinear) least squares regression, using the SIC to determine included regressors and the appropriate value for q. Specifically, a forward stepwise linear regression is performed first, with regressors added one at a time until no additional regressor can be added to improve the SIC. The linear regression coefficients are thereafter fixed. Next a single hidden unit is added (i.e., q is set to 1), and regressors are selected one by one for *connection* to the first hidden unit, until the SIC can no longer be improved. Then a second hidden unit is added and the process repeated, until four hidden units have been tried, or the SIC for q hidden units exceeds that for $q - 1$ hidden units. This ANN model selection procedure is begun anew each time the data window moves forward one period. A different set of regressors and a different number of hidden units connected to different inputs may therefore be chosen at each point in time. This procedure simulates a fairly sophisticated real time ANN forecasting implementation. One should expect the ANN models to have SIC values superior to (i.e. smaller than) those of the linear models, as the ANN model can choose any of the linear models as a special case.

It should be stressed, though, that the estimation strategy outlined here is one of a plethora of available techniques for estimating neural network models. Indeed, a whole host of more sophisticated techniques which rely on various types of cross validation, etc. are now also available, often in "canned" computer packages. For references to other estimation strategies, the reader need look no further than any recent book or periodical on neural networks (of which there are many).

4. Further Empirical Issues

4.1 Data Transformations

Many macroeconomic data are seldom available to the applied econometrician in raw format. For example, quarterly and monthly data often consists of aggregated quantities, averaged prices, or even interpolated data. Also, seasonally observed data are frequently only available after some form of seasonal adjustment has been applied. In brief, many seasonal correction methods apply sequences of outlier removal and moving average filtering techniques to discretely measured data. Such correction methods may consist of the following steps. First, a trend is estimated and removed. Second, outliers are removed and two-sided moving average filters are applied. Third, Steps 1 and 2 are repeated in a number of iterations. In any step, irregular data are treated according to the functional form of the estimated trend and seasonal factors. Perhaps not surprisingly, in many cases these types of procedures may result in seasonally adjusted time series which have properties quite different for the original unadjusted series. For example, Ghysels, Granger, and Siklos (1996) show that seasonal adjustment may introduce nonlinearity into an otherwise linear process.

On the other hand, seasonal adjustment may actually reduce the relevance of switching regimes, say, leading to a finding of "less" nonlinearity. This arises as sequences of moving average filters clearly smooth away the effects of structural shifts. Since many nonlinear models are designed to model regime switches, one may expect that such models are less useful for seasonally adjusted data. Intuitively, the application of a sequence of moving average filters implies that a current, say recessionary, observation is replaced by some weighted average of past and future observations, some of which may lie within expansionary phases. This in turn suggests that the seasonally adjusted observations dampen the oscillatory affect that economic expansions and contractions have on economic variables, for example. In terms of switching regime models, one may expect that seasonal adjustment decreases the probability of switching regimes (transition probability), and thus the probability that one stays within the same regime increases. Using simulations and empirical examples, Franses and Paap (1996) confirm the conjecture for the Markov switching model of Hamilton (1989). In summary, we recommend that extreme caution be taken when using seasonally adjusted data, as the incorrect use of such data may lead to quite misleading findings concerning the presence of nonlinearity, for example.

4.2 Outliers

A typical feature of economic time series is that some observations are irregular. For example, the May 1968 strike in France resulted in a very obvious downward spike. Taken one step further, one may, in some cases, wish to classify two successive negative valued observations of GDP growth as outliers, if in the middle of an extremely long expansionary phase, for example. One nice feature of nonlinear models is that they may be explicitly designed to jointly describe the different dynamics associated with expansions and recessions. However, since nonlinear models already tend to be constructed using a substantial number of parameters one may wish to stipulate that parameters are not used to fit only one or two (outlying) observations. If such a strategy is not followed, difficulties at the estimation stage may arise. Further, in such cases, out-of-sample forecasts may be poor relative to forecasts constructed using simpler more parsimonious models. Given these considerations, it becomes relevant to test for nonlinearity in the presence of outliers. Van Dijk, Franses and Lucas (1996a, b) propose tests for nonlinearity and ARCH in the presence of outliers. Their tests are based on the generalized M-estimator, where possibly irregular data points are given less weight. Extensive Monte Carlo evidence in these two papers shows that these types of robust tests statistics have good size properties, and suffer little from diminished power. It is further shown that standard tests for nonlinearity and ARCH can have empirical sizes ranging from 20% to 100% if not corrected for outliers!

The application of these robust tests for nonlinearity to the industrial production data used by Anderson and Teräsvirta (1992) reveals that for Austria, Belgium, and the U.S.A. there actually appears to be no nonlinearity after all. Upon further inspection of the data, it appears that the nonlinearity found using standard nonrobust tests is attributable to a small number of outlying observations. For financial data, Franses and Van Dijk (1997) document that one may also often find ARCH because of neglected outliers. In particular, they examine 22 weekly and monthly exchange rate series, as well as 13 stock market indices for samples of length 5 years. Their main result is that spurious GARCH is found over 50% of the time. Monte Carlo experiments are also run, providing further evidence that Frances and Van Dijk's (1997) results are indeed driven by outliers.

In another related study, Franses and Draisma (1997) find that hidden layers in ANN activation functions are frequently activated by only a single observation. This suggests that the hidden unit is used to fit only one observation, which may lead to the problems discussed at the beginning of the section. From a practical perspective, we recommend careful investigation into the possibility that findings of nonlinearity are driven by one or few irregular observations. This can be done quite easily by using the robust nonlinearity tests discussed above, for example. However, there is a clear tradeoff between how many outliers are admitted into the DGP and the usefulness of nonlinear models for capturing asymmetries and nonlinearities which may characterize a data series. In particular, allowing too few outliers also has drawbacks, as one may argue that certain so called outliers are precisely those data which are not well treated using linear models, and which we wish to model using nonlinear models.

4.3 Model Selection

Probably one of the most difficult (and currently unresolved) issues in empirical modelling is that of model selection. Once we leave the linear world, there are a virtually endless collection of nonlinear models available, making comparison of all nonlinear alternatives essentially impossible. Indeed, comparing only a few nonlinear alternatives is already, in many cases, quite time consuming. Broadly speaking, there are two strategies which one may wish to consider in such cases. The first involves fitting a linear model and applying nonlinearity tests which have power in the direction in which the investigator is interested. One drawback of this approach is that many of the currently available nonlinearity tests are portmanteau tests, have some power in many directions, but little power in any particular direction. Other tests are powerful in the direction of interest, but also have power against selected other alternatives. An example here is the STAR (smooth transition autoregressive) test (see Granger and Teräsvirta (1993) and the references contained therein), which has power against ARCH, while the ARCH-LM test has power against STAR. A second possible strategy is just to fit one's "favorite" nonlinear model, and investigate coefficient restrictions (such as zero restrictions). However, it is often difficult to exactly identify a nonlinear model, leading to some difficulty when implementing this strategy, in many cases. Also, when estimated parameters have population coefficients equal to zero, some of the estimation techniques used in practice break down. An applied strategy which is often used in place of the above strategies (see also Sections 2 and 3 above) is to use forecast based comparison, or to use the SIC and AIC, for example. These criteria are quite popular, even though much remains unknown with respect to their small sample and consistency properties, in a nonlinear context. However, the SIC and AIC are based on mean-square-error loss, and when somewhat complex nonlinear models (such as ANN and STAR) models are examined, it is not clear if such loss criteria are relevant. For example, in many cases asymmetric loss may be of interest. Further, as competing nonlinear models often have very different characteristics (and thus strengths and weaknesses), one must be careful to compare either (i) all features of the competing models, or (ii) only those features which are of particular interest. Boswijk and Franses (1997) suggest that one time series characteristic which is often of interest is the impulse response function (IRF), and they propose using IRFs as an aid to model selection. For example, seemingly very different models may have similar IRFs, while models which seem alike in many respects may have very different IRFs. While IRFs cannot be used to form a final decision when comparing models, they can serve as an aid, and may be used, in addition to the various other model selection criteria discussed above, when selecting among competing models.

In summary, when selecting among nonlinear models, it is sensible to first select model selection criteria based on carefully constructed loss functions. Individual end users generally require models which are useful for many different purposes, from forecasting mean to maximizing profits. Second, it is reasonable to assume than any single model selection criteria only sheds light on a small "part" of the overall picture. Using a number of model selection criteria may thus be useful when comparing models. Third, there are an infinite variety of models which may be compared. When using nonlinear models, it is perhaps sensible to begin with some basic benchmark linear models, and then consider a small set of alternative nonlinear models which are of particular interest, given economic theory and other considerations.

Acknowledgements

Some parts of the discussion in this chapter are summaries of more detailed discussions which can be found in "An Introduction to Stochastic Unit Root Processes," "A Model Selection Approach to Assessing the Information in the Term Structure Using Linear Models and Artificial Neural Networks," "A Model Selection Approach to Real Time Macroeconomic Forecasting Using Linear Models and Artificial Neural Networks," and "Further Developments in the Study of Cointegrated Variables" (see references). Swanson, thus, wishes to thank Clive W.J. Granger and Halbert White, who co-authored all of these papers. Swanson also wishes to thank the Research and Graduate Studies Office at Pennsylvania State University for financial support during the writing of this article.

Note

1. This is an approximation of the true variance of x_t, since $E[\pi_{t,j}^2] \neq \exp(2j\phi)$ for small j.

References

Bishop, C. M. (1995). *Neural Networks for Pattern Recognition.* Oxford: Claredon Press.

Boswijk, H. P. and P. H. Franses (1997). "Common Persistence in Nonlinear Autoregressive Models," Econometric Institute Report 9702, Erasmus University.

Brandt, A. (1986). "The Stochastic Equation $Y_{n+1} = A_n Y_n + B_n$ with Stationary Coefficients," *Advances In Applied Probability*, 18, 211-220.

Brock, W. A., W. D. Dechert, J. A. Scheinkman, and B. LeBaron (1996). "A Test for Independence Based on the Correlation Dimension," *Econometric Reviews*, 15, 197-235.

Brock, W. A., D. A. Hsieh, and B. LeBaron (1991). *Nonlinear Dynamics, Chaos, and Instability.* Cambridge: MIT Press.

Burns, A. F. and W. C. Mitchell (1946). *Measuring Business Cycles.* New York: Columbia University Press.

Carroll, S. M. and B. W. Dickinson (1989). "Construction of Neural Nets Using the Radon

Transform," in *Proceedings of the International Joint Conference on Neural Networks.* Washington DC. New York: IEEE Press, 607-611.

Chen, X. and N. R. Swanson (1997). "Semiparametic ARX Neural Network Models With an Application to Forecasting Inflation," working paper, Department of Economics, Pennsylvania State University.

Corradi, V., N. R. Swanson, and H. White (1997). "Testing for Stationarity Ergodicity and for Comovement Between Nonlinear Discrete Time Markov Processes," working paper, Department of Economics, Pennsylvania State University.

Croushore, D. (1993). "Introducing: The Survey of Professional Forecasters," *The Federal Reserve Bank of Philadelphia Business Review*, November-December, 3-15.

Cybenko, G. (1989). "Approximation by Superpositions of a Sigmoid Function," *Mathematics of Control Signals and Systems*, 2, 303-314.

Diebold, F. X. and R. S. Mariano (1995). "Comparing Predictive Accuracy," *Journal of Business and Economic Statistics*, 13, 253-263.

Diebold, F. X. and J. A. Nason (1990). "Nonparametric Exchange Rate Prediction," *Journal of International Economics*, 28, 315-332.

Dorsey, R. E., J. D. Johnson, and M. V. van Boening (1994). "The Use of Artificial Neural Networks for Estimation of Decision Surfaces in First Price Sealed Bid Auctions," in Cooper, W. W. and A. Whinston, (eds.) *New Directions in Computational Economics.* Boston: Kluwer, 19-40.

Dropsy, V. (1992). "Exchange Rates and Neural Networks," working paper No. 1-92, Department of Economics, California State University Fullerton.

Dutta, S. and S. Shekhar (1989). "Bond Rating: A Non-Conservative Application of Neural Networks," in *Proceedings of the IEEE International Conference on Neural Networks*, San Diego. New York: IEEE Press, 443-450.

Engle, R. F. (1982). "Autoregressive Conditional Heteroskedasticity With Estimates of the Variance of U.K. Inflation," *Econometrica*, 50, 987-1008.

Engle, R. F. and S. J. Brown (1986). "Model Selection for Forecasting," *Applied Mathematics and Computation*, 20, 313-327.

Fair, R. C. and R. J. Shiller (1990). "Comparing Information in Forecasts from Econometric Models," *American Economic Review*, 80, 375-389.

Franses, P. H. (1996). *Periodicity and Stochastic Trends in Economic Time Series.* Oxford: Oxford University Press.

Franses, P. H. and D. Van Dijk (1997). "Do We Often Find ARCH Because of Neglected Outliers?" Economic Institute Report 9706, Erasmus University.

Franses, P. H. and R. Paap (1996). "Does Seasonal Adjustment Change Inference from Markov Switching Models?" Econometric Institute Report 9615, Erasmus University.

Franses, P. H. and G. Draisma (1997). "Recognizing Changing Seasonal Patterns Using Artificial Neural Networks," *Journal of Econometrics*, forthcoming.

Franses, P. H. and P. Van Homelen (1997). "On Forecasting Exchange Rates Using Neural Networks," *Applied Financial Economics*, forthcoming.

Funahashi, K. (1989). "On the Approximate Realization of Continuous Mappings by Neural Networks," *Neural Networks,* 2, 183-192.

Ghysels, E., C. W. J. Granger, and P. L. Siklos (1996). "Is Seasonal Adjustment a Linear or Nonlinear Data Filtering Process?" *Journal of Business and Economic Statistics*, 14, 374-386.

Granger, C. W. J. (1993). "Strategies for Modeling Nonlinear Time Series Relationships," *The Economic Record*, 69, 233-238.

Granger, C. W. J. and A. P. Andersen (1978). *An Introduction to Bilinear Time Series Models.* Göttingen: Vandenhoeck and Ruprecht.

Granger, C. W. J. and N. R. Swanson (1996). "Further Developments in the Study of Cointegrated Variables," *Oxford Bulletin of Economics and Statistics,* 58, 537-553.

Granger, C. W. J. and N. R. Swanson (1997). "An Introduction to Stochastic Unit Root Processes," *Journal of Econometrics*, 80, 35-62.

Granger, C. W. J. and T. Teräsvirta (1993). *Modelling Nonlinear Relationships*. New York: Oxford University Press.

Guyton, D. A., N. -F. Zhang and R. V. Foutz (1986). "A Random Parameter Process for Modeling and Forecasting Time Series," *Journal of Time Series Analysis*, 7, 105-115.

Hall, R. E. (1978). "Stochastic Implications of the Life Cycle-Permanent Income Hypothesis: Theory and Evidence," *Journal of Political Economy*, 96, 971-987.

Hamilton, J. D. (1989). "A New Approach to the Economic Analysis of Nonstationary Time Series and the Business Cycle," *Econometrica*, 57, 357-384.

Hamilton, J. D. (1994). *Time Series Analysis*. New Jersey: Princeton University Press.

Henriksson, R. D. and R. C. Merton (1981). "On Market Timing and Investment Performance. II. Statistical Procedures for Evaluating Forecasting Skills, " *Journal of Business*, 54, 513-533.

Hornik, K., M. Stinchcombe, and H. White (1989). "Multilayer Feedforward Networks are Universal Approximators," *Neural Networks*, 2, 359-366.

Hornik, K., M. Stinchcombe, and H. White (1990). "Universal Approximation of an Unknown Mapping and its Derivatives Using Multilayer Feedforward Networks," *Neural Networks*, 3, 551-560.

Journal of Applied Econometrics (1992). "Special Issue on Nonlinear Dynamics and Econometrics," 7, December, New York: Wiley.

Keane, M. P. and D. E. Runkle (1990). "Testing the Rationality of Price Forecasts," *American Economic Review*, 80, 714-735.

Koopmans, L. H. (1974). *The Spectral Analysis of Time Series*. New York: Academic Press.

Kuan, C. -M. and T. Liu (1995). "Forecasting Exchange Rates Using Feedforward and Recurrent Neural Networks," *Journal of Applied Econometrics*, 10, 347-364.

Kuan, C. -M. and H. White (1994). "Artificial Neural Networks: An Econometric Perspective," *Econometric Reviews*, 13, 1-91.

Laroque, G. and B. Salanie (1994). "Estimating the Canonical Disequilibrium Model: Asymptotic Theory and Finite Sample Properties," *Journal of Econometrics*, 62, 165-210.

Lee, T. -H., H. White, and C. W. J. Granger (1993). "Testing for Neglected Nonlinearity in Time Series Models: A Comparison of Neural Network Methods and Alternative Tests," *Journal of Econometrics*, 56, 269-290.

Leitch, G. and J. E. Tanner (1991). "Economic Forecast Evaluation: Profits Versus the Conventional Error Measures," *American Economic Review*, 81, 580-590.

Leybourne, S. J., B. P. M. McCabe, and A. R. Tremayne (1994). "Can Economic Time Series be Differenced to Stationarity?" working paper, University of Nottingham.

Luukkonen, R., P. Saikkonen, and T. Teräsvirta (1988a). "Testing Linearity Against Smooth Transition Autoregressive Models," *Biometrika*, 75, 491-499.

Luukkonen, R., P. Saikkonen, and T. Teräsvirta (1988b). "Testing Linearity in Univariate Time Series Models," *Scandinavian Journal of Statistics*, 15, 161-175.

Mishkin, F. S. (1988). "The Information in the Term Structure: Some Further Results," *Journal of Applied Econometrics*, 3, 307-314.

Mizrach, B. (1992). "Multivariate Nearest-Neighbor Forecasts of EMS Exchange Rates," *Journal of Applied Economics*, 7, 151-164.

Moody, J. and J. Utans (1991). "Principled Architecture Selection for Neural Networks: Applications to Corporate Bond Rating Predictions," in J. E. Moody, S. J. Hanson, and R. P. Lippmann, (eds.) *Advances in Neural Information Processing Systems* 4. San Mateo: Morgan Kaufman, 638-690.

Neftci, S. N. (1984). "Are Economic Time Series Asymmetric Over the Business Cycles?" *Journal of Political Economy*, 92, 306-328.

Pesaran, M. H. and A. G. Timmerman (1992). "A Simple Nonparametric Test of Predictive

Performance, *Journal of Business and Economic Statistics*, 10, 461-465.
Pesaran, M. H. and A. G. Timmerman (1994a). "The Use of Recursive Model Selection Strategies in Forecasting Stock Returns," working paper, Department of Economics, University of Cambridge.
Pesaran, M. H. and A. G. Timmerman (1994b). "A Generalization of the Non-Parametric Henriksson-Merton Test of Market Timing," *Economic Letters*, 44, 1-7.
Potter, S. M. (1995). "A Nonlinear Approach to US GNP," *Journal of Applied Econometrics* , 10, 109-125.
Pourahmadi, M. (1986). "On Stationarity of the Solution of a Doubly Stochastic Model," *Journal Of Time Series Analysis*, 7, 123-131.
Pourahmadi, M. (1988). "Stationarity of the Solution of $X_t = A_t X_{t-1} + \epsilon_t$ and Analysis of Non-Gaussian Dependent Random Variables," *Journal of Time Series Analysis*, 9, 225-230.
Ripley, B. D. (1994). "Neural Networks and Related Methods for Classification," *Journal of the Royal Statistical Society Series B*, 56, 409-456.
Rothman, P. (1992). "The Comparative Power of the TR Test Against Simple Threshold Models," *Journal of Applied Econometrics*, 7, 187-195.
Rumelhart, D. E. and J. L. McClelland (1986). *Parallel Distributed Processing: Explorations in the Microstructures of Cognition*. Cambridge: MIT Press.
Steckler, H. O. (1991). "Macroeconomic Forecast Evaluation Techniques," *International Journal of Forecasting*, 7, 375-384.
Steckler, H. O. (1994). "Are Economic Forecasts Valuable?" *Journal of Forecasting*, 13, 495-505.
Swanson, N. R. (1996). "Forecasting Using First Available Versus Fully Revised Economic Time Series Data," *Studies in Nonlinear Dynamics and Econometrics*, 1, 47-64.
Swanson, N. R. and H. White (1995). "A Model Selection Approach to Assessing the Information in the Term Structure Using Linear Models and Artificial Neural Networks," *Journal of Business and Economic Statistics*, 13, 265-275.
Swanson, N. R. and H. White (1997). "A Model Selection Approach to Real Time Macroeconomic Forecasting Using Linear Models and Artificial Neural Networks, *Review of Economics and Statistics*, 79, 540-550.
Teräsvirta, T. and H. M. Anderson (1992). "Characterizing Nonlinearities in Business Cycles Using Smooth Transition Autoregressive Models," *Journal of Applied Econometrics*, 7, 119-136.
Tjøstheim, T. (1986). "Some Doubly Stochastic Time Series Models," *Journal of Time Series Analysis*, 7, 51-72.
Tong, H. (1983). *Threshold Models in Non-Linear Time Series Analysis*. New York: Springer-Verlag.
Tong, H. (1990). *Non-Linear Time Series: A Dynamical System Approach*. Oxford: Oxford University Press.
Trippi, R. and E. Turbau (1993). *Neural Networks in Finance and Investing*. Chicago: Probus Publishing Company.
Van Dijk, D., P. H. Franses, and A. Lucas (1996a). "Testing for Smooth Transition Autoregression in the Presence of Outliers," Econometric Institute Report 9622, Erasmus University.
Van Dijk, D., P. H. Franses, and A. Lucas (1996b). "Testing for ARCH in the Presence of Additive Outliers," Econometric Institute Report 9659, Erasmus University.
Weiss, A. A. (1996). "Estimating Time Series Models Using the Relevant Cost Function," *Journal of Applied Econometrics*, 11, 539-560.
White, H. (1988). "Economic Prediction Using Neural Networks: The Case of IBM Daily Stock Returns," in *Proceedings of the IEEE International Conference on Neural Networks*, San Diego. New York: IEEE Press, 451-458.
White, H. (1989). "Learning in Artificial Neural Networks: A Statistical Perspective," *Neural Computation*, 1, 425-464.
White, H. (1990). "Connectionist Nonparametric Regression: Multilayer Feedforward Networks

Can Learn Arbitrary Mappings," *Neural Networks*, 3, 535-549.

Zarnowitz, V. and P. Braun (1992). "Twenty-Two Years of the NBER-ASA Quarterly Economic Outlook Surveys: Aspects and Comparisons of Forecasting Performance," *NBER* working paper No. 3965.

5 UNIT-ROOT TESTS AND EXCESS RETURNS

Marie-Josée Godbout
Bank of Canada
Simon van Norden
École des Hautes Études Commercial

1. Introduction

When introducing students to the modern theory of financial markets, it is common to characterize the behavior of the log of asset prices as martingales, and their excess returns as being serially uncorrelated or even unpredictable. This is consistent with Fama's (1970) characterization of weak, semi-strong, and strong market efficiency. To be sure, there is an extensive literature documenting the deviations of asset prices from this characterization. Despite this, most would accept the proposition that asset prices contain a unit root in their time-series representation and that excess returns do not. Put another way, the stylized fact is that asset prices are integrated of order one (I(1)) and excess returns are integrated of order zero (I(0)). However, a small number of prominent recent papers have presented evidence that appears to reject this characterization in a surprising way. They show that, according to some tests, some excess returns appear to contain a unit root.

Of the four papers of which we are aware, three provide evidence from foreign exchange markets; Evans and Lewis (1993) (herinafter EL93), Evans and Lewis (1995) (herinafter EL95) and Crowder (1994).[1] All three begin with the premise that the log of the spot exchange rate s_t is I(1). Therefore, if it is not cointegrated with the log of the *k*-

period forward rate f_t, or, more specifically, if its cointegrating vector is not [1,-1], then excess returns to holding foreign exchange for k periods $x_t^k = s_{t+k} - f_t^k$ will be I(1).[2]

There is extensive published evidence that such returns appear to be I(0) and that spot and forward rates seem to be cointegrated with cointegrating vector [1, -1]. (For example, see Phillips, McFarland, and McMahon (1996) or the survey by Brenner and Kroner (1995)). However, Evans and Lewis carefully argue that the published evidence is not conclusive, noting the difficulties inherent in determining the asymptotic behavior of a series from a finite sample.[3] For example, if the permanent trend in x_t accounts for only a small fraction of its sample variance, even over the space of a decade or more, then unit-root tests on excess returns may incorrectly tend to reject the null hypothesis of a unit root, a fact documented in EL93 via Monte Carlo experiments.

More importantly, however, Evans and Lewis are careful not to interpret their results as literal evidence of unit roots in excess returns.[4] Instead, they argue that the presence of infrequent changes in regime that are rationally anticipated by the market (i.e., a "peso problem") will lead to very persistent "trends" in excess returns. Their simulations lead them to conclude that such changes in regime can cause tests to reject the null hypothesis of stationary excess returns in realistic sample sizes, even when they are truly stationary in large enough samples. Their preferred interpretation of the evidence they present is that there are highly persistent trends in excess returns that are consistent with regime-switching behavior in exchange rates.

On the other hand, Crowder (1994) interprets the results of unit root and cointegration tests more narrowly. He extends Baillie and Bollerslev's (1989) work on the apparent cointegration of nominal spot exchange rates, noting that such cointegration implies that changes in some foreign exchange rates must be predictable. Such predictability is commonly attributed to the existence of a risk premium. However, if the spot and forward exchange rates are not cointegrated, markets can only be economically efficient if the risk premium is I(1), whereas if spot exchange rates are cointegrated, then economic efficiency implies that the risk premium must be I(0). Since the risk premium cannot be both, Crowder interprets this as evidence against the economic efficiency of foreign exchange markets.

This paper reconsiders these new results about the persistence of excess returns. Specifically, we examine the literature on finite-sample problems in the application of cointegration tests and show how these may have affected the above conclusions. Section 1 begins by critically reviewing the above papers in more detail and explains why we believe that their conclusions may not be as sound as they appear. Section 2 attempts to replicate previously published results based on tests for the number of cointegrating vectors in a system of variables and applies a series of simulation experiments to assess their robustness. Section 3 does the same for tests of parameter restrictions on the cointegrating vector. Section 4 summarizes our conclusion and suggests possible extensions.

2. Literature Survey

We begin with a review of the empirical evidence presented by Evans and Lewis, which is based on tests of cointegrating relationships between spot and forward exchange rates.

These authors also present a model which predicts persistent trends in excess returns, based on the presence of regime-switching behavior in exchange rates. Such models are not entirely new and follow a well-established line of research on "peso problems" and related phenomena. However, in our minds the most striking contribution of EL93 and EL95 is the new empirical evidence they present on the apparent persistence of these trends. For reasons of brevity, we therefore limit our review to the evidence they present on these trends.

As noted in the introduction, Crowder (1994) presents similar evidence. However, his results are sensitive to the lag length selected, and he does no simulations to check the finite sample behavior of his tests. Moreover, his article is immediately followed by a rebuttal by Baillie and Bollerslev (1994). For these reasons, we think Evans and Lewis' results are the most convincing seen to date and we limit ourselves to their work. However, we should note that Crowder's methodology is essentially the same as that in EL93. Therefore, we suspect that the results we present on their work could apply equally to Crowder's.

Both of the studies by Evans and Lewis use the same data set, taken from Bekaert and Hodrick (1993).[5] This consists of log spot exchange rates against the U.S. dollar for the British pound, the German mark and the Japanese yen, and the three corresponding 1- or 3-month log forward exchange rates. The data are monthly and cover the period 1975 to 1989. Excess returns are taken to be the difference between the log forward rate and the log spot rate corresponding to the same settlement data as the forward rate.

1.1 Evans and Lewis (1993)

1.1.1 Summary

If each spot rate is cointegrated with its respective forward rate, then a system of three spot rates and three forward rates should have three cointegrating vectors in the system. If there are fewer than three cointegrating vectors, then at least one of the excess return series (i.e., the difference between the spot rate and its respective forward rate) must have a unit root. EL93 use the Johansen (1988) maximum-likelihood tests for the number of cointegrating vectors in such a three-exchange rate system, and they never find significant evidence of more than two cointegrating vectors.[6]

An alternative explanation for their failure to find evidence of three cointegrating vectors might be that the Johansen (1988) tests simply lack the power to detect them. EL93 address this problem with a series of Monte Carlo experiments (documented in Evans and Lewis (1992)). Simulating six-variable systems with three cointegrating vectors, they find that a lack of power in the Johansen test cannot explain their results. More precisely, they find that the value of the test statistic for the null hypothesis that there are no more than two cointegrating vectors lies well below the 99 per cent critical value of its simulated empirical distribution under the alternative hypothesis. Therefore, in their experiment where three cointegrating vectors exist, they find test statistics larger than those in the true data with a probability of over 99 per cent. They conclude that "...there is strong evidence of statistically significant trends in forward rates relative to spot rates."[7]

1.1.2 Critique

Since Evans and Lewis (EL) are dealing with a sample size of 180 observations, finite sample problems associated with cointegration tests might affect the results. However, the size-distortion problem commonly associated with this test should make the maximum-likelihood (ML) tests find evidence of too many cointegrating vectors, not too few.[8] Furthermore, EL attempted to compensate for these finite-sample effects with their simulation experiments. Despite this, we think that there are grounds for questioning the EL93 results.

First, contrary to the conclusion of EL93, EL95 (discussed in more detail below) state that these same spot rates are cointegrated with forward rates. Specifically, they say:

> Before running the regressions, we used the Johansen (1988) procedure to check that the pairs of individual spot and forward rates were cointegrated. The same procedure was also used to test for the number of trends in the vectors of three spot and three forward rates used in the system estimation. We could not eject the hypothesis that all these vectors contained three trends.[9]

EL95 do not reference their earlier work, so it is not obvious how these results are to be reconciled with those of EL93.

More importantly, however, we are concerned that the data-generating process (DGP) used in the simulations in EL93 and Evans and Lewis (1992) might give unrepresentative results. Their DGP is specified as $\bar{z}_t = [\bar{z}_{1t}', \bar{z}_{2t}']'$, where[10]

$$\begin{aligned} \bar{v}_t &= \bar{v}_{t-1} + \bar{\epsilon}_t & E_{t-1}(\bar{\epsilon}_t \cdot \bar{\epsilon}_t') &= Q_t \\ \bar{z}_{1t} &= \bar{v}_t + \Lambda \cdot \bar{u}_{1t} & E_{t-1}(\bar{u}_{1t} \cdot \bar{u}_{1t}') &= \Sigma_{1t} \\ \bar{z}_{2t} &= \Gamma \cdot \bar{z}_{1t} + \bar{u}_{2t} & E_{t-1}(\bar{u}_{2t} \cdot \bar{u}_{2t}) &= \Sigma_{2t} \end{aligned} \tag{1}$$

$\bar{z}_t$ is a d dimensional vector with m stochastic trends (d-m cointegrating vectors), Γ is the matrix of cointegrating vectors, $\bar{u}_{1t}, \bar{u}_{2t}, \bar{\epsilon}_t$ are mutually uncorrelated innovations of dimension m, d-m, and d.

Although EL vary the number of cointegrating relationships (m) in their simulations, we are concerned here with the case where we have three pairs of cointegrated series (so $d = 6$, $m = 3$.) They also considered both homoskedastic and heteroskedastic innovation processes, finding that they gave very similar results. We will restrict our attention to the homoskedastic case. They further assume that each of these three matrices is diagonal, and that both Γ and Λ are the identity matrix. Together, this implies that (1) is equivalent to three unrelated bivariate systems of the form

$$\begin{aligned} v_t^i &= v_{t-1}^i + \epsilon_t^i & E_{t-1}(\epsilon_t^i \cdot \epsilon_t^i) &= q^i \\ z_{1t}^i &= v_t^i + u_{1t}^i & E_{t-1}(u_{1t}^i \cdot u_{1t}^i) &= \sigma_1^i \\ z_{2t}^i &= z_{1t}^i + u_{2t}^i & E_{t-1}(u_{2t}^i \cdot u_{2t}^i) &= \sigma_2^i \end{aligned} \tag{2}$$

One can think of each of these systems as corresponding to the spot and forward rate for a particular currency i. Each system has two integrated moving-average IMA (1,1) variables that are jointly cointegrated with cointegrating vector [1,1]. To see that both series are IMA(1,1), note that

$$\Delta z_{1t}^{i} = \epsilon_t^i + \Delta u_t^i = e_t^i + \theta_1 \cdot e_{t-1}^i, \text{ where } \theta_1 = -\sigma_1^i/(\sigma_1^i + q^i) \tag{3}$$

for some innovation sequence e_t^i. The same logic applies to z_{2t}^i, except that its moving-average (MA) parameter will be $\theta_2 = -(\sigma_1^i + \sigma_2^i)/(\sigma_1^i + \sigma_2^i + q^i)$. The process is then calibrated so that $\sigma_1^i = q^i = E(\Delta s^i \cdot \Delta s^i)$ and $\sigma_2^i = E(\Delta f^i \cdot \Delta f^i)$, where s^i and f^i are the log spot and forward exchange rates for currency i. If their first differences have roughly equal variances, then the MA parameters will be $\theta_1 = -0.5$ and $\theta_2 = -0.67$.

Evans and Lewis offer no justification for choosing this particular DGP or parameterization. It has at least one unrealistic feature: the presence of a strong negative MA component in both spot and forward exchange rates, when both are thought to closely approximate a martingale; for example, see Meese and Rogoff (1983). This difference could plausibly affect the outcome of their simulations. It is well understood that substantial MA components will require long autoregressive (AR) lags in the cointegration test to serve as a suitable approximation of the short-run dynamics. However, Evans and Lewis (1992) consider only the case of one and three AR lags. It is also understood that failure to adequately capture negative MA dynamics can make series appear to be more stationary than they are in reality. This in turn might cause cointegration tests on the simulated data to find fewer stochastic trends than are truly present, which would tend to exaggerate the power of the cointegration test. We are concerned that this, rather than the presence of very persistent trends in excess returns, might be the reason that EL93 find much less evidence of cointegration in the true data than in their simulation experiments with three cointegrating vectors.

We investigate the latter problem below with our own simulation experiments. However, we think that some of the Monte Carlo work presented by Evans and Lewis (1992) is consistent with the reinterpretation of their results suggested above. Table B in Evans and Lewis (1992) shows that when there is no cointegration, asymptotic 95 per cent critical values would detect cointegration more than 50 per cent of the time in their three-variable system and more than 20 per cent of the time in their six-variable system. Such size distortion in the simulation experiment could exaggerate the estimated power of the test in the actual data. Furthermore, small increases in the number of lags used in their test have important effects on the simulation results. For example, their Table B reports that in a six-variable system with three cointegrating vectors and homoskedastic errors, increasing the number of lags from one to three lowers the 1 per cent critical value from 63.99 to 48.26. This is consistent with the hypothesis that the combination of an important MA component and few lags are driving their results.

1.2 Evans and Lewis (1995)

1.2.1 Summary

EL95 use the same data set as EL93 but, as discussed in endnote 9, work under the assumption that spot and forward rates are cointegrated. Instead of testing for the number of cointegrating vectors, EL simply test whether spot and forward rates move one-for-one in the long run (i.e., whether the cointegrating vector is [1, -1].)[11] EL95 use the dynamic ordinary least squares (DOLS) estimator proposed in Stock and Watson (1993) to test the cointegrating vectors. Tests of a single cointegrating vector require only ordinary least squares (OLS) estimation of

$$s_{t+k} = \alpha_0 + \alpha_1 \cdot f_t + \sum_{n=-3}^{3} \beta_n \cdot \Delta f_{t-n} + u_t \tag{4}$$

Hypotheses about α_1 can then be tested with the usual t-statistics, or if Newey-West standard errors are used, with a χ^2 (1) statistic. Testing the joint hypothesis that $\alpha_1 = 1$ for each of three exchange rates simply requires seemingly unrelated regression (SUR) estimation of three OLS equations (one for each currency), and the usual Wald statistic has a asymptotic $\chi^2(3)$ distribution under the null. EL95 show that t-tests for both 1- and 3-month forward rates allow forward rates allow them to reject the hypothesis that the cointegrating vector is [1,-1] for some currencies, and that the joint test always rejects with a marginal significance level of less than 0.1 per cent.

To address concerns about the finite-sample of their estimator, Table 3 in EL95 reports the results of a Monte Carlo experiment that examines the finite-sample properties of their test statistics under the null hypothesis that the cointegrating vectors are all [1, -1]. They conclude that linear DGPs cannot explain their rejection of the [1, -1] cointegrating vector, based on 1000 replications.

1.2.2 Critique

Stock and Watson (1993) find that the DOLS estimator tends to overstate the true precision of the estimated cointegrating vectors, and that this is worsened when non-parametric variance-covariance (VCV) estimators such as Newey-West are used. This may give grounds for believing that the marginal significance levels at which EL95 reject the cointegrating vector restriction are exaggerated. However, it does not explain why their Monte Carlo experiment would not detect such a problem.

EL begin their Monte Carlo experiment by estimating (4) and saving the residuals. They then bootstrap a new series of residuals and simulate an artificial series for s_{t+k} under the restriction that $\alpha_1 = 1$.[12] These artificial series are then used to determine the distribution of the DOLS test statistics under the null hypothesis that $\alpha_1 = 1$. This experiment therefore changes $\{s_t\}$ while holding $\{f_t\}$ constant. It also imposes the assumption that (4) is well specified.

The key to the validity of the Stock-Watson test is that the differences of the explanatory variable in (4) must produce errors that are orthogonal to the level of the explanatory variable; otherwise, estimates of the cointegrating vector will be biased. The design of the EL95 Monte Carlo guarantees that the generated errors will be independent of the forward rate, so that such bias cannot arise. Evans and Lewis justify this by noting that: "As discussed in Stock and Watson (1993), the residuals from this equation are independent of the entire sequence of the right-hand side variable and so can be treated as strictly exogenous."[13] This is true asymptotically if enough leads and lags of the differences are included. The extent to which this may hold in a finite sample for a given number of lags is not clear.

The conjecture that the endogeneity of the forward rate can account for EL95's rejections of the [1, -1] cointegrating vector between spot and forward rates is consistent with at least one other piece of evidence. EL95 present additional simulations of a regime-switching model, which manages to account for the large Stock-Watson test statistics they find in the true data. However, a potentially important difference between their simulations of (4) and their regime-switching model simulations is that, in the latter, spot and forward rates are modelled simultaneously instead of the forward rate being fixed. Therefore, the difference that they document in the behavior of their linear and non-linear models might not be evidence of regime-switching in exchange rates; it could instead be an artifact of finite-sample problems in the Stock-Watson test with endogenous regressors.

1.3 Conclusion

For the reasons given above, we think that the conclusions presented by Evans and Lewis may be premature, and that their results deserve closer scrutiny. In the following sections, we re-evaluate the results in EL93 and EL95 with the aid of further simulation experiments designed to test the hypotheses advanced above. Specifically, we examine whether the simulation results in EL93 on the power of the cointegration tests are sensitive to their assumptions on the exogeneity of the forward rate.

One additional aspect of this literature should be noted. Evans and Lewis use cointegration tests indirectly, to provide evidence of the importance of regime-switching in financial markets. There is no dispute over whether one can find evidence of regime-switching in financial markets and there is no shortage of direct empirical work on this. Apparently, their intended contribution is to show that this regime-switching gives rise to very persistent trends in excess returns as evidenced by the cointegration tests.

Viewed from this perspective, we think their work leaves several important questions unanswered. Is a cointegration test a more meaningful measure of persistence than autocorrelation coefficients? How is it superior to previous measures of return persistence? Is it particularly sensitive to the effects of regime-switching?

There is some evidence on the last question, but it does not tend to support the conclusions of Evans and Lewis. Evans (1991)[14] found that regime-switching tended to make explosive series appear stationary and to make non-cointegrated series appear to be cointegrated. Evans and Lewis are claiming the opposite, based not on the theoretical properties of cointegration tests, but instead solely on the Monte Carlo evidence they present. We think that this is another reason for re-examining the robustness of this evidence. We do this in the next two sections.

2. Evidence from Tests for the Number of Cointegrating Vectors

2.1 Reproduction

The first step in analyzing Evans and Lewis' (EL) results is to try to reproduce them using the same data set from Bekaert and Hodrick (1993). Table 1 shows that while we are not able to replicate EL93's results precisely, our statistics are quite close and produce the same conclusions when using asymptotic critical values with the maximum eigenvalue

Table 1. Replicating Evans and Lewis (1993): 1-Month Forward Rate Johansen (1988) Test Results (One Lag)

Number of cointegrating vectors	Maximum Eigenvalue			Trace Statistic				
	Godbout & van Norden	Evans & Lewis 1993	5% Critical Value[a]	Godbout & van Norden	Evans & Lewis 1993	5% Critical Value[a]	Reinsel & Ahn	Linear Combination Test
0	58.65	53.13	36.36	139.18	*118.00*	82.49	134.49	131.95
1	38.54	36.83	30.04	80.54	*64.91*	59.46	77.82	77.64
2	20.10	15.20	23.80	40.99	*28.09*	39.89	39.61	40.14
3	12.88	---[b]	17.89	20.93	---[b]	24.31	20.22	20.62
4	7.56	---[b]	11.44	8.05	---[b]	12.53	7.78	7.97
5	0.49	---[b]	3.84	0.49	---[b]	3.84	0.47	0.49

Notes: [a] Taken from Osterwald-Lenum (1992, Table 0). [b] Results not reported.

statistic.[15] However, using the Trace statistic, Reinsel and Ahn's scaled trace statistics or Pitarakis and Gonzalo's linear combination test (LCT) we obtain test statistics roughly equal to their 95 per cent asymptotic critical values for the test of the third cointegrating vector, whereas EL93 found evidence of only two cointegrating vectors. Given that their Monte Carlo evidence suggests that the asymptotic critical values are very conservative in this case, this weak evidence of a third cointegrating vector may not be convincing by itself. Furthermore, we find that this evidence vanishes if we add more lags to the Johansen test. Although not reported above, we also repeated these tests using the 3-month forward rates considered by EL93. In that case, regardless of the test statistic or the number of lags used, we always found exactly two significant cointegrating vectors.

2.2 Experimental Design

We would now like to repeat EL93's simulations to determine whether these results are consistent with the presence of three cointegrating vectors. However, for the reasons explained in the previous section, we will use a different DGP to perform these simulations. For each pair of spot and forward exchange rates, we begin by estimating the following bivariate system via ordinary least squares:[16]

$$\begin{bmatrix} s_t \\ f_t \end{bmatrix} = \begin{bmatrix} \alpha & 1-\alpha \\ \beta & 1-\beta \end{bmatrix} \cdot \begin{bmatrix} s_{t-1} \\ f_{t-1} \end{bmatrix} + \begin{bmatrix} \epsilon_{st} \\ \epsilon_{ft} \end{bmatrix} \tag{5}$$

The restriction that the sum of the coefficients in each equation equals one ensures that the simulated spot and forward rates will be cointegrated with cointegrating vector [1, -1]. We do not impose any weak exogeneity restrictions, which in this case amount to restricting $\alpha = 1$ or $\beta = 1$. (Uncovered interest rate parity would imply $\alpha = 0$ for 1-period forward rates.) Since Evans and Lewis use few lags in their cointegration tests, we deliberately keep the dynamics of the system as simple as possible.

Having estimated (5) for each of the three currencies, we estimate the variance-covariance matrix of the six residual series. This estimate is then used to calibrate a mean-zero multivariate normal distribution, from which we make independent and identically distributed draws to simulate new residuals. These are then used in (5) to generate new data series of the same length as the original data. Johansen (1988) test statistics (using a single lag) are then calculated and the number of cointegrating vectors significant at the 95 per cent level tabulated. This was repeated for 1000 simulated data sets.

We feel that (5) is at least as plausible as DGP as that of EL93, since (5) relies on estimated rather than exogenously fixed parameters, and is consistent with the near-martingale properties of spot and forward exchange rates (whereas EL93's is not.) However, we also investigated a richer DGP, which added lagged differences of spot and forward rates to each equation in (5). We found that it produced similar conclusions, as explained below.

2.3 New Simulation Results

The results are very different from those reported by EL93. Although three cointegrating vectors are present by construction, in 1000 simulations we literally never find significant evidence of three such vectors, regardless of the test statistic used. Significant evidence of even two cointegrating vectors, such as we find in the true data, is relatively rare, occurring in less than 1 per cent of the trials. If we focus on the tests of the simple null hypothesis of less than two cointegrating vectors, we found that the test statistics were below their 95 per cent critical values in over 99 per cent of the simulations.[17] This means that the Johansen test would have effectively no power to detect the presence of three cointegrating vectors for the DGP we consider. Furthermore, in over 97 per cent of the trials we found that the simulated Lambda-max statistic was less than that produced by the true data. This suggests that, if anything, the true data show greater evidence of three cointegrating vectors than our artificial DGP.

Although not reported here, we similarly tested the data for the 3-month forward exchange rates. Again, the simulations literally never found evidence of three or more cointegrating vectors, despite their being present by construction. Again, we can reject the null of two or fewer cointegrating vectors in less than 1 per cent of the trials, and the true data give a Lambda-max statistic greater than the vast majority of the trials (in this case, 96 per cent).[18]

Based on the assumption that the DGP we used is at least as plausible as that considered by EL93, it seems that failure to find evidence of three cointegrating vectors may simply be due to the low power of the cointegration test. This in turn suggests that it is premature for EL93 to conclude that this is evidence of permanent of very persistent trends in excess returns.

3. Evidence from Tests of the Value of the Cointegrating Vector

3.1 Reproduction

Table 2 shows the results of our attempt to reproduce the results reported in EL95. The first three lines show the estimates of α_1 in the OLS regression.

$$s_{t+k} = \alpha_0 + \alpha_1 f_t + v_{t+k} \tag{6}$$

Regardless of whether 1-month or 3-month forward rates are used, we always manage to reproduce the results in EL95 to within an accuracy of 0.2 per cent, and to reproduce them to the limits of published accuracy in half of the cases. While these estimates of α_1 are all close to 1.0, they are known to be biased downwards in small samples. For comparison, we also report unbiased estimates (α_1 from (4)) produced by the same estimation procedure used to test the parameter estimate. Shown in the third line (under "Corrected"), they are considerably closer to 1.0. Rejecting the hypothesis that they are equal to 1.0 implies that we can discriminate between an estimate of 0.997 and a value of 1.0.

In reporting tests of the hypothesis that $\alpha_1 = 1$, we report both the calculated p-value of the test (to allow comparison with EL95) as well as the corresponding χ^2 test statistic. We are able to replicate their conclusions, although we cannot precisely replicate their results.[19] We always obtain very strong rejections of the joint hypothesis that $\alpha_1 = 1$ for all three currencies, regardless of the method used to calculate the standard errors and regardless of whether 1-month or 3-month forward rates are used.

We also used the Johansen and Juselius (1990) methodology to test restrictions on the cointegrating vectors. We tested both two-variable systems, consisting of the spot and forward rate for a given currency, as well as the full six-variable system. The former tests were done under the maintained hypothesis of one cointegrating vector, while the latter assumed three cointegrating vectors. In all cases, we were able to reject the restrictions on the cointegrating vectors at the 1 per cent significance level using asymptotic critical values. Given the problems associated with size distortion in the ML tests for cointegration, we cannot say how reliable such a conclusion is in finite samples.

3.2 Experimental Design

To investigate the sensitivity of the simulation results to the design of the experiment, we performed two separate experiments. First, we simulated the data using the method used by EL95 for their linear DGP. Then we simulated the data using the same DGP as in Section 2.2, with one minor change. Instead of generating the errors from a multivariate normal distribution, we followed EL95 and used random draws from the residuals of (5) to simulate new series. In all experiments, we performed 1000 replications.

Table 2. EL95 Estimation of Cointegrating Vectors: Jan. 1975 to Dec. 1989 (Heteroskedasticity and Serial-Correlation Robust Results)

Term	Statistic	Standard errors	Source	Pound	Mark	Yen	Joint
k=1	$\hat{\alpha}_1$		EL95	0.971	0.986	0.991	
			GvN 96	0.973	0.986	0.991	
			Corrected	0.999	0.998	0.997	
	Test $\hat{\alpha} = 1$	Heteroskedasticity-consistent	EL95	0.209	0.010	<0.001	<0.001
			GvN 96	0.463	0.013	<0.001	0.001
			χ^2	0.54	**6.12**	**16.87**	**16.20**
		HAC robust (Newey-West)	EL95	0.463	0.013	<0.001	<0.001
			GvN 96	0.628	0.082	0.011	0.001
			χ^2	0.24	3.03	**6.48**	**15.97**
k=3	$\hat{\alpha}_1$		EL95	0.909	0.946	0.967	
			GvN 96	0.908	0.948	0.969	
			Corrected	0.992	0.996	0.993	
	Test $\hat{\alpha} = 1$	Heteroskedasticity-consistent	EL95	0.021	0.052	<0.001	<0.001
			GvN 96	0.021	0.052	<0.001	<0.001
			χ^2	**5.33**	3.78	**15.91**	**32.82**
		HAC robust (Newey-West)	EL95	0.297	0.057	0.052	<0.001
			GvN 96	0.145	0.195	0.014	<0.001
			χ^2	2.12	1.68	**6.04**	**32.49**
		HAC robust (flat kernel)	GvN 96	0.243	0.284	0.051	<0.001
			χ^2	1.36	1.15	3.82	**32.14**

Notes: **Heteroskedasticity-consistent** indicates *p*-values calculated using heteroskedasticity-robust standard errors. **HAC robust** indicates *p*-values are calculated using standard errors that allow for heteroskedasticity and serial correlation using the correction suggested by Newey and West (1987). **HAC robust (flat kernel)** indicates *p*-values are calculated using standard errors that allow for heteroskedasticity and serial correlation using an unweighted kernel in place of the weighted kernel suggested by Newey and West (1987). This is arguably better suited to the case where MA serial correlation is induced by overlapping observations (as with 3-month forward rates). **Corrected** shows the estimated parameter corrected for the downward bias in finite samples using the same Stock and Watson (1993) estimator used for the subsequent hypothesis tests. χ^2 shows the test statistics upon which the GvN96 *p*-values are based. Under the null, these are distributed $\chi^2(1)$ for individual currencies and $\chi^2(3)$ for the joint test.

Table 3. Percentage of Trials Where $\alpha_1 = 1$ Not Rejected at the 5 Per Cent Level When True in (4) (Based on 1000 Trials)

Forward (months)	Standard errors	Critical Values	DGP	Yen	Mark	Pound	Joint
1	Heteroskedasticity-consistent	95%	(4)	92.9	94.4	94.8	94.4
			(5)	41.8	53.1	39.0	12.3
		Data based	(4)	52.1	98.6	100.	100.
			(5)	14.6	64.9	71.4	31.2
	HAC robust (Newey-West)	95%	(4)	91.4	94.0	93.8	94.5
			(5)	60.2	72.1	57.5	12.3
		Data based	(4)	34.9	90.0	98.0	100.
			(5)	15.3	66.1	69.8	30.9
3	Heteroskedasticity-consistent	95%	(4)	92.9	94.4	94.8	94.4
			(5)	32.9	47.1	32.0	8.6
		Data based	(4)	97.3	94.4	100.	100.
			(5)	39.4	46.6	61.4	43.5
	HAC robust (Newey-West)	95%	(4)	91.4	94.0	93.8	94.5
			(5)	50.7	65.6	50.2	8.6
		Data based	(4)	83.1	79.5	97.7	100.
			(5)	39.2	47.5	59.7	43.1
	HAC robust (flat kernel)	95%	(4)	89.8	92.5	91.6	94.7
			(5)	59.2	75.3	59.1	8.6
		Data based	(4)	71.0	69.4	91.6	100.
			(5)	39.5	48.0	58.9	42.9

Notes: **Heteroskedasticity-consistent** indicates *p*-values calculated using heteroskedasticity-robust standard errors. **HAC robust** indicates *p*-values are calculated using standard errors that allow for heteroskedasticity and serial correlation using the correction suggested by Newey and West (1987). **HAC robust (flat kernel)** indicates *p*-values are calculated using standard errors that allow for heteroskedasticity and serial correlation using an unweighted kernel in place of the weighted kernel suggested by Newey and West (1987). This is arguably better suited to the case where MA serial correlation is induced by overlapping observations (as with 3-month forward rates). **95%** indicates that critical values are from the 95th percentile of the asymptotic distribution. **Data based** indicates that the critical values are the actual values of the test statistics shown in Table 2.

3.3 Simulation Results

Simulation results for tests of the hypothesis that the cointegrating vector between spot and 1-month forward rates is [1,-1] are shown in Table 3. The rows labeled (4) show the results of repeating EL95's simulation experiment, while those labeled (5) use the same DGP considered previously. If the tests are correctly sized, then using the 95 per cent asymptotic critical values, we should not reject the true null hypothesis in close to 95 per cent of the trials. This is what we find using EL95's simulation experiment, regardless of the currency, the horizon of the forward rates, and the method used to construct the standard errors; we cannot reject the null hypothesis in 89 to 95 per cent of the trials.

However, this conclusion is very fragile to the design of the experiment. When we use (5) as our DGP, we find much more evidence of size distortion. For individual currencies, we accept the correct null hypothesis in only 32 to 75 per cent of trials. The problem is more severe for the joint test, where the true null is accepted in only 12.3 per cent of trials with the 1-month forward rates, and 8.6 per cent with the 3-month forward rates. This is consistent with the hypothesis put forward in Section 1.2; that the endogeneity of the forward rate might cause significant finite-sample problems with Stock-Watson tests that would not be detected by EL95's simulation experiment.[20]

Of course, the presence of size distortion in the Stock-Watson test does not rule out the possibility that the cointegrating vector may be different from 1.0.; it simply means that asymptotic critical values cannot be used to test this restriction. However, since our simulation experiment using (5) gives us the approximate finite-sample distribution of the test statistic under the null, we can use this to determine the marginal significance level of the test statistics from the true data (i.e., those reported in Table 2). The results are shown in Table 3 in the rows labeled "data based" critical values. The values shown are the fraction of times the simulated statistics are smaller than the test statistics from the true data. A value greater than x per cent in this row would lead us to reject (conditional on the chosen DGP) the null hypothesis of $\hat{\alpha}_1 = 1$ at the $100 - x$ per cent significance level.

Based on tests for individual currencies, we find that the true test statistics have marginal significance levels that are always more than 25 per cent (i.e., the numbers in the table are always less than 75 per cent), and that the joint tests have marginal significance levels of just over 50 per cent (i.e., the numbers in the table are always less than 50 per cent.). Therefore, this DGP would imply that there is no evidence to reject the null hypothesis at any plausible level of significance. This strongly contrasts with the conclusion we would draw had we used the EL95 simulation method, which suggests that the joint tests were always significant at the 1 per cent level.

4. Conclusion and Directions for Research

We find that EL's conclusion of very persistent trends in returns in foreign exchange markets is premature, at least on the basis of the evidence from cointegration tests. We have suggested that the design of the Monte Carlo experiments in EL93 was flawed and that these flaws would tend to exaggerate the apparent power of the ML tests to detect three cointegrating vectors. Our simulations show that EL93's simulation evidence is very sensitive to specification of the DGP, and that less restrictive dynamics reverse their results. We find that the simulation results EL95 report on the finite-sample size of their tests are very sensitive to the experimental design. We show that a less restrictive design reverses their conclusions and suggests that their ability to reject the restrictions on the cointegrating vectors can be attributed to size distortion. This also casts doubt on the conclusions of Crowder (1994), whose test results are sensitive to the lag length used and are not checked for finite-sample distortions. Comparing our own test statistics with their estimated finite-sample distributions, we find no evidence that would lead us to reject the hypothesis that spot and forward exchange rates are cointegrated and that they move one-for-one in the long run. If anything, the results from our Johansen tests suggest that the evidence of cointegration found is to be expected given the apparently low power of the test.

To be convincing, future research in this area will have to pay more careful attention to the reliability of the tests used to investigate the long-run properties of the data. However, given the fundamental problems of distinguishing between stationary and non-stationary behavior in finite samples, we feel this line of research would also benefit from a more careful explanation of what is to be learned from such investigation. It is hard for us to understand why cointegration tests are a compelling way to test for the importance of regime-switching, or why they are a better measure of the persistence of excess returns than more conventional methods.

6. Appendix: Maximum-Likelihood Test for Cointegration

While there are many alternative tests for cointegration, Gonzalo (1994) suggests that the maximum-likelihood (ML) system estimation approach performs better than both single-equation and alternative multivariate methods in detecting cointegration.[21] This approach is also among the best-known and the most widely applied in empirical work. The starting point of these tests is a VAR specification for the $n x 1$ vector of I(1) variables, namely,

$$X_t = \mu + A_1 X_{t-1} + \ldots + A_k X_{t-k} + u_t \tag{A1}$$

where u_t is assumed to be an independent and identically distributed Gaussian process. Note that we can rewrite (A1) as

$$\Delta X_t = \mu + \Gamma_1 \Delta X_{t-1} + \ldots + \Gamma_{k-1} \Delta X_{t-k+1} + \Pi X_{t-k} + u_t \tag{A2}$$

$$\begin{aligned} \Gamma_j &= -(I - A_1 - \ldots - A_t) \\ \Pi &= -(I - A_1 - \ldots - A_k) \end{aligned} \qquad j = 1, \ldots, k \tag{A3}$$

By rewriting (A1) into (A2) we are able to summarize the long-run information in X_t by the long-run impact matrix, Π; it is the rank of this matrix that determines the number of cointegrating vectors. Note that under the null hypothesis of $r (0 < r < n)$ cointegrating vectors, Π can be factored as $\Pi = \alpha\beta^T$, where α and β are nxr matrices. Therefore under the null we can write the process for X_t as

$$\Delta X_t = \mu + \sum_{i=1}^{k-1} \Gamma_i \Delta X_{t-1} + \alpha\beta^T X_{t-k} + u_t \tag{A4}$$

Johansen and Juselius (1990) demonstrate that β, the cointegrating vectors, can be estimated as the eigenvectors associated with the r largest, statistically significant eigenvalues found by solving the problem

$$|\lambda S_{kk} - S_{k0} S_{00}^{-1} S_{0k}| = 0 \tag{A5}$$

where S_{00} represents the residual moment matrix from a regression of $\Delta X_{t-1}, \ldots, \Delta X_{t-k+1}; S_{kk}$ is the residual moment matrix from a regression of X_{t-k} on ΔX_{t-k+1}; and S_{0k} is the cross-moment matrix. These eigenvalues readily permit the formation of likelihood ratio tests to determine the value of r. Johansen and Juselius propose two tests with differing assumptions about the alternative hypothesis: (i) the Trace statistic tests the restriction $r \leq q$ $(q<n)$ against the completely unrestricted model $r \leq n$; and (ii) the λ^{max} statistic makes the alternative more precise by specifying that only one additional cointegrating vector exists $(r \leq q+1)$. The log-likelihood ratio test statistics are formed as

$$\text{Trace} = -T \sum_{i=1+1}^{n} ln(1-\hat{\lambda}_i) \tag{A6}$$

$$\lambda^{max} = -T\, ln(1-\hat{\lambda}_{q+1}) \tag{A7}$$

The asymptotic critical values are non-standard and are tabulated in Osterwald-Lenum (1992). They depend on whether a constant is included in (A1), whether the restriction is imposed that any cointegrating vectors must also annihilate drift, and whether drift is present when it is allowed for. The test without a constant was originally proposed by Johansen (1988), the others by Johansen and Juselius (1990).

The critical values for these tests are based on their asymptotic distribution under the null hypothesis of interest. Recently, a number of authors have begun to examine how reliable this asymptotic approximation is in finite samples, and some have suggested modifications to the ML test.[22] Contributions to this research include Toda (1994, 1995), Gregory (1994), Gonzalo and Pitarakis (1994), Reinsel and Ahn (1988), and Cheung and Lai (1993). The importance of this problem will depend on the specific application. Edison, Gagnon, and Melick (1994) examine the size and power of ML cointegration tests in the context of testing purchasing power parity in the post-Bretton Woods period, and find important size distortions. Hendry (1995) examines the ML tests in the context of Canadian money demand and reaches similar conclusions. Ho and Sørenson (1994) review Durlauf's (1989) evidence on cointegration in U.S. sectoral value-added across one-digit standard industrial classification sectors and suggest that neither asymptotic critical values nor the finite-sample corrections suggested by Reinsel and Ahn (1992) or by Cheung and Lai (1993) are reliable in very short samples (40-50 observations). Godbout and van Norden (1997) show similar evidence from studies of cointegration in international financial markets. Edison and Melick (1995) find evidence of significant size distortions in their study of real interest rate parity. In arriving at these conclusions, these studies all use a simulation methodology similar to ours.

Notes

1. Evans and Lewis (1994) present similar evidence for bond markets. For compactness, we will limit our attention to the results from the foreign exchange markets, although our discussion may have implications for research on bond markets as well.
2. The timing distinction is irrelevant for the purposes of defining the cointegrating relationship, since (s_t, f_t^k) will be cointegrated with vector [1, -1] if and only if (s_{t+k}, f_t^k) is cointegrated with cointegrating vector [1,-1].
3. Evans and Lewis reference Campbell and Perron (1991), but more extensive discussions of this problem may be found in Blough (1994) and Cochrane (1991).
4. While this is true of Evans and Lewis (1993, 1995), it is less true of Evans and Lewis (1992, 1994).
5. These data were generously provided by Robert Hodrick and are more fully documented in Bekaert and Hodrick (1993). We used the average of the bid and ask prices throughout. Evans and Lewis noted that their conclusions were the same when the bid spot and ask forward prices were used instead.
6. See Evans and Lewis (1993, Table 1). Note that using the Johansen Trace test and 1-month forward rates, they find significant evidence of only one cointegrating vector. The Appendix provides more details on the construction of these tests and others.
7. Evans and Lewis (1993, p. 1011).
8. References to the literature on the size-distortion problem are given in the Appendix.
9. Evans and Lewis (1995, footnote 18). One possibility is that the failure to reject the hypothesis of three cointegrating vectors simply means that they could not find evidence of four or more cointegrating vectors.
10. The following based on Evans and Lewis (1992), Appendix, pages iii-iv.
11. As mentioned in endnote 9, EL do not explicitly discuss how to reconcile the assumption that spot and forward rates are cointegrated with the results in EL95. First, their assumption that spot and forward rates are cointegrated is a necessary condition for excess returns to be stationary. Therefore, this assumption should seem innocuous when (as they do) one tries to find evidence to reject the null hypothesis of stationarity. Second, the method that they use to test restrictions on the cointegrating vector is robust to the absence of cointegration, so that their inferences will be asymptotically valid whether or not cointegration is present. See Stock and Watson (1993).
12. In the case where $k>1$, EL first fit an AR$(k-1)$ model to the residuals, then bootstrap the AR residuals to obtain a new series of residuals for (4). They also try generating errors from ARCH processes.
13. Evans and Lewis (1995, pp. 739-40).
14. Evans (1991) is by *George* Evans. *Martin* Evans coauthored Evans and Lewis (1992, 1993, 1995).
15. The accuracy of our statistical programs was checked by verifying the results of our own code against that produced by commercially available programs by Johansen, Juselius and Hansen, and by Phillips and Ouliaris.
16. Data series were demeaned prior to estimation of (5).
17. This is consistent with the results reported in the previous sentence. Using the sequential determination procedure suggested by Johansen, failure to reject the null hypothesis of one cointegrating vector leads to the conclusion of no more than one cointegrating vector, even if we can reject the null of two cointegrating vectors. Put another way, this implies that in those rare cases where we could reject the null of two cointegrating vectors, we were always unable to reject the null of either zero or one such vectors.
18. These conclusions were robust to variations in the number of lags in the DGP and in the cointegration test. We never found significant evidence of three cointegrating vectors in as much as 1 percent of the trials for any case we examined. Tests for the null hypothesis of two cointegrating vectors were never rejected in more than 2 per cent of the trials, and the true Lambda-max statistics were always greater than at least 78 per cent of their simulated counterparts.
19. With the 1-month forward rate data, we found that our results using the heteroskedasticity-robust standard errors were identical to the results EL reported for the serial-correlation-robust standard errors, suggesting a typographical error in the reporting of their results.
20. Additional simulations suggested that the conclusions were robust to minor changes in the number of lags used in (4) to construct the test statistics, the number of lags used to construct the Newey-West standard errors, the use of heteroskedasticity-robust or conventional standard errors, a reversal of the roles of spot and forward rates in the Stock-Watson regression, and the use of bootstrapped or multivariate normal innovations.
21. See also Watson (1995), especially Section 3.d.
22. In addition, see the literature survey in Ho and Sørenson (1996). Gonzalo and Lee (1998) give examples of several "pitfalls" — cases where the size of the ML test for cointegration approaches 1 asymptotically.

References

Baillie, R. T. and T. Bollerslev (1989). "Common Stochastic Trends in a System of Exchange Rates," *Journal of Monetary Economics*, 44, 167-181.

Baillie, R. T. and T. Bollerslev (1994). "The Long Memory of the Forward Premium," *Journal of International Money and Finance*, 13, 565-571.

Bekaert, G. and R. J. Hodrick (1993). "On Biases in the Measurement of Foreign Exchange Risk Premiums," *Journal of International Money and Finance*, 12, 115-138.

Brenner, R. J. and K. F. Kroner (1995). "Arbitrage, Cointegration and Testing the Unbiasedness Hypothesis in Financial Markets," *Journal of Quantitative and Financial Analysis*, 30, 23-42.

Campbell, J. and P. Perron (1991). "Pitfalls and Opportunities: What Macroeconomists Should Know About Unit Roots," in O. Blanchard and S. Fischer (eds.) *NBER Macroeconomics Annual*. Cambridge: MIT Press.

Cheung, Y. W. and K. S. Lai (1993). "Finite-Sample Sizes of Johansen's Likelihood Ratio Tests for Cointegration," *Oxford Bulletin of Economics and Statistics*, 55, 313-328.

Cochrane, J. H. (1991). "A Critique of the Application of Unit Root Tests," *Journal of Economic Dynamics and Control*, 15, 275-284.

Crowder, W. J. (1994). "Foreign Exchange Market Efficiency and Common Stochastic Trends," *Journal of International Money and Finance*, 13, 551-564.

Durlauf, S. (1989). "Output Persistence, Economic Structure and the Choice of Stabilization Policy," *Brookings Papers on Economic Activity*, 69-137.

Edison, H. J., E. Gagnon, and W. R. Melick (1994). "Understanding the Empirical Literature on Purchasing Power Parity: the Post-Bretton Woods Era," Board of Governors of the Federal Reserve System International Finance Discussion Papers No. 465.

Engle, R. F. and C. W. J. Granger (1987). "Cointegration and Error Correction: Representation, Estimation and Testing," *Econometrica*, 55, 251-276.

Evans, M. D. D. and K. Lewis (1992). "Trends in Excess Returns in Currency and Bond Markets," NBER working paper No. 4116.

Evans, M. D. D. and K. Lewis (1993). "Trends in Excess Returns in Currency and Bond Markets," *European Economic Review*, 37, 1005-1019.

Evans, M. D. D. and K. Lewis (1994). "Do Stationary Risk Premia Explain it All? Evidence from the Term Structure," *Journal of Monetary Economics*, 3, 285-318.

Evans, M. D. D. and K. Lewis (1995). "Do Long-Term Swings in the Dollar Affect Estimates of the Risk Premia?" *Review of Financial Studies*, 8, 709-742.

Fama, E. F. (1970). "Efficient Capital Markets: A Review of Theory and Empirical Work," *Journal of Finance*, 25, 383-417.

Godbout, M. and S. van Norden (1997). "Reconsidering Cointegration in International Finance: Three Case Studies of Size Distortion in Finite Samples," working paper No. 97-1, Bank of Canada.

Gonzalo, J. (1994). "A Comparison of Five Alternative Methods of Estimating Long Run Equilibrium Relationships," Journal of Econometrics, 60, 203-233.

Gonzalo, J. and T. Lee (1998). "Pitfalls in Testing for Long Run Relationships," *Journal of Econometrics*, 86, 129-154.

Gonzalo, J. and J. Pitarakis (1994). "Comovements in Large Systems," discussion paper No. 9465, CORE.

Gregory, A. (1994). "Testing for Cointegrating in Linear Quadratic Models," *Journal of Business and Economic Statistics*, 12, 347-360.

Hendry, S. (1995). "Long-Run Demand for M1," working paper No. 95-11, Bank of Canada.

Ho, M. S. and B. E. Sørenson (1996). "Finding Cointegration Rank in High Dimensional Systems Using the Johansen Test: An Illustration Using Data Based Monte Carlo Simulations," *Review of Economics and Statistics*, 78, 726-732.

Johanson, S. (1988). "Statistical Analysis of Cointegration Vectors," *Journal of Economic Dynamics and Control*, 12, 231-254.

Johansen, S. and K. Juselius (1990). "Maximum Likelihood Estimation and Inference on Cointegration with Applications to the Demand for Money," *Oxford Bulletin of Economics and Statistics*, 52, 169-210.

Meese, R. A. and K. Rogoff (1983). "Empirical Exchange Rate Models of the Seventies: Do They Fit Out of Sample?" *Journal of International Economics*, 14, 3-24.

Newey, W. K. and K. D. West (1987). "A Simple, Positive Semi-Definite, Heteroskedasticity and Autocorrelation Consistent Covariance Matrix," *Econometrica*, 55, 703-708.

Osterwald-Lenum, M. (1992). "A Note with Quantiles of the Asymptotic Distribution of the Maximum Likelihood Cointegration Rank Test Statistics," *Oxford Bulletin of Economics and Statistics*, 54, 461-472.

Phillips, P. C. B., J. McFarland, and P. C. McMahon (1996). "Robust Tests of Forward Exchange Market Efficiency with Empirical Evidence from the 1920s," *Journal of Applied Econometrics*, 11, 1-22.

Reinsel, G. C. and S. H. Ahn (1988). "Asymptotic Distribution of the Likelihood Ratio Test for Cointegration in the Nonstationary Vector AR Case," technical report, Department of Statistics, University of Wisconsin.

Reinsel, G. C. and S. H. Ahn (1992). "Vector Autoregressive Models with Unit Roots and Reduced Rank Structure; Estimation, Likelihood Ratio Tests, and Forecasting," *Journal of Time Series Analysis*, 13, 353-375.

Stock, J. H. and M. W. Watson (1993). "A Simple Estimator of Cointegrating Vectors in Higher Order Integrated Systems," *Econometrica*, 61, 783-820.

Toda, H. Y. (1994). "Finite Sample Properties of Likelihood Ratio Tests for Integration When Linear Trends are Present," *Review of Economics and Statistics*, 76, 66-79.

Toda, H. Y. (1995). "Finite Sample Performance of Likelihood Ratio Tests for Cointegrating Ranks in Vector Autoregressions," *Econometric Theory*, 11, 1015-1032.

Watson, M. W. (1995). "Vector Autoregressions and Cointegration," in Engle, R.F. and D. McFadden (eds.), *Handbook of Econometrics*, Vol. 4. New York: North Holland.

6 ON THE INHERENT NONLINEARITY OF FREQUENCY DEPENDENT TIME SERIES RELATIONSHIPS

Hui Boon Tan
Universiti Putra Malaysia
Richard Ashley
Virginia Polytechnic Institute and State University

1. Introduction

Macroeconomic theory suggests that many of the most important aggregate relationships depend on frequency. That is, these relationships are different in sign and/or intensity for high frequency fluctuations than for low frequency fluctuations. For example, the Permanent Income Hypothesis is essentially synonymous with the proposition that the consumption-income relationship is strong at low frequencies and weak at high frequencies. And the growth rate in the supply of money is thought to be directly related to the growth rate in output, but only at high frequencies; whereas the opposite pattern is thought to obtain for the relationship between the growth rate in the supply of money and the inflation rate. And so forth.

Such relationships have been studied by many authors, using a number of different techniques, including Friedman (1957), Engle (1974, 1978), Lucas (1980), Mills (1982), Geweke (1982, 1986), Summers (1983, 1986), Cochrane (1989), Phillips (1991), Thoma (1992, 1994), Lee (1994), Corbae, *et al.* (1994), Tan and Ashley (1997), and – no doubt – many others.

Below we argue on fundamental grounds that frequency dependent time series relationships of the sort envisioned by macroeconomic theory cannot be captured by typical linear modelling methods – vector ARMA models and the like. These frequency dependent relationships are inherently and intrinsically *nonlinear* in nature. Thus, in positing significant frequency dependence in macroeconomic relationships, theorists are also asserting that nonlinear dynamic modelling is crucial for understanding these relationships.

This result also implies that there is something fundamentally wrong with using the usual cross spectral analysis (or Geweke's (1982) elegant elaboration of it) to analyze the frequency dependence of time series relationships. In the context of a transparently simple example, we provide a new interpretation of the frequency dependence results these linear methods yield.

Finally, we illustrate the point using the Tan and Ashley (1997) procedure for detecting and modelling frequency dependence in estimated regression coefficients. Here we apply this technique to a linear bivariate model, first studied by Jaditz and Sayers (1994), relating fluctuations in the growth rate of the U.S. Producer Price Index (PPI_t) to future fluctuations in the growth rate of the Consumer Price Index (CPI_t). We find that the coefficient on lagged PPI_t in this model is reasonably stable over time but quite unstable across frequencies, implying that the bivariate model is seriously mis-specified. This result provides a rationale for the otherwise inexplicably poor postsample forecasting performance of the linear bivariate model.

2. Linear Models and Frequency Dependent Relationships

The following simple example demonstrates how and why linear causal processes with constant coefficients are in general incapable of modelling frequency dependent relationships of the sort posited by macroeconomic theory.

Consider a pair of time series, x_t and y_t, generated by the bivariate moving average process:

$$\begin{aligned} x_t &= u_t + \phi u_{t-1} + \rho v_{t-1} \\ y_t &= v_t \\ \begin{pmatrix} u_t \\ v_t \end{pmatrix} &\sim \text{NIID}\left[\begin{pmatrix} 0 \\ 0 \end{pmatrix}, \begin{pmatrix} \sigma_u^2 & 0 \\ 0 & \sigma_v^2 \end{pmatrix}\right] \end{aligned} \tag{1}$$

where σ_u^2 and σ_v^2 are both strictly positive and the parameters ϕ and ρ are fixed constants not depending on u_{t-k} or v_{t-k} for any value of k.

This linear, causal model clearly exhibits what Geweke (1982) calls "unidirectional feedback from y_t to x_t" for all non-zero values of ρ. Note, however, that the relationship between x_t and y_t posited in this model does *not* depend on frequency in the sense macroeconomists use this phrase. In particular, the effect of, say, a positive fluctuation in y_{t-1} on x_t does not depend on whether y_{t-2}, y_{t-3}, y_{t-4}, etc. were also positive – it is just ρy_{t-1}

regardless. Thus, a fluctuation in y_{t-1} that is part of a low frequency (i.e., smooth, persistent) fluctuation in y has precisely the same effect on x_t as does one that is part of a high frequency fluctuation.[1] Indeed, the only way that the effect of a fluctuation in y_{t-1} could possibly depend on the extent to which this value for y_{t-1} is part of a pattern of similar values for y_{t-2}, y_{t-3}, etc., would be for the parameter ρ to be a function of y_{t-2}, y_{t-3}, etc. But if ρ depends on past values of y_t, then equation 1 represents a *nonlinear* process. Thus, meaningful frequency dependence in the strength of a relationship can never occur in a linear model such as this.

However, if that is the case, then what is the significance of the frequency variation typically found in examining the relationship between a pair of time series using the gain function, as in Engle (1974, p. 10)? And what about the frequency dependence of the measure of the strength of linear feedback relationships given by Geweke (1982)? Since Geweke's approach is closely related to cross spectral analysis, applying his approach to the bivariate moving average process given above as equation 1 sheds light on both of these questions.

Recognizing that equation 1 is a special case of Geweke's general vector MA model and making the appropriate substitutions into his equation (3.6) for $f_{Y\to X}(\omega)$, Geweke's measure of the strength of the linear feedback from y to x at frequency ω is

$$f_{Y\to X}(\omega) = \frac{1}{2}\ln\left\{\frac{[\sigma_u^2(1+\phi^2+2\phi\cos(\omega)) + \rho^2\sigma_v^2]^2}{[\sigma_u^2(1+\phi^2+2\phi\cos(\omega))]^2}\right\} \tag{2}$$

for this process. This measure clearly *does* depend on the frequency ω so long as $\rho^2\sigma_v^2$ is non-zero . And $\rho^2\sigma_v^2$ will be strictly positive so long as y_t has positive variance and does affect future values of x_t.

Geweke's measure, $f_{Y\to X}(\omega)$, is plotted against ω in Figure 1 for $\sigma_v^2 = \sigma_u^2 = 1$, $\rho = 0.6$, and $\phi = \pm 0.6$. Since we have just seen that, in actuality, there is no frequency dependence in the strength of this relationship, what is the meaning of this clearly observable frequency dependence in $f_{Y\to X}(\omega)$? The sign of ϕ is obviously crucial in determining the form of the frequency dependence in $f_{Y\to X}(\omega)$. In fact, a fluctuation in y_{t-1} merely augments u_t, the current innovation in x_t. Evidently, the frequency dependence of Geweke's measure is actually quantifying the degree to which an innovation in y_t causes smooth, persistent fluctuations (low frequencies dominant) as opposed to rough, transient fluctuations (high frequencies dominant) in x_t, not the degree to which the intensity of the relationship between x_t and y_t varies across frequencies.

3. Frequency-Dependent Relationships – A Simple Example

Thus, techniques for assessing the frequency dependence of time series relationships based on linear causal models, such as the vector MA model used by Geweke, are actually quantifying something else. In contrast, the band spectrum regression model introduced in Engle (1974), the system spectral regression model of Phillips (1991), and the blockwise time domain spectral regression model of Tan and Ashley (1997) are all frameworks in which bona fide frequency dependence in relationships can be detected and

Figure 1. $f_{Y\to X}(\omega)$, Geweke's Measure of the Strength of the Linear Feedback from y to x at frequency ω, for the process given in equation 1.

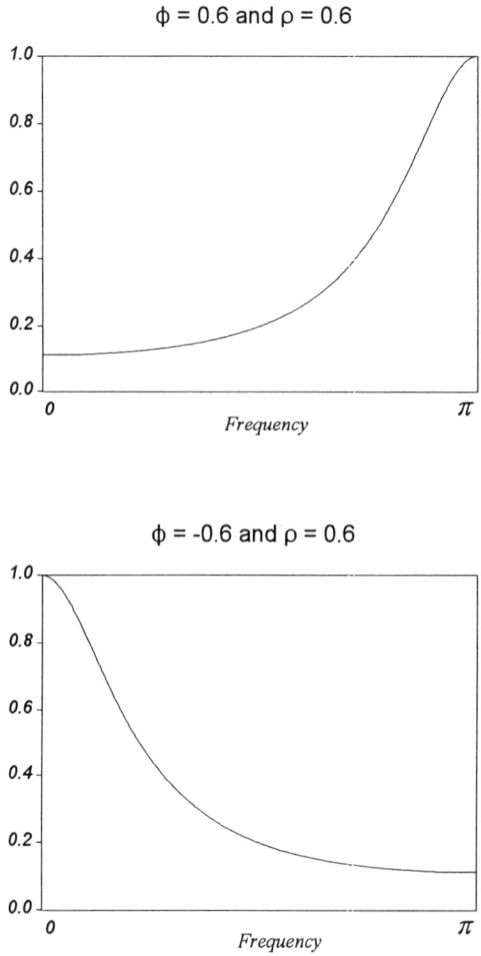

Notes: This figure plots the expression for $f_{Y\to X}(\omega)$ in equation 2 with $\sigma_v^2 = \sigma_u^2 = 1$, $\rho = 0.6$, and $\phi = \pm 0.6$.

modeled. The relative merits of each of these methods are discussed and compared in Tan and Ashley (1997); here we provide an illustrative example using our own method simply because it is so straightforward.

Following Jaditz and Sayers (1994), we examine the postsample forecasting performance of a bivariate model for CPI_t, the monthly growth rate in the Consumer Price Index, based on its own past and on past values of PPI_t, the growth rate in the Producer Price Index. Estimating a bivariate linear model for this relationship over a 60 month fitting period from November, 1982 to October, 1987 yields:[2]

$$CPI_t = \underset{(.000439)}{.00203} + \underset{(.135)}{.235}\, CPI_{t-1} + \underset{(.069)}{.195}\, \mathrm{PPI}_{t-1} + e_t, \quad \bar{R}^2 = .281 \tag{3}$$

This estimated model looks fairly reasonable, although perhaps a bit weaker than one might expect. As Jaditz and Sayers found, however, its one-step-ahead forecasting performance over the ensuing 80 month postsample period (from November 1987 to June 1994) is quite dismal. Even following Jaditz and Sayers' procedure of updating the parameter estimates each month using a 60 month moving window, the MSE of the forecasts from equation 3 is 35.4% *higher* than that of an AR(1) model for CPI_t.

It is commonplace to observe a time series regression model forecast substantially less well than it fits, but this case is puzzling. For one thing, as Jaditz and Sayers point out in some detail, the "cost-push" inflation theory on which this relationship is based has strong empirical support from other data sets. Moreover, the model specification itself makes sense and is clearly not seriously over-parameterized. Time plots of the series provide no explanation either; both series are more variable in the first part of the postsample period (and less variable in the second part) than is observed in the fitting period, but not egregiously so. Moreover, Jaditz and Sayers test the stability of the coefficient on PPI_{t-1} using a Chow test over the entire data set and are unable to reject the null hypothesis that this coefficient is stable at the 5% level of significance. Consequently, the extremely poor postsample forecasting performance of equation 3 seems paradoxical.

Since parameter stability nevertheless seemed suspect to us, we first tested for more subtle forms of (time domain) instability in the coefficient on PPI_{t-1} using the "stabilogram" test proposed in Ashley (1984). This test is equivalent to doing a Chow test over multiple subperiods by including a dummy variable on the coefficient in question for each subperiod and then testing the hypothesis that the explanatory variable enters with the same coefficient in each subperiod.

Figure 2 plots the stabilograms obtained by partitioning the entire data set (T = 140 months) into either three or seven approximately equal subperiods. In each stabilogram the middle line is the estimated coefficient on the dummy variable for that subperiod; the top and bottom lines form an estimated 95% confidence interval for the population coefficient. The seven-subperiod specification was chosen *a priori* because it partitions the sample period into the first three subperiods and the postsample period into the remaining four subperiods; the three-subperiod specification was chosen *a posteriori* because it fits the data best using the Schwarz criterion.

The corresponding estimated regression equations are, for the three subperiod model:

$$\begin{aligned} CPI_t = {} & \underset{(0.0003)}{0.0024} + \underset{(0.095)}{0.220}\, CPI_{t-1} + \underset{(0.063)}{0.236}\, D1_t PPI_{t-1} + \underset{(0.055)}{0.069}\, D2_t PPI_{t-1} \\ & + \underset{(0.055)}{0.069}\, D2_t PPI_{t-1} + \underset{(0.047)}{0.050}\, D3_t PPI_{t-1} + e_t \end{aligned} \tag{4}$$

with standard errors in parentheses and $F(2,135) = 3.637$ (p-value = 0.029), and, for the

Figure 2. 3 Subperiod and 7 Subperiod Time Domain Stabilograms for $PPI_{t\ 1}$ Coefficient from Regression Equations 4 and 5.

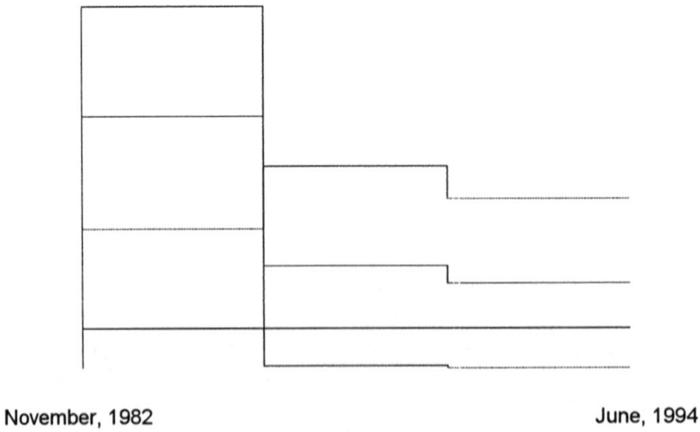

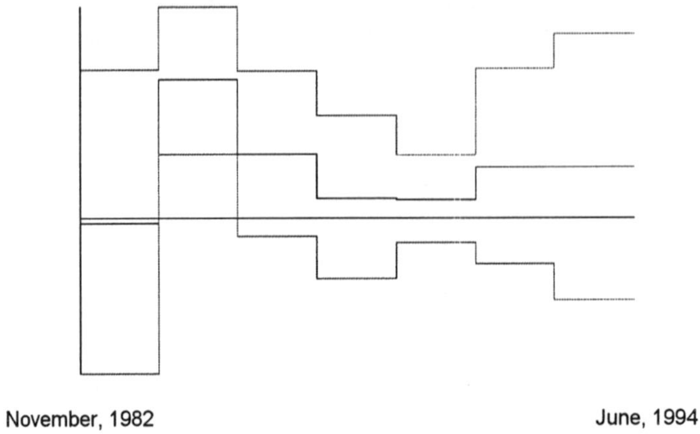

Notes: These figures are time plots of the estimated coefficients of the dummy variables in equations 4 and 5: $D1_tPPI_{t-1},\ldots,D3_tPPI_{t-1}$ for equation 4 and $D1_tPPI_{t-1},\ldots,D7_tPPI_{t-1}$ for equation 5. The upper and lower lines in each plot delineate estimated 95% confidence intervals. Both of these regressions used 140 observations - from November, 1982, to June, 1994 ; they differ in how the data set was partitioned.

seven subperiod model:

$$
\begin{aligned}
CPI_t = {} & \underset{(0.0003)}{0.0023} + \underset{(0.098)}{0.267}\,CPI_{t-1} - \underset{(0.163)}{0.012}\,D1_t PPI_{t-1} + \underset{(0.079)}{0.292}\,D2_t PPI_{t-1} \\
& + \underset{(0.088)}{0.136}\,D3_t PPI_{t-1} + \underset{(0.088)}{0.041}\,D4_t PPI_{t-1} + \underset{(0.046)}{0.040}\,D5_t PPI_{t-1} \\
& + \underset{(0.105)}{0.108}\,D6_t PPI_{t-1} + \underset{(0.142)}{0.106}\,D7_t PPI_{t-1} + v_t
\end{aligned}
\tag{5}
$$

with standard errors in parentheses and $F(6,131) = 1.531$ (p-value = 0.1729). For the three subperiod model, $D1_t$ is a dummy variable which is one in the first third of the sample of 140 observations, $D2_t$ is a dummy variable for the second third of the sample, and $D3_t$ is a dummy variable which is one for the remaining observations. Similarly, for the seven subperiod model, $D1_t$ is a dummy variable which is one for the first twenty periods, $D2_t$ is a dummy variable for the second twenty periods, and so forth.

These time domain stabilogram results indicate that the coefficient on PPI_{t-1} probably *is* a bit unstable; in particular, it appears to be larger than usual in the twenty-month subperiod from July 1984 to February 1986. The null hypothesis that all three subperiods in the three-subperiod model have equal coefficients can be rejected at the 3% level. That is hardly convincing, however, in view of the downward bias in this significance level due to the fact that regression equations were actually estimated for the 2, 3, 4 ... 10 subperiod models. In fact, the test is significant at only the 17% level for the seven-subperiod model we chose *a priori*.

Moreover, the estimated parameter variation does little to explain the poor postsample forecasting performance of equation 3. True, there is evidence that equation 3 is probably overstating the coefficient on PPI_{t-1} relative to its value in the postsample period, but it's not as if the sign of the coefficient is fluctuating. In fact, over the 80 month postsample period, the estimated coefficient on PPI_{t-1} is $.059 \pm .04$; so the coefficient on PPI_{t-1} is still positive in the postsample period, albeit not significantly so. In the end, we must concur with Jaditz and Sayers in concluding that time domain parameter variation is not the reason that equation 3 forecasts so very poorly.

To investigate the stability of the coefficient on PPI_{t-1} in the *frequency* domain using the method proposed by Tan and Ashley (1997), it is convenient to re-write equation 3 as

$$Y = X\beta + \epsilon \tag{6}$$

where Y is a 60-element vector containing the sample data on CPI_t and X is a 60×3 dimensional matrix containing the sample data on CPI_{t-1} and PPI_{t-1} in its second and third columns, so that β_3 is the coefficient on PPI_{t-1}. We pre-multiply equation 6 by the orthogonal 60×60 dimensional matrix A, first suggested by Harvey (1978), with (j,t)th element:

$$a_{j,t} = \begin{cases} \left(\frac{1}{60}\right)^{\frac{1}{2}} & \text{for } j = 1 \\ \left(\frac{2}{60}\right)^{\frac{1}{2}} cos\left[\frac{\pi j(t-1)}{60}\right] & \text{for } j = 2, 4, 6, \ldots 58 \\ \left(\frac{2}{60}\right)^{\frac{1}{2}} sin\left[\frac{\pi (j-1)(t-1)}{60}\right] & \text{for } j = 3, 5, 7, \ldots 59 \\ \left(\frac{1}{60}\right)^{\frac{1}{2}} (-1)^{t+1} & \text{for } j = 60 \end{cases} \tag{7}$$

to obtain the frequency domain regression equation,

$$\begin{aligned} AY &= AX\beta + A\epsilon \\ Y^{**} &= X^{**}\beta + \epsilon^{**} \end{aligned} \tag{8}$$

where $Y^{**} = AY$, etc. Note that Y^{**}, X^{**}, and ϵ^{**} are all real-valued, the β vector is unchanged, and that (since A is orthogonal) ϵ^{**} is distributed the same as ϵ. However, since the even numbered rows of A are the weights in the finite cosine transform and the odd numbered rows of A are the weights in the finite sine transform, the $j = 1 \ldots 60$ "observations" in equation 8 now correspond to 31 different frequencies rather than to 60 time periods; Table 1 summarizes which frequency corresponds to each observation. There are fewer distinct frequencies in equation 8 than time periods in equation 6 because there is both a sine and a cosine transform for each of the frequencies in the interval $(0, \pi)$.

Looking closely at the form of equation 7, the transformations corresponding to the low frequency (small j index) rows of A are extracting the slowly-varying components of whatever column vector they multiply, whereas the high frequency rows of A are extracting the quickly-varying components; see Tan and Ashley (1997) for details.

The parameter vector β can be estimated via OLS regression applied to either equation 6 or equation 8 – it makes no difference. But when the stabilogram test is applied to equation 8 to test the stability of β_3 across a set of subperiods, the "subperiod" corresponding to each dummy variable now represents a band of frequencies. Thus, the test is now examining the stability of the coefficient on PPI_{t-1} across different frequency bands. The only difference in the way the test is applied is that some attention must be paid to the partitioning of the sample into subsamples so that both observations corresponding to each frequency always wind up in the same subsample.

Estimating regression models with the observations partitioned into 2, 3, ... 10 approximately equal "subperiods" (i.e., frequency bands) we find that the two-band model has the smallest Schwarz criterion. The stabilogram for this two-band model using the 60 sample observations is plotted in Figure 2 along with the analogous two-band

Table 1. Frequency Corresponding to Each "Observation" in Equation 8.

"Observation" Number	Frequency
1	0
2	$2\pi\{1/60\}$
3	$2\pi\{1/60\}$
4	$2\pi\{2/60\}$
5	$2\pi\{2/60\}$
6	$2\pi\{3/60\}$
7	$2\pi\{3/60\}$
...	
58	$2\pi\{29/60\}$
59	$2\pi\{29/60\}$
60	π

stabilogram obtained using only the 80 postsample observations. Note that the estimated pattern of parameter variation across frequencies looks significant and is, itself, fairly stable across the two periods – the relationship between CPI_t and PPI_{t-1} is apparently strong and positive at low frequencies, but weak and/or negative at high frequencies in both the sample and postsample periods. The corresponding estimated regression models are, for the two band frequency domain stabilogram model with sample period observations:

$$\underset{}{Y_j^{**}} = \underset{(0.0004)}{0.0022} + \underset{(0.135)}{0.161X_{j2}^{**}} + \underset{(0.076)}{0.274D_j^1} - \underset{(0.132)}{0.056D_j^2} + \nu_j \qquad (9)$$

with standard errors in parentheses and $F(1, 156) = 4.935$ (p-value = 0.0304), and for the two band frequency domain stabilogram model with postsample period observations:

$$Y_j^{**} = \underset{(0.0004)}{0.0024} + \underset{(0.126)}{0.239X_{j2}^{**}} + \underset{(0.041)}{0.117D_j^1} - \underset{(0.065)}{0.133D_j^2} + \nu_j \qquad (10)$$

with standard errors in parentheses and $F(1, 76) = 13.057$ (p-value = 0.00054). In each

Figure 3. Two-Band Frequency Domain Stabilograms for $PPI_{t\;1}$ Coefficient in Regression Equations 9 and 10.

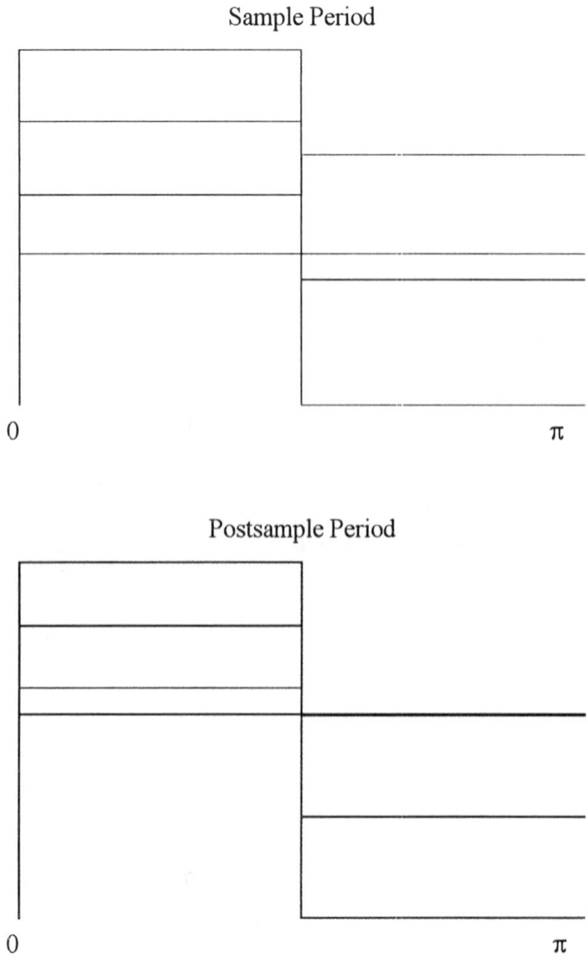

Notes: These figures are plots of the estimated coefficients on the dummy variables D_j^1, and D_j^2, in equations 9 and 10 versus the frequency index *j*. This index runs from 1 to 60 for equation 9 and from 1 to 80 for equation 10; in both cases this interval corresponds to the frequency interval from 0 to π. The upper and lower lines in each plot delineate estimated 95% confidence intervals. Equation 9 used the 60 observations from November, 1982 to October, 1987; equation 10 used the 80 postsample observations from November, 1987 to June, 1994.

equation D_j^1 equals X_{j3}^{**} for each j, which corresponds to one of the lowest half of the frequencies and is zero otherwise ; whereas D_j^2 equals X_{j3}^{**} for each j which corresponds to one of the remaining frequencies and is zero otherwise.

Thus, the null hypothesis that the CPI_t - $PPI_{t\text{-}1}$ coefficient is stable across frequencies can be rejected in both the sample and postsample periods. The sample period rejection at the 3% level is not all that convincing, but the rejection at the .05% level using the postsample data seems compelling.

Jaditz and Sayer's paradoxical result now becomes explicable: the time domain bivariate model fails at postsample forecasting because it is mis-specified, leading to a mildly unstable CPI_t - $PPI_{t\text{-}1}$ coefficient in the time domain and a dramatically unstable CPI_t - $PPI_{t\text{-}1}$ coefficient in the frequency domain.

Finally, it is instructive to transform this model for a frequency dependent $CPI - PPI$ relationship back into the time domain. Since the transformation matrix, A, is orthogonal, this backtransformation requires nothing more involved than pre-multiplying the regression equation by the transpose of A. Rewriting equation 8 to explicitly include dummy variables allowing the coefficient on β_k to vary across m (frequency domain) subsamples yields the regression model

$$Y^{**} = X^{**}_{\{k\}}\beta_{\{k\}} + D\gamma + \nu \tag{11}$$

where $X^{**}_{\{k\}}$ is X^{**} with its kth column removed and $\beta_{\{k\}}$ is the β vector with its kth component removed. D is the $T{\times}m$ matrix $[D^1 \ldots D^m]$,, with D_j^s equal to X_{jk}^{**} if observation j is in frequency band s and D_j^s equal to zero otherwise; equations 8 and 9 are estimated versions of this equation, with $k = 3$ and $m = 2$. Pre-multiplying by A^T, this becomes:

$$\begin{aligned} A^TY^{**} &= A^TX^{**}_{\{k\}}\beta_{\{k\}} + A^TD\gamma + A^T\nu \\ Y &= X_{\{k\}}\beta_{\{k\}} + [A^TD^1 \ldots A^TD^m]\gamma + \tilde{\nu} \end{aligned} \tag{12}$$

Equation 12 can be viewed as a generalization of the original model (equation 6) in which the data on the kth explanatory variable has been decomposed ("filtered") into m disjoint frequency components so that each can potentially enter the regression with a different coefficient. Figure 4 illustrates for a three-band ($m = 3$) model how the three frequency domain dummy variable series (D^1, D^2, and D^3) and the concomitant (filtered) time domain series (A^TD^1, A^TD^2, A^TD^3) look; the equally weighted sum of the latter three series precisely reproduces PPI_t.

Returning to the estimated two-band model of equation 9, these coefficient estimates imply that a positive fluctuation in $PPI_{t\text{-}1}$ which is part of a recent pattern of similar fluctuations will have a significant positive influence on CPI_t. This is because such a fluctuation in $PPI_{t\text{-}1}$ will have a positive impact on the period t element of A^TD^1 – the low frequency component of PPI – which enters the regression with a significant positive coefficient. In contrast, an isolated fluctuation in $PPI_{t\text{-}1}$ will primarily impact

Figure 4. Frequency Domain Dummy Variables for PPI_t and Time Domain Equivalents.

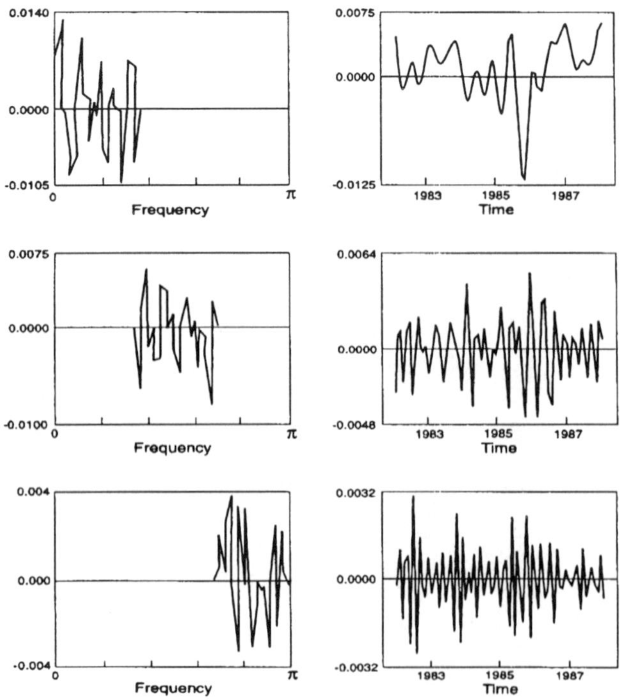

$A^T D^2$, which enters the regression with a weaker (and negative) coefficient.

4. Conclusions

The results reported in Section 2 indicate that linear processes (vector MA, VARMA, etc.) cannot be used to model frequency dependent time series relationships: the frequency dependence that they appear to exhibit is actually quantifying something else.

As illustrated in Section 3, authentic frequency dependence in the intensity and/or sign of a time series relationship can be conveniently detected and modeled using the technique given by Tan and Ashley (1997). In many cases this, or one of the alternative techniques due to Engle (1974) or Phillips (1991), will provide satisfactory results.

However, where the non-causal nature of the resulting models is awkward, nonlinear dynamic modelling is necessary. In this sense, one can conclude that the frequency dependence posited by economic theorists to characterize many of the most crucial macroeconomic relationships – consumption–income, money–income, inflation rate–unemployment rate, inflation rate–interest rate, and so forth – is actually a call for nonlinear modelling of these relationships.

Notes

1. Since $y_t = v_t$ is assumed to be serially uncorrelated in this model, its spectrum is flat. Thus, "low frequency" and "high frequency" fluctuations in y_t are equally likely, but that is irrelevant; all that matters here is that the effect of any particular fluctuation in y_{t-1} on x_t is *independent* of whether or not that fluctuation is part of a pattern of recent fluctuations in y_{t-2}, y_{t-3}, etc.
2. Earlier data were not used because of nonstationarity problems in the 1981-82 period. The figures in parentheses are estimated standard errors; the fitting errors appear to be reasonably homoskedastic and non-autocorrelated.

References

Ashley, R. A. (1984). "A Simple Test for Regression Parameter Instability," *Economic Inquiry*, 253-267.

Cochrane, J. H. (1989). "The Return of the Liquidity Effect: A Study of the Short-Run Relation Between Money Growth and Interest Rates," *Journal of Business and Economic Statistics*, 7, 75-83.

Corbae, D., S. Ouliaris, and P. C. B. Phillips (1994), "A Reexamination of the Consumption Function Using Frequency Domain Regressions," *Empirical Economics*, 19 595-609.

Engle, R. F. (1974). "Band Spectrum Regression," *International Economic Review*, 15, 1-11.

Engle, R. F. (1978). "Testing Price Equations for Stability Across Spectral Frequency Bands," *Econometrica*, 46, 869-881.

Friedman, M. (1957). *A Theory of the Consumption Function*. Princeton: Princeton University Press.

Geweke, J. (1982). "Measurement of Linear Dependence and Feedback Between Multiple Time Series," *Journal of the American Statistical Association*, 77, 304-314.

Geweke, J. (1986). "The Superneutrality of Money in the United States: An Interpretation of the Evidence," *Econometrica*, 54, 1-21.

Harvey, A. C. (1978). "Linear Regression in the Frequency Domain," *International Economic Review*, 19, 507-512.

Jaditz, T. and C. L. Sayers (1994). "Predicting Inflation," working paper, Bureau of Labor Statistics Division of Price and Index Number Research.

Lee, K. (1994). "Causality of Nominal Money Growth at Various Frequency Bands: An Empirical Analysis Using Time Series Domain Chebyshev Filters," working paper, Department of Economics, University of Missouri.

Lucas, R. E. (1980). "Two Illustrations of the Quantity Theory of Money," *American Economic Review*, 70, 1005-1014.

Mills, T. C. (1982). "Signal Extraction and Two Illustrations of the Quantity Theory of Money," *American Economic Review*, 72, 1162-68.

Phillips, P. C. B. (1991). "Spectral Regression for Cointegrated Time Series," in Barnett, W., J. Powell, and G. Tauchen (eds.) *Nonparametric and Semiparametric Methods in Economics and Statistics*. Cambridge: Cambridge University Press.

Summers, L. H. (1983), "The Non-Adjustment of Interest Rates: a Study of the Fisher Effect," in Tobin, J. (ed.) *Macroeconomics, Prices and Quantities: Essays in Memory of Arthur M. Okun*. Washington, DC: Brookings Institution.

Summers, L. H. (1986). "Estimating the Long-Run Relationship Between Interest Rates and Inflation (a Response to McCallum)," *Journal of Monetary Economics*, 18, 7-86.

Tan, H. B. and R. A. Ashley (1997). "An Elementary Method for Detecting and Modelling Regression Parameter Variation Across Frequencies With an Application to Testing the Permanent Income Hypothesis," *Macroeconomic Dynamics*, forthcoming.

Thoma, M. A. (1992). "The Effects of Inside and Outside Money on Industrial Product Across Spectral Frequency Bands," *Review of Economics and Statistics*, 74, 737-741.

Thoma, M. A. (1994). "The Effects of Money Growth on Inflation Rates Across Spectral Frequency Bands," *Journal of Money, Credit, and Banking*, 26, 218-230.

7 STATIONARITY TESTS WITH MULTIPLE ENDOGENIZED BREAKS

Junsoo Lee
Vanderbilt University
University of Central Florida

1. Introduction

Since the pioneering work of Perron (1989), allowing for a structural break in testing for a unit root has been an important issue. He suggests that unit root tests would be biased toward accepting a false null hypothesis of a unit root if a time series exhibits stationary fluctuations around a trend containing a structural break. Since the break point used by Perron is given exogenously rather than determined from the data, subsequent literature has incorporated an endogenous break point. This approach includes Zivot and Andrews (1992), Christiano (1992), and Bannerjee, *et al*. (1992), among others.

The tests with a stationary null hypothesis rather than a unit root hypothesis have been considered by several authors including Kwiatkowski, *et al.* (1992, hereafter KPSS). These stationarity tests are also affected by the presence of a structural break, as demonstrated by Lee, *et al.* (1997) who show that stationarity tests ignoring an existing break diverge. Thus, stationarity tests ignoring a break when it exists tend to reject the null hypothesis of stationarity in favor of the false alternative unit root hypothesis. This result calls for the need of allowing for a structural break for stationarity tests as well.

In this chapter, we accordingly present modified stationarity tests allowing for multiple structural breaks. In particular, we consider an endogenizing testing procedure for which *both* the number of breaks and their placements are endogenously determined from the data. Previous studies assume the existence of a fixed number of structural breaks. Finding the true number of breaks correctly is crucial for precise statistical inference. It can be also shown that stationarity tests diverge when the number of breaks is under-specified, and they exhibit a loss of power when the number of breaks is over-specified. This finding is a generalization of that of Lee, *et al.* (1997). A correct estimation of the number of breaks is also critical for unit root tests. While Lumsdaine and Papell (1997) allow for two breaks, they assume *a priori* that the number of breaks is fixed.

Existing endogenized break models for a unit root have employed a criterion of minimizing t-statistics to identify a break point. Lee (1996) considers a minimum stationarity test that employs an endogenizing procedure similar to that employed for unit root tests. He finds that while the minimum stationarity test performs reasonably well in identifying a break point, it can not be used for determining the number of breaks. The minimum statistics can be used only when the number of breaks is given *a priori*. Recent developments in testing for the number of breaks include Bai and Perron (1998), Bai (1995, 1997), Yao (1988), and Nunes, *et al.* (1995). These works have made important contributions regarding the issue of determining the number of breaks and break points. This paper incorporates the Bayes information criterion to determine the number of breaks and break points, just as some of the above mentioned works do. The problem considered in this paper, however, differs from that considered in these works. We are concerned with the problem of testing for stationarity around an unknown number of breaks at unknown break points, against a unit root alternative hypothesis.

The paper is organized as follows. In Section 2, we provide modified stationarity tests allowing for exogenous multiple breaks whose placements are assumed to be known. Section 3 extends these to the endogenized break case. We also discuss statistical properties of the estimators of the number of breaks and break points. Section 4 provides simulation results on the performance of the tests. Section 5 gives a summary. Throughout this paper, "$\rightarrow$" indicates weak convergence at $T \rightarrow \infty$.

2. Test Statistics with Known Break Points

As in KPSS, we consider an unobserved components representation in which a time series y_t, $t = 1,2,...,T$, is decomposed into the sum of a random walk and a stationary error:

$$y_t = r_t + \epsilon_t, \tag{1}$$

where

$$r_t = r_{t-1} + u_t,$$

and where ϵ_t are assumed stationary, and u_t are *iid* $(0, \sigma_u^2)$. KPSS consider the testing regression, $y_{t_2} = \alpha + e_t$, to test the null hypothesis of stationarity around a level ($r_0 = \alpha$), $H_0: \sigma_u^2 = 0$. The initial value r_0 is treated as fixed and serves the role of an intercept. When a deterministic trend term $\gamma \cdot t$ ($\gamma \neq 0$) is added to the above model, the null hypothesis is stationarity around the trend function.

To allow for structural breaks, we consider the following models, using different specifications for r_t:

$$(M_1)^* \quad r_t = r_{t-1} + \sum_{j=1}^{R} \delta_j \cdot B_{jt} + u_t \tag{2a}$$

$$(M_2)^* \quad r_t = \gamma + r_{t-1} + \sum_{j=1}^{R} \delta_j \cdot B_{jt} + u_t \tag{2b}$$

$$(M_3)^* \quad r_t = \gamma + r_{t-1} + \sum_{j=1}^{R} \delta_j^* \cdot D_{jt} + u_t \tag{2c}$$

$$(M_4)^* \quad r_t = \gamma + \sum_{j=1}^{R} \delta_j \cdot B_{jt} + \sum_{j=1}^{R} \delta_j^* \cdot D_{jt} + u_t \tag{2d}$$

where $B_{jt} = 1$ for $t = T_{Bj} + 1$ and zero otherwise; $D_{jt} = 1$ for $t \geq T_{Bj} + 1$ and zero otherwise; and T_{Bj} stands for the time period when the jth structural change occurs. R is the number of structural breaks.

We then consider the following corresponding models:

$$(M_1) \quad y_t = \alpha + \sum_{j=1}^{R} \delta_j \cdot D_{jt} + e_t \tag{3a}$$

$$(M2) \quad y_t = \alpha + \gamma \cdot t + \sum_{j=1}^{R} \delta_j \cdot D_{jt} + e_t \tag{3b}$$

$$(M_3) \quad y_t = \alpha + \gamma \cdot t + \sum_{j=1}^{R} \delta_j^* \cdot DT_{jt} + e_t \tag{3c}$$

$$(M_4) \quad y_t = \alpha + \gamma \cdot t + \sum_{j=1}^{R} \delta_j \cdot D_{jt} + \sum_{j=1}^{R} \delta_j^* \cdot DT_{jt} + e_t \tag{3d}$$

where $DT_{jt}^* = t - T_{Bj}$ for $t \geq T_{Bj} + 1$ and zero otherwise.

Perron (1989) previously considered these models. M_1 describes a stationary process with a one time shift in the level. M_2 is a model in which the trend function has a shift in the intercept. M_3 is a piecewise model with a change in the slope of the trend function, but without any sudden change in the level at the time of the break. Lastly, M_4 allows for a sudden change in the level followed by a change in the slope of the trend function.

We consider M_1 and M_4 only, since M_2 and M_3 are special cases of M_4 and the asymptotic distributions of the tests from them are not separately identified. Consequently, the DGP and the testing regressions are either as M_1 in (3a) or M_4 in (3d). The level stationarity test statistic $\tilde{\eta}_\mu$ is defined as:

$$\tilde{\eta}_\mu = T^{-2}\sum_{t=1}^{T}\tilde{S}_t^2 / \tilde{s}^2(l), \tag{4}$$

where

$$\tilde{S}_t = \sum_{j=1}^{t}\tilde{e}_j,$$

and $\tilde{e}_t$ are residuals from the regression (3a); and where $\tilde{s}^2(l)$ is the estimate of the long-run variance,

$$\sigma^2 = \lim_{T\to\infty} T^{-1}\left(\Sigma e_t\right)^2.$$

It is constructed via non-parametric estimation,

$$\hat{s}^2(l) = \hat{\Gamma}_0 + 2\Sigma w_j \hat{\Gamma}_j,$$

by choosing a truncation lag parameter l and a set of weights, w_j, $j = 1,\ldots,l$, where $\hat{\Gamma}_j$ is the jth sample autocovariance of the residuals $\tilde{e}_t$. The null hypothesis implies level stationarity around the multiple structural breaks. The alternative hypothesis implies an integrated process with dummy variables, $y_t = y_{t-1} + \delta B_t + \epsilon_t$, where $B_t = [B_{1t},\ldots,B_{Rt}]'$ and $\delta = [\delta_1,\ldots,\delta_R]'$. The trend stationarity test statistic $\tilde{\eta}_\tau$ is obtained from (3d) in the same manner.

Assumption:
(a) The innovations ϵ_t *in (3a) - (3d) satisfy the strong mixing regularity conditions of Phillips and Perron (1988, p. 336).*
(b) The break points T_{Bj} *are known a priori, and* $T_{Bj}/T \to \lambda_j$ *as* $T\to\infty$.

To examine the asymptotic distributions of the test statistics, we let $[bT]$ stand for the nearest integer to bT for $b \in (0,1)$. We first show that the following result on the convergence of partial sum processes is established.

Proposition 1: *Let* $\tilde{S}_{[bT]}$ *be the partial sum of the residuals from the regression (3a). Then,*

$$\frac{1}{\sqrt{T}}\tilde{S}_{[bT]} \rightarrow \begin{cases} \sqrt{\lambda_1-\lambda_0}\;\sigma V_1\left(\dfrac{b-\lambda_0}{\lambda_1-\lambda_0}\right) & \text{for } 0<b\leq\lambda_1 \\ \sqrt{\lambda_j-\lambda_{j-1}}\;\sigma V_j\left(\dfrac{b-\lambda_{j-1}}{\lambda_j-\lambda_{j-1}}\right) & \text{for } \lambda_{j-1}<b<\lambda_j \\ \sqrt{\lambda_{R+1}-\lambda_R}\;\sigma V_{R+1}\left(\dfrac{b-\lambda_R}{\lambda_{R+1}-\lambda_R}\right) & \text{for } \lambda_R<b\leq 1 \end{cases} \tag{5}$$

where V_i *are independent Brownian bridges;* $v_i(b_i) = W(b_i) - b_i W(1)$, *for* $i = 1,\ldots,R+1$, *with*

$$b_j = \frac{b-\lambda_{j-1}}{\lambda_j-\lambda_{j-1}}$$

defined on the (0,1) region for $\lambda_0=0$, $\lambda_{R+1}=1$, *and* $\lambda_j\neq\lambda_{j-1}$, *and* W *is a Brownian motion.*

For the $\tilde{\eta}_\tau$ statistic, the partial sum of the residuals is obtained from the regression (3d), and its asymptotic distribution is the same as the above expression in (5) except that the term $V_i(b_i)$ is replaced by the second level Brownian bridge,

$$V_i^{(2)}(b_i) = W(b_i) + (2b_i-3b_i^2)W(1) + (-6b_i+6b_i^2)\int_0^1 W(b_i)\,db_i. \tag{6}$$

This expression appears in KPSS (1992, p. 167). The next theorem gives the asymptotic distributions of the $\tilde{\eta}_\mu$ and $\tilde{\eta}_\tau$ statistics under the null hypothesis.

Theorem 1: *Suppose that Assumptions (a) and (b) hold and the DGP is subject to R structural breaks. Under the null hypothesis of stationarity around the breaks,* $\sigma_u^2 = 0$,

$$\tilde{\eta}_\mu \rightarrow \sum_{j=1}^{R+1}(\lambda_j-\lambda_{j-1})^2\int_0^1 V_i(b_i)^2\,db_i. \tag{7}$$

The limiting distribution of the $\tilde{\eta}_\mu$ test statistic is the same as the above expression, except that $V_i(b_i)$ is replaced with $V_i^{(2)}$ in (6). The limiting distribution is the weighted sum of $R+1$ independent terms. Each of them can be obtained from the regression using the pre- or post sub-sample between jth and $(j+1)$th structural break. When $R = 1$, the symmetry of the distribution is easily observed, since we can interchange λ and $(1-\lambda)$. Symmetry is also possible with $R>1$. For instance, when $R = 2$, a set of $\lambda = (0.25, 0.5)'$ and $\lambda = (0.5, 0.75)'$ gives the same asymptotic distribution. The asymptotic distribution of the statistics $\tilde{\eta}_\mu$ and $\tilde{\eta}_\tau$ can be simulated accordingly. Critical values are calculated via Monte Carlo simulation, using the sample size of 2,000 and 5,000 replications. We provide them in Table 1 for $\lambda = 0.1, 0.2, 0.3, 0.4$, and 0.5, when $R = 1$, and for $\lambda_1, \lambda_2 = 0.1, 0.2, \ldots, 0.9$, $\lambda_1 \neq \lambda_2$, when $R = 2$. Critical values for three or more breaks are not provided, but they can be simulated in the same manner.[1]

Table 1. Upper Tail Critical Values for $\tilde{\eta}_\mu$ and $\tilde{\eta}_\tau$ with One Break

	$\tilde{\eta}_\mu$			$\tilde{\eta}_\tau$		
λ	90%	95%	99%	90%	95%	99%
0.1	0.288	0.381	0.601	0.098	0.122	0.179
0.2	0.230	0.303	0.484	0.079	0.098	0.143
0.3	0.188	0.245	0.388	0.065	0.079	0.113
0.4	0.160	0.200	0.300	0.056	0.066	0.090
0.5	0.153	0.189	0.269	0.053	0.062	0.081

We now consider the asymptotic distributions of our test statistics under the alternative hypothesis that $\sigma_u^2 > 0$. KPSS show that under the alternative, the numerator is $O_p(T^2)$ and the denominator is $O_p(lT)$, so that the test statistics are $O_p(T/l)$. Since $T/l \to \infty$ as $T \to \infty$, the tests are consistent. The order of divergence of our test statistics is the same as that of the KPSS statistics, so our tests are also consistent. The only difference is that the aysmptotic distributions of our statistics are again expressed in terms of a weighted sum of $R+1$ independent terms.

Theorem 2: *Suppose that Assumptions (a) and (b) hold and the DGP is subject to R structural breaks. Let* $\tilde{\eta}_\mu$ *and* $\tilde{\eta}_\tau$ *be the statistics in (4). Under the alternative hypothesis of a unit root,* $\sigma_u^2 > 0$,

$$(l/T)\tilde{\eta}_\mu^* \to \frac{\sum_{j=1}^{R+1} (\lambda_j - \lambda_{j-1})^4 \int_0^1 \left(\int_0^{bi} \underline{W}(s_i) ds_i \right)^2 db_i}{K \left[\sum_{j=1}^{R+1} (\lambda_j - \lambda_{j-1}) \int_0^1 \underline{W}(s_i)^2 ds_i \right]} \tag{8}$$

and the asymptotic distribution of $(1/T)\tilde{\eta}_\tau$ *is obtained by replacing the demeaned Wiener process* $\underline{W}(s_i)$ *in the above expression with the demeaned and detrended Wiener process* $\underline{W}(s)^*$. *The term K is a constant (equation (24), KPSS, p. 168), and the expressions for* $\underline{W}(s_i)$ *and* $\underline{W}(s_i)^*$ *appear in KPSS (equations (21) and (25), pp. 168-169).*

3. Test Statistics with Unknown Break Points

In practical estimation, one hardly knows the break points T_{Bj}; thus Assumption (b) is seldom satisfied. This violation of the assumption calls for a new procedure to determine unknown break points endogenously from the data. The estimation scheme in this paper is to choose the number of breaks and their placements using the Bayes information criterion. Let $\lambda = (\lambda_1, \ldots, \lambda_R)'$, with $\lambda_i \neq \lambda_j$. We consider a double minimum,

$$\{\tilde{R}, \tilde{\lambda}\} = \underset{0 \le R \le R^*}{argmin}\ \underset{\lambda \in \Lambda}{argmin}\left[T \log \tilde{\sigma}_R^2(R, \lambda) + (k + R(k+1))(\log T)\right] \tag{9}$$

where k is the number of regressors, R^* is the pre-specified upper bound for R, and Λ is a closed subset of (0, 1); and where $\tilde{\sigma}_R^2(R, \lambda)$ is the estimate of the usual error variance in (3) associated with fixed R and λ. Here, $\tilde{R}$ and $\tilde{\lambda}$ can be estimated by a grid search on the region (0, 1), excluding the two end points. As a practical matter, the region can be restricted on [0.15, 0.85] to achieve efficiency in calculation. The above Bayes information criterion was advocated by Yao (1988) and Nunes, *et al.* (1995). The endogenized test statistic is defined as the $\tilde{\eta}_\mu$ test which uses $\{\tilde{R}, \tilde{\lambda}\}$,

$$\tilde{\eta}_\mu^* = \tilde{\eta}_\mu(\tilde{R}, \tilde{\lambda}). \tag{10}$$

The trend stationarity test $\tilde{\eta}_\tau^*$ is obtained in a similar manner.

Here, $\tilde{\lambda}$ has a well-defined probability distribution over Λ, and the true value λ is included in the sample space of $\tilde{\lambda}$. The same is true for $\tilde{T}_{Bj}$ defined over (2,...,T-1). Then the asymptotic distribution of the statistic $\tilde{\eta}_\mu$ is given by:

$$\int_0^1 \Phi\,[\tilde{\eta}_\mu(\tilde{\lambda}) | \tilde{\lambda} = \lambda] P(\tilde{\lambda} = \lambda)\, d\lambda,$$

where $\Phi\,[\tilde{\eta}_\mu(\tilde{\lambda}) | \tilde{\lambda} = \lambda]$ is the asymptotic distribution of $\tilde{\eta}_\mu^*$ conditional on $\tilde{\lambda} = \lambda$, and $P(\tilde{\lambda} = \lambda)$ is the probability that $\tilde{\lambda} = \lambda$. The asymptotic distribution of $\tilde{\eta}_\tau^*$ is obtained in the same manner. As an extreme case, suppose that $P(\tilde{\lambda} = \lambda) = 1$, and the break points are correctly estimated. This presumption holds when Assumption (c) is satisfied.

Table 2. Upper Tail Critical Values for $\tilde{\eta}_\mu$ and $\tilde{\eta}_\tau$ with Two Breaks.

		$\tilde{\eta}_\mu$			$\tilde{\eta}_\tau$		
λ_1	λ_2	90%	95%	99%	90%	95%	99%
0.1	0.2	0.222	0.302	0.482	0.078	0.097	0.142
0.1	0.3	0.175	0.231	0.370	0.063	0.076	0.111
0.1	0.4	0.144	0.186	0.288	0.050	0.060	0.086
0.1	0.5	0.128	0.157	0.231	0.044	0.053	0.072
0.1	0.6	0.130	0.158	0.236	0.044	0.051	0.070
0.1	0.7	0.146	0.188	0.282	0.052	0.062	0.085
0.1	0.8	0.172	0.227	0.369	0.062	0.076	0.110
0.1	0.9	0.219	0.285	0.460	0.077	0.095	0.144
0.2	0.3	0.182	0.238	0.383	0.060	0.074	0.111
0.2	0.4	0.139	0.179	0.290	0.048	0.060	0.083
0.2	0.5	0.114	0.141	0.217	0.040	0.047	0.064
0.2	0.6	0.104	0.125	0.181	0.036	0.042	0.056
0.2	0.7	0.113	0.139	0.208	0.040	0.047	0.066
0.2	0.8	0.142	0.184	0.285	0.048	0.059	0.085
0.2	0.9	0.175	0.223	0.365	0.064	0.078	0.113
0.3	0.4	0.146	0.188	0.307	0.050	0.060	0.083
0.3	0.5	0.116	0.145	0.230	0.039	0.047	0.066
0.3	0.6	0.097	0.117	0.160	0.034	0.039	0.051
0.3	0.7	0.097	0.116	0.159	0.034	0.039	0.050
0.3	0.8	0.112	0.144	0.218	0.040	0.047	0.066
0.3	0.9	0.148	0.191	0.299	0.051	0.061	0.086
0.4	0.5	0.129	0.159	0.229	0.045	0.052	0.070
0.4	0.6	0.104	0.126	0.190	0.036	0.042	0.054
0.4	0.7	0.097	0.116	0.162	0.034	0.039	0.050
0.4	0.8	0.104	0.128	0.178	0.037	0.042	0.054
0.4	0.9	0.126	0.153	0.228	0.044	0.051	0.070
0.5	0.6	0.126	0.157	0.226	0.044	0.051	0.071
0.5	0.7	0.110	0.137	0.206	0.040	0.047	0.063
0.5	0.8	0.113	0.141	0.207	0.040	0.047	0.065
0.5	0.9	0.126	0.159	0.223	0.044	0.050	0.069
0.6	0.7	0.138	0.177	0.265	0.050	0.060	0.084
0.6	0.8	0.137	0.173	0.269	0.048	0.059	0.081
0.6	0.9	0.147	0.188	0.291	0.050	0.061	0.084
0.7	0.8	0.180	0.235	0.362	0.061	0.075	0.106
0.7	0.9	0.177	0.234	0.376	0.061	0.076	0.108
0.8	0.9	0.221	0.301	0.474	0.077	0.097	0.133

Assumption:
(c) $\tilde{T}_{Bj} \to T_{Bj}$, *and* $\tilde{\lambda}_j \to \lambda_j$, *as* $T \to \infty$.

Then the null distribution of $\tilde{\eta}_\mu^*$ approaches asymptotically that of $\tilde{\eta}_\mu$, which assumes fixed break points.

Corollary 1: *Suppose that Assumptions (a) and (c) hold. Then, the asymptotic null distribution of* $\tilde{\eta}_\mu^*$ *converges to that of* $\tilde{\eta}_\mu$.

The above result implies that the asymptotic distribution of the $\tilde{\eta}_\mu^*$ statistic approaches that of the $\tilde{\eta}_\mu$ statistic.[2] This result, however, hinges on the consistency of $\tilde{\lambda}$. For the maximum likelihood estimate of λ, say $\tilde{\lambda}^*$, Bai (1997) shows that

$$T(\tilde{\lambda}^* - \lambda) = O_p\left(\frac{1}{d'd}\right),$$

where d denotes the magnitude of structural breaks, and $T^{1/2-b}\|b\| \to \infty$ for $b \in (0, \frac{1}{2})$, in the regression model allowing for structural breaks. Similar reasoning can be found from the confidence interval of $\tilde{\lambda}$. Picard (1985) and Bai (1995) show that the confidence interval of λ is given as $\tilde{\lambda}^* \pm q_{\alpha/2}\tilde{\sigma}^2(d'd)^{-1}$, where $q_{\alpha/2}$ is the $\alpha/2$ percentile defined in their papers, under some regularity conditions involving stationarity. A smaller value of d results in a wider confidence interval. These results also hold in our case. In the MLE, $\tilde{\lambda}^*$ is obtained as the value minimizing the sum of squared residuals, while $\tilde{\lambda}$ is obtained in (10). In addition, a relevant order of convergence T^2 is applied to the stationarity statistic. Thus, $\tilde{\lambda}$ is also T-consistent, but only when d is big enough.[3] Therefore, $\tilde{\lambda}^*$ or $\tilde{\lambda}$ is not expected to be correctly estimated in the finite sample if d is small. It should be noted that although the asymptotic distribution of the test statistic $\tilde{\eta}_\mu^*$ or $\tilde{\eta}_\tau^*$ is invariant to d, the probability $P(\tilde{\lambda} = \lambda)$ depends on it in finite samples. Incorrect placement of λ will affect the performance of the endogenized stationarity tests. When d is big enough, $P(\tilde{\lambda} = \lambda)$ is close to one. When d is small, $P(\tilde{\lambda} = \lambda)$ departs from one. Accordingly, estimated statistics tend to differ from their asymptotic values. This phenomenon is not critical, but affects finite sample performance of the tests, as we see in the next section.

Nonetheless, break points cannot be consistently estimated under the unit root alternative. Intuitively speaking, this is so because the distribution of stationarity tests is not dependent on λ under the unit root alternative; see Lee, *et al.* (1997). Nunes, *et al.* (1996) provide evidence that the ML estimate of a break point for an integrated process is imprecise. The same is true for the endogenized stationarity tests under the alternative hypothesis of a unit root, or for usual endogenized unit root tests under the null hypothesis of a unit root. This difficulty of estimating λ for the integrated process, however, does not negate the endogenized stationarity tests, since stationarity statistics diverge under the alternative, regardless of good or bad estimates of λ. Thus, stationarity tests will have power to reject the null of stationarity if the DGP implies the unit root alternative hypothesis. Only estimated break points may not be reliable under the alternative hypothesis.

We next look at the effects of possible mis-specification of the number of breaks.

Lee, *et al.* (1997) have shown that usual stationarity tests ignoring a break when it exists would be biased toward rejecting the true null of stationarity (see Theorem 1, Lee, *et al.* (1997)). Their result can be easily generalized. Let $\tilde{R}$ denote the number of breaks in the testing regression.

> **Corollary 2:** *Suppose that the true data generating process implies (1) including R structural breaks. And suppose that the number of breaks is under-specified for the* $\tilde{\eta}^*_\mu$ *or* $\tilde{\eta}^*_\tau$ *test such that* $0 \le \tilde{R} < R$. *Then*
>
> $$\tilde{\eta}^*_\mu = O_p(T/l).$$

This result suggests that stationarity tests diverge from the null distribution when the number of breaks is under-specified. Thus, they tend to reject the true null hypothesis of stationarity too often. This bias problem of the stationarity tests ignoring an existing break corresponds to the property of the unit root tests which tend to accept the false unit root null hypothesis in the presence of a break, as Perron (1989) and Amsler and Lee (1995) showed.

While the above result holds under the null hypothesis of stationarity, the effect of mis-specifying the number of breaks under the alternative is different. Lee, *et al.* (1997) suggest that the asymptotic distribution of the stationarity test ignoring a structural break when it exists is not affected under the alternative hypothesis of a unit root.

> **Corollary 3:** *Suppose that Assumption (a) holds and the DGP is subject to an unknown number of structural breaks. Under the alternative hypothesis of a unit root,* $\sigma^2_u > 0$, *the distributions of the statistics* $\tilde{\eta}_\mu$, $\tilde{\eta}_\tau$, $\tilde{\eta}^*_\mu$, *and* $\tilde{\eta}^*_\tau$ *are asymptotically invariant to the inclusion of exclusion of a finite number of breaks.*

This result indicates that the existence of breaks matters mainly under the null hypothesis for stationarity tests, while it does not affect the power of the tests under the alternative hypothesis of a unit root.

4. Simulation Results

In this section we provide Monte Carlo simulation results on the performance of the endogenized stationarity tests. Pseudo-*iid* N(0,1) random numbers were generated using the Gauss procedure RNDNS, and all calculations were conducted using version 3.2.12 of Gauss. The DGP used in the simulations has the form of (3a) or (3d). The initial values (y_0 and ϵ_0) are assumed to be zero, and we assume $\sigma^2_\epsilon = 1$. All simulation results are calculated using 2,000 replications, and the size (rejection frequencies when $\sigma^2_u = 0$) or power (rejection frequencies when $\sigma^2_u > 0$) of the tests are evaluated using the upper 5% critical value. We focus on the $\tilde{\eta}^*_\mu$ test, because the behavior of the $\tilde{\eta}^*_\tau$ test is very similar.

Table 3. Performance of Endogenized Stationarity Tests

Panel A: Under the Null

Exp	DGP				$\hat{R}$(%)					
	R	λ	δ	R^*	0	1	2	3	4	Size
A	0	-	-	4	99	1	0	0	0	0.047
B	1	0.5	1	4	21	69	10	0	0	0.348
	1	0.5	2	4	0	100	0	0	0	0.062
	1	0.5	3	4	0	100	0	0	0	0.051
C	2	0.25, 0.5	1, 1	4	0	92	7	1	0	0.522
	2	0.25, 0.5	2, 2	4	0	4	90	7	0	0.114
	2	0.25, 0.5	3, 3	4	0	0	100	0	0	0.053

Panel B: Under the Alternative

Exp	DGP				$\hat{R}$(%)					
	R	λ	δ	R^*	0	1	2	3	4	Power
4C	0	-	-	4	1	11	33	36	19	0.867
	1	0.5	1	4	1	10	35	35	19	0.866
	1	0.5	2	4	1	9	37	35	18	0.859
	1	0.5	3	4	1	12	33	39	15	0.875
4D	2	0.25, 0.5	1, 1	4	1	12	35	35	17	0.863
	2	0.25, 0.5	2, 2	4	1	11	34	37	17	0.864
	2	0.25, 0.5	3, 3	4	0	10	34	35	21	0.860

The parameter λ indicating break points can be estimated from a grid search. A more efficient method would be to use a dynamic programming algorithm, as demonstrated by Bai and Perron (1998). We adopt a similar procedure.[4] We impose the restriction that λ lies on the interval [0.15, 0.85] to exclude values of λ near the end points. Since the $\tilde{\eta}_\mu^*$ test depends on λ, we use the asymptotic critical values (with T = 2,000) corresponding to the estimated number of breaks and estimated break points, for each replication. We let R, R^*, and $\hat{R}$ denote the true number of breaks in the DGP, the maximum number of breaks in the simulation, and the estimated number of breaks, respectively.

We first examine how well the $\tilde{\eta}_{\mu}^{*}$ test determines the number of breaks and their placements, and how these estimates affect the performance of the test under the null hypothesis. We consider a maximum of four breaks ($R^* = 4$). The results are given in Table 3. The columns of $\hat{R}$ provide the relative frequency of the estimated number of breaks. Experiment A examines the case of $R = 0$, so that the DGP implies no break. The results indicate that the test detects the number of breaks fairly well. The relative frequency of choosing no break ($\hat{R} = 0$) when there is no break ($R = 0$) is almost 100%. The size of the test in this situation is close to its nominal size of 5%.

Experiment B assumes the presence of one break ($R = 1$) in the DGP. We examine the case that $\lambda = 0.5$. The results with other values of λ are similar, and thus are omitted. We examine the effect of using different magnitudes of a structural break; $d =$ 1, 2, and 3. When the magnitude of a break is small ($d = 1$), the relative frequencies of the results that $\hat{R} = 0$, 1, and 2 are 21.5%, 68.5%, and 10.0%, respectively. Since this is the case where the extent of a structural break is minimal, a precise estimation of the number of breaks would not be easy. The size of the test in this case in 34.8%. The over-rejection might be driven by the cases in which the number of breaks is underestimated ($\hat{R} = 0$, when $R = 1$), as Corollary 2 may well indicate.

To visualize the distribution of their placements, we provide relative frequencies of $\tilde{\lambda}$ in Figure 1a for the cases of $\hat{R} = 1$ and in Figure 1b for the cases of $\hat{R} = 2$, respectively. The estimated values of $\tilde{\lambda}$ center around 0.5 for the cases of $\hat{R} = 1$, while they show no clear pattern for the cases of $\hat{R} = 2$. When the magnitude of a break is big enough that $d = 2$ or 3, the procedure detects the number of breaks very closely. The frequency of correctly estimating the number of breaks (choosing $\hat{R} = 1$, when $R = 1$) is almost 100%. There is no significant size distortion in this situation; the size is close to its nominal value, which confirms the finding in Corollary 1. Figures 1c and 1d show the distribution of the estimated values of $\tilde{\lambda}$. They clearly center around $\lambda = 0.5$, and they collapse closely to 0.5 when $d = 3$.

Experiment C considers the case in which the true number of breaks is two ($R = 2$). We consider values of $\lambda = (\lambda_1, \lambda_2)$ to be (0.25, 0.5). We also examine the effects of different magnitudes of structural breaks, and consider values of $d = (d_1, d_2)$ to be (1, 1), (2, 2) and (3, 3). The simulation results are very similar to the cases of $R = 1$. When $d = (1,1)$, the number of breaks is often underestimated. The relative frequency of choosing $\hat{R} = 1$ is 92.1%, and accordingly the test exhibits over-rejections of the null hypothesis. Figure 2a shows that the estimated values of $\tilde{\lambda}$ center around either 0.25 or 0.5. When $d = (2,2)$, however, the relative frequency of estimating the number of breaks correctly ($\hat{R} = 2$) becomes higher. Then the performance of the test improves, reducing size distortions. When $d = (3,3)$, the test performs very well. The relative frequency of estimating the number of breaks correctly ($\hat{R} = 2$) is almost 100%, and the size of the test is close to its nominal value. Figures 2c and 2d show the surface of estimated values of $\tilde{\lambda}$, which converges to λ as d increases.

We also investigated the effects of autocorrelated errors. Their effects on the endogenized tests are about the same as those on the tests assuming exogenous breaks. While it is hard to distinguish net effects of autocorrelated errors, the performance of

Figure 1. Empirical Distribution of Estimated Break Points (R = 1, λ = 0.5)

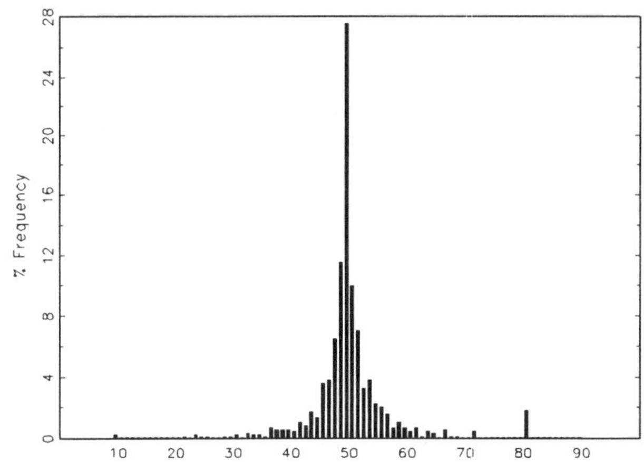

1a: $\hat{R} = 1, d = 1$

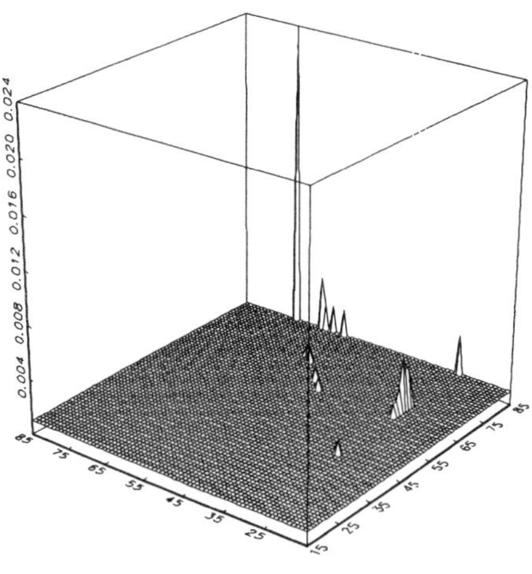

1b: $\hat{R} = 2, d = 1$

Figure 1, Continued

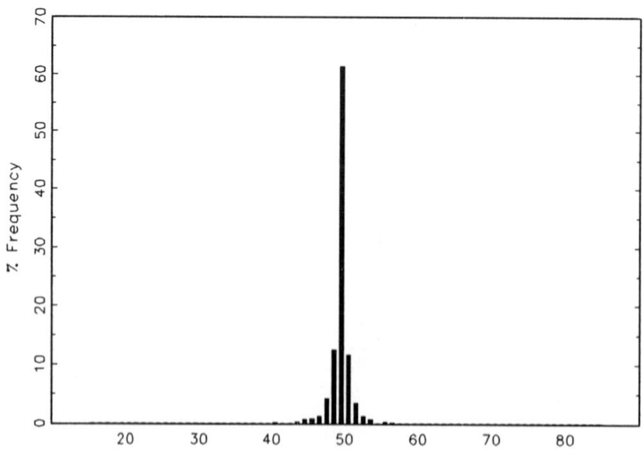

1c: $\hat{R}=1, d=2$

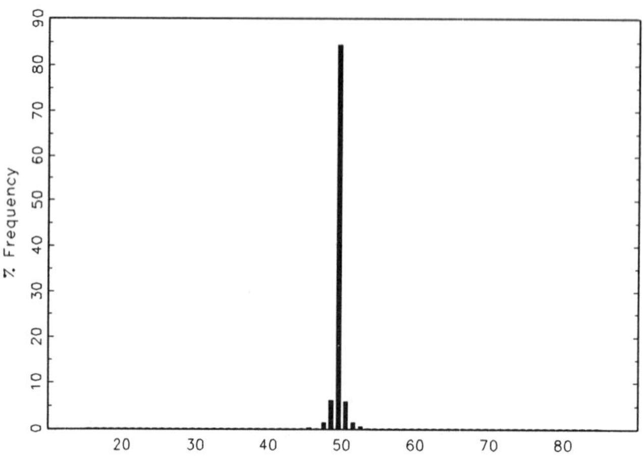

1d: $\hat{R}=1, d=3$

Figure 2. Empirical Distribution of Estimated Break Points (R = 2, λ = 0.25, 0.5)

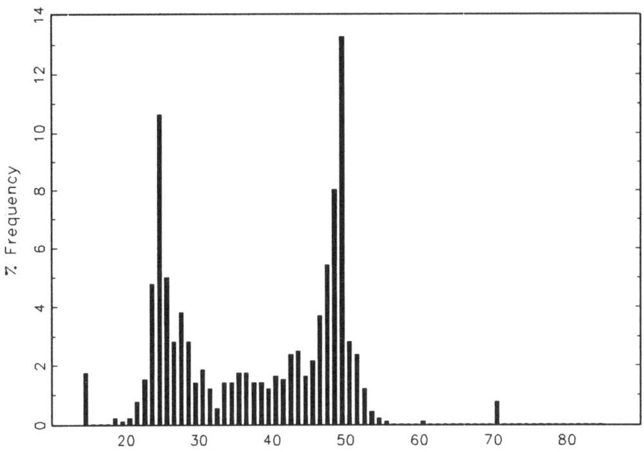

2a: $\hat{R} = 1, d = (1,1)$

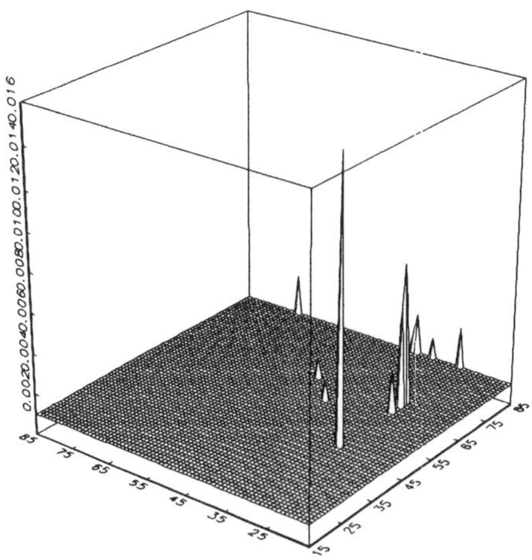

2b: $\hat{R} = 2, d = (1,1)$

Figure 2, Continued

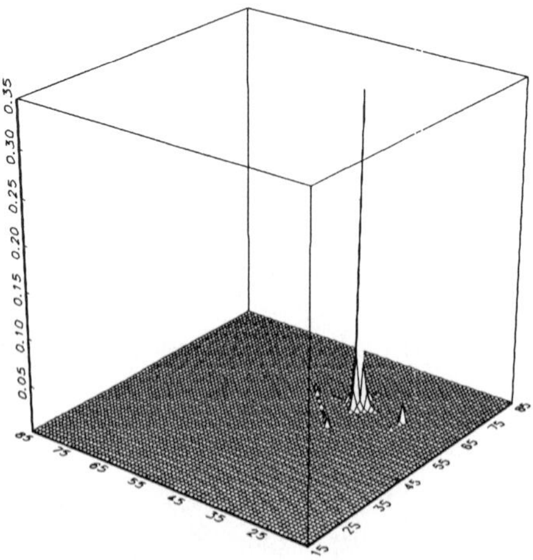

2c: $\hat{R}=2, d=(2,2)$

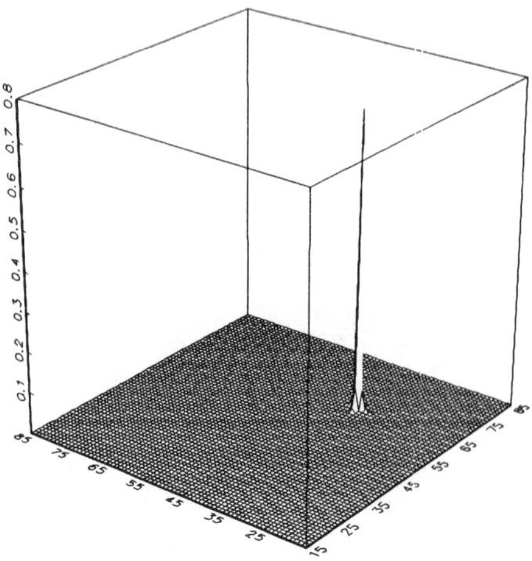

2d: $\hat{R}=2, d=(3,3)$

the testing procedures in estimating R and λ does not appear to be affected by them, and thus we do not require a separate discussion of the simulation results.

Lastly, we investigate the performance of the $\tilde{\eta}_\mu^*$ test under the alternative hypothesis of a unit root. In Experiments D, E, and F, we examine the cases where the true number of breaks is 0, 1, and 2, respectively ($R=0, R=1$, and $R=2$). The results are given in Panel (b) of Table 3. Selections of the number of breaks under the alternative are very different from the cases under the null hypothesis. The number of breaks is not correctly estimated under the null hypothesis. The number of breaks is not correctly estimated under the alternative hypothesis of a unit root. This result is expected, since the distribution of the stationarity statistic does not depend on λ under the alternative hypothesis, as Corollary 3 indicates. Nonetheless, the power of the stationarity test does not depend on whether the number of breaks is correctly estimated. Even when the number of breaks is over-fitted, or under-fitted, there is no loss of power. In addition, power does not depend on d. Therefore, neither the number of breaks nor their placements can carry any significant meaning when the stationarity null hypothesis is rejected in favor of the alternative hypothesis of a unit root.

5. Conclusions

In this chapter we proposed endogenized stationarity tests and investigated their performance. We have shown that the proposed testing procedures using the Bayes information criterion performs reasonably well under the stationarity null hypothesis in estimating the unknown number of breaks and their placements endogenously from the data, when the magnitude of structural breaks is big enough. Under the alternative hypothesis of a unit root, neither the number of breaks nor their placements are precisely estimated. But this difficulty does not negate the value of the endogenized stationarity test, because they still have power to reject the null hypothesis, regardless of possibly incorrect estimates of the number of breaks and their placements.

Appendix

Proof of Proposition 1 and Theorem 1: We first consider the case of one structural break. The proof can be easily extended to cases of multiple breaks. We start with the level stationarity test $\tilde{\eta}_\mu$. Without loss of generality, we express equation (3a) as: $y_t = \alpha_1 D_{1t} + \alpha_2 D_{2t} + e_t$, where $D_{1t} = 1$ for $t \le T_B$ and zero otherwise, and $D_{2t} = 1$ for $t > T_B$ and zero otherwise. Denoting $\boldsymbol{x}_t = [D_{1t}, D_{2t}]'$, $\boldsymbol{X} = (\boldsymbol{x}_1', \ldots, \boldsymbol{x}_T')'$, $\boldsymbol{e} = (e_1, \ldots, e_T)'$, and $P = diag[T^{1/2}, T^{1/2}]$, we express the partial sum of the residuals as:

$$T^{-1/2}\tilde{S}_{[bT]} = T^{-1/2}\sum_{j=1}^{[bT]}[e_j - (PD_{1j}\alpha_1, PD_{2j}\alpha_2)(PX'XP)^{-1}PX'\boldsymbol{e}].$$

Here,

$$T^{-1/2}\sum_{j=1}^{bT} e_j \to \sigma W(b),\ (P\boldsymbol{X}'\boldsymbol{X}P) \to \begin{bmatrix} \lambda & 0 \\ 0 & 1-\lambda \end{bmatrix},\ P\boldsymbol{X}'\boldsymbol{e} \to \begin{pmatrix} \sigma W(\lambda) \\ \sigma W(1-\lambda) \end{pmatrix},$$

$$T^{-1}\sum_{j=1}^{[bT]} D_{1j} \to \begin{cases} \lambda \text{ for } 0 \le b \le \lambda \\ b \text{ for } \lambda \le b \le 1 \end{cases}, \text{ and } T^{-1}\sum_{j=1}^{[bT]} D_{2j} \to \begin{cases} 0 \text{ for } 0 \le b \le \lambda \\ b-\lambda \text{ for } \lambda < b \le 1 \end{cases}.$$

For $b \le \lambda$,

$$\sigma^{-1}T^{-1/2}\tilde{S}_{[bT]} \to W(b) - (b/\lambda)W(\lambda) = \sqrt{\lambda}\,W(b/\lambda) - (b/\sqrt{\lambda})W(1)$$

$$= \sqrt{\lambda}\left[W(b/\lambda) - (b/\lambda)W(1)\right] \equiv \sqrt{\lambda}\,V(b_1),$$

where $V(b_1) \equiv W(b_1) - b_1 W(1)$ and $b_1 \equiv b/\lambda$ defined on [0,1].

For $b>\lambda$,

$$\sigma^{-1}T^{-1/2}\tilde{S}_{[bT]} \to W(b) - W(\lambda) - \frac{b-\lambda}{1-\lambda}W(1-\lambda) = \sqrt{1-\lambda}\left[W(b-\lambda) - \frac{b-\lambda}{1-\lambda}(b/\sqrt{\lambda})W(1)\right]$$

$$= \sqrt{1-\lambda}\left[W(b_2) - b_2 W(1)\right] \equiv \sqrt{1-\lambda}\,V(b_2),$$

where $V(b_2) \equiv W(b_2) - b_2 W(1)$ and $b_2 \equiv \dfrac{b-\lambda}{1-\lambda}$ defined on [0,1].

Since $T^{-2}\sum_{t=1}^{T}\tilde{S}_t^2 = T^{-2}\sum_{t=1}^{[\lambda T]}\tilde{S}_t^2 + T^{-2}\sum_{t=[\lambda T+1]}^{T}\tilde{S}_t^2$, we obtain:

$$T^{-2}\sum_{t=1}^{T}\tilde{S}_t^2 \to \sigma^2\int_0^{\lambda}\left[\sqrt{\lambda}V(b_1)\right]^2 db + \sigma^2\int_{\lambda}^{1}\left[\sqrt{1-\lambda}V(b_2)\right]^2 db$$

$$= \lambda\sigma^2\int_0^1\left[\sqrt{\lambda}V(b_1)\right]^2 db_1 + (1-\lambda)\sigma^2\int_0^1\left[\sqrt{1-\lambda}V(b_2)\right]^2 db_2$$

$$= \lambda^2\sigma^2\int_0^1\left[V(b_1)\right]^2 db_1 + (1-\lambda)^2\sigma^2\int_0^1\left[V(b_2)\right]^2 db_2.$$

In the presence of multiple breaks, the above expression is given as the weighted sum of $R+1$ terms, with weights being the fractions λ_j of the time periods between breaks. For the trend stationarity test $\tilde{\eta}_\tau$, we alternatively let $\boldsymbol{x}_t = [D_{1t}, DT_{1t}, D_{2t}, DT_{2t}]'$, where $DT_{it} = D_{it}\cdot t$ for $i = 1,2$, and $P = diag[T^{1/2}, T^{3/2}, T^{1/2}, T^{3/2}]$. Following similar procedures used for the $\tilde{\eta}_\mu$ test, we can show for the $\tilde{\eta}_\tau$ test that

$$\sigma^{-1}T^{-1/2}\tilde{S}_{[bT]}\rightarrow\sqrt{\lambda}V_2(b_1) \text{ for } b\leq\lambda, \text{ and } \sigma^{-1}T^{-1/2}\tilde{S}_{[bT]}\rightarrow\sqrt{1-\lambda}V_2(b_2) \text{ for } b>\lambda.$$

The denominator of the statistic is a consistent estimator of the long-run variance σ^2; see Phillips and Perron (1988).

Proof of Theorem 2: We also provide proof for the case of one structural break. The proof can be similarly extended to cases of multiple breaks. As indicated in KPSS, under the alternative, both the numerator and the denominator diverge. This is also true for our tests. When the DGP implies M_1^* under the alternative, the partial sum of the residuals for the level stationarity test is:

$$T^{-3/2}\tilde{S}_{[bT]} = T^{-3/2}\sum_{j=1}^{[bT]}[r_j - \boldsymbol{x}_j'P(P\boldsymbol{X}'\boldsymbol{X}P)^{-1}P\boldsymbol{X}'r + e_j - \boldsymbol{x}_j'(\boldsymbol{X}'\boldsymbol{X})^{-1}\boldsymbol{X}'\boldsymbol{e} +$$

$$B_j\delta_1 - \boldsymbol{x}_j'(\boldsymbol{X}'\boldsymbol{X})^{-1}\boldsymbol{X}'\boldsymbol{B}\delta_1],$$

where $\boldsymbol{x}_t = [D_{1t},D_{2t}]'$, $\boldsymbol{B} = (B_1,\ldots,B_T)'$, $\boldsymbol{e} = (e_1,\ldots,e_T)'$, and $P = diag[T^{1/2},T^{1/2}]$ in this case. Except for the first two terms on the right hand side, all terms are $o_p(1)$. We want to show that the first two terms are asymptotically distributed as a sum of two independent terms expressed by the demeaned Wiener process. Since

$$T^{-3/2}\sum_{t=1}^{T}r_tD_{1t}\rightarrow\sigma_\mu\int_0^\lambda W(s)ds, \text{ and } T^{-3/2}\sum_{t=1}^{T}r_tD_{2t}\rightarrow\sigma_\mu\int_\lambda^1 W(s)ds,$$

for $b\leq\lambda$, we obtain:

$$T^{-3/2}\tilde{S}_{[bT]}\rightarrow\sigma_\mu\int_0^b W(s)ds - \sigma_\mu(b/\lambda)\int_0^\lambda W(s)ds = \lambda^{3/2}\sigma_\mu\left[\int_0^{b_1}W(s_1)ds_1 - b_1\int_0^1 W(s_1)ds_1\right]$$

where $b_1 \equiv b/\lambda$ and $s_1\equiv s/\lambda$ are defined on [0,1]. For $b>\lambda$,

$$T^{-3/2}\tilde{S}_{[bT]}\rightarrow\sigma_\mu\int_\lambda^b W(s)ds - \frac{b-\lambda}{1-\lambda}\sigma_\mu\int_\lambda^1 W(s)ds$$

$$= (1-\lambda)^{3/2}\sigma_\mu\left[\int_0^{b_2}W(s_2)ds_2 - b_2\int_0^1 W(s_2)ds_2\right],$$

where $b_2 \equiv \dfrac{b-\lambda}{1-\lambda}$ and $s_2 \equiv \dfrac{s-\lambda}{1-\lambda}$ are defined on [0,1]. We have used the results

$$\lambda^{-1/2}W(s) = W(s/\lambda) \text{ and } (1-\lambda)^{-1/2}W(s-\lambda) = W[(s-\lambda)/(1-\lambda)]$$

in the above derivation. Denoting $\underline{W}(s_i) = W(s_i)ds - b_i\int_0^1 W(s_i)ds_i$, $i = 1, 2$, we can show

$$T^{-4}\sum_{t=1}^{T}\tilde{S}_t^2 = T^{-1}\sum_{t=1}^{\lambda T}(T^{-3/2}\tilde{S}_t)^2 + T^{-1}\sum_{t=\lambda T+1}^{T}(T^{-3/2}\tilde{S}_t)^2$$

$$\to \sigma_\mu^2\int_0^\lambda\left[\int_0^{b_1}\lambda^{3/2}\underline{W}(s_1)\right]^2 db + \sigma_\mu^2\int_\lambda^1\left[\int_0^{b_2}(1-\lambda)^{3/2}\underline{W}(s_2)\right]^2 db$$

$$= \lambda^4\sigma_\mu^2\int_0^1[\underline{W}(s_1)]^2 db_1 + (1-\lambda)^4\sigma_\mu^2\int_0^1[\underline{W}(s_2)]^2 db_2.$$

Adopting the results from equation 23 of KPSS, it can be shown that the denominator follows

$$(lT)^{-1}s(l)^2 \to K\sigma_\mu^2[\lambda\int_0^1\underline{W}(s_1)^2 ds_1 + (1-\lambda)\int_0^1\underline{W}(s_2)^2 ds_2].$$

Thus, the $\tilde{\eta}_\mu$ statistic is $O_p(T/l)$, too. The proof for the trend stationarity test $\tilde{\eta}_\tau$ can be obtained in the same manner.

Notes

1. A Gauss program to evaluate critical values is available from the author upon request.

2. The above result implies that the asymptotic distribution of the $\tilde{\eta}_\mu^*$ statistic is not free of λ, since the $\tilde{\eta}_\mu$ statistic depends on λ. This property contrasts with that of the endogenized unit root test. Zivot and Andrews (1992) have shown that the asymptotic distribution of their minimum t-statistic follows the distribution of the minimum order statistic of t-statistics. Then the asymptotic distribution is free of the nuisance parameter λ. An intuitive reason is that while the null hypothesis of a unit root test does not imply λ, the null hypothesis of the stationarity test implies λ.

3. This result does not necessarily mean that the estimate of break points (T_{Bj}) is T-consistent; the result is applied to $\tilde{\lambda}$, but not to $\tilde{T}_{Bj}$.

4. Details of the use of the dynamic programming method is illustrated in Bai and Perron (1998, Appendix). The Gauss coded program used in this paper in searching for break points based on the dynamic programming technique is available from the author upon request.

References

Amsler, C. and J. Lee (1995). "An LM Test for A Unit Root in the Presence of a Structural Change," *Econometric Theory*, 11, 359-368.

Bai, J. (1997). "Estimating Multiple Breaks one at a Time," *Econometric Theory*, 13, 315-352.

Bai, J. (1995). "Least Absolute Deviation Estimation of s Shift," *Econometric Theory*, 11, 403-436.

Bai, J. and P. Perron (1998). "Estimating and Testing Linear Models with Multiple Structural Changes," *Econometrica*, 66, 47-78.

Banerjee, A., R. L. Lumsdaine, and J. H. Stock (1992). "Recursive and Sequential Tests of the Unit Root and Trend Break Hypothesis: Theory and International Evidence," *Journal of Business and Economic Statistics*, 10, 271-288.

Christiano, L. J. (1992). "Searching for a Break in GNP," *Journal of Business and Economic Statistics*, 10, 237-250.

Kwiatkowski, D., P. C. B. Phillips, P. Schmidt, and Y. Shin (1992). "Testing The Null Hypothesis Of Stationarity Against The Alternative of A Unit Root: How Sure Are We That Economic Time Series Have a Unit Root?," *Journal of Econometrics*, 54, 159-178.

Lee, J., C. Huang, and Y. Shin (1997). "On Stationarity Tests in the Presence of Structural Breaks," *Economics Letters*, 55, 165-172.

Lee, J. (1996). "Minimum Statistics Testing for Stationarity in the Presence of a Structural Break," working paper, Department of Economics, Vanderbilt University.

Lumsdaine, R. L. and D. H. Papell (1997). "Multiple Trend Breaks and the Unit Root Hypothesis," *Review of Economics and Statistics*, 79, 212-218.

Nunes, L., C. Kuan and P. Newbold (1995). "Spurious Break," *Econometric Theory*, 11, 736-749.

Nunes, L., P. Newbold and C. Kuan (1996). "Spurious Number of Breaks," *Economics Letters*, 50, 175-178.

Perron, P. (1989). "The Great Crash, the Oil Price Shock, and the Unit Root Hypothesis, *Econometrica*, 57, 1361-1401.

Phillips, P. C. B. and P. Perron (1988). "Testing for a Unit Root in Time Series Regression, *Biometrika*, 75, 335-346.

Picard, D. (1985). "Testing and Estimating Change-Points in Time Series," *Advances in Applied Probability*, 176, 841-867.

Yao, Y. C. (1988). "Estimating the Number of Change-points via Schwarz' criterion," *Statistics and Probability Letters*, 6, 181-189.

Zivot, E. and D. W. K. Andrews (1992). "Further Evidence on the Great Crash, the Oil-Price Shock, and the Unit-Root Hypothesis," *Journal of Business and Economic Statistics*, 11, 3, 251-270.

8 NONLINEAR EVOLUTION IN UK STOCK RETURNS AND VOLUME

Chris Brooks
University of Reading
Melvin J. Hinich
University of Texas at Austin
Michael J. Smith
University of Reading

1. Introduction

Testing for nonlinearity in financial asset returns has become an important area of econometric research in the last decade as a consequence of the rapid development in the econometric methodologies necessary to compute the tests, and as a consequence of the increasingly widely held view that the world is probably not linear. A common finding is that there is substantial evidence in favor of some nonlinear structure (see, for example, Scheinkman and LeBaron (1989), Hsieh (1991), or Hsieh (1989) for applications to stock returns), although there still exists a lively debate as to whether members of the GARCH family of models (Bollerslev *et al.* (1992) provide a broad survey of these) are able to capture all of the important nonlinear features of the data (see Brooks (1996)). The motivations for this line of inquiry are at least twofold. First, a finding of nonlinear dependence implies that expenditure of research effort to include nonlinear models as plausible functional forms is probably a worthwhile exercise. Second, nonlinear dependence may also imply hitherto unrevealed forecastability in the series, although the evidence to date on the relative forecasting performance of nonlinear time series models is largely negative (see Ramsey (1996) for a discussion of and suggested explanations for this phenomenon).

At the same time, a literature has developed considering the relationship between the price of a stock (or rather, the returns of a stock) and stock trading volume. Data on the volume of transactions in financial markets have been available for many years, although research interest in volume has been expressed only relatively recently, presumably since researchers are unsure of exactly what to do with volume data. Gallant *et al.* (1992) argue, however, that one may glean more information about how financial markets work by studying the joint dynamics of price and volume than by considering only a series of prices. Karpoff (1987) provides a survey of early research into the relationship between prices and trading volume; his main conclusion is that volume is positively contemporaneously related to the absolute size of the price change (although we argue that there is no reason why this relationship need be contemporaneous). A recent development in this literature is due to Hiemstra and Jones (1994), who use a linear and nonlinear Granger causality test on NYSE return and volume data. They find strong evidence of bi-directional nonlinear causality, which they cite as evidence against limiting the class of relationships of interest to the linear set. There have also been a number of theoretical investigations into this relationship. Campbell *et al.* (1993), for example, propose a model where a set of "noise" traders cause changes in trading volume which market makers will only absorb if their expected return is higher. Other explanations for a price-volume relationship include the mixture of distributions model of Epps and Epps (1975) and the sequential information arrival approach of Copeland (1976).

The purpose of this chapter is to employ two recently proposed tests for univariate and multivariate nonlinearity in order to consider the inter-temporal relationship between high frequency returns and volumes of stocks traded on the London Stock Exchange (LSE). The vast majority of studies to date on this relationship have used much lower frequencies of data (such as daily, weekly, or even monthly), and hence the stiff data requirements of the tests require these researchers to have long time histories, often spanning decades. This type of analysis brings with it the implicit assumption that the parameters of any model estimated, and more generally the underlying data generating process, is constant for the entire sample period. This is unlikely to be an appropriate assumption in the case of a financial market where the number and composition of transactors in the markets, as well as the microstructures and even the assets which are traded, are likely to change beyond recognition from one decade to the next. Potential arbitrage opportunities implied by causality tests are perhaps unlikely to persist for more than a few hours, let alone several years. Thus studies which observe a causal relationship between series, or nonlinear dependence within a series may simply be picking up a pattern which existed some time ago, but which has long since disappeared. This is always a potential (but usually ignored) problem when using old vintages of data in combination with newer data. Moreover, the behavior of complex nonlinearity tests in the face of this is not well known. Our approach is to use tests which have good small sample properties (see Hinich (1996) for Monte Carlo evidence) over short horizons of data which should mitigate this problem.

The remainder of this paper is organized as follows. Sections 2 and 3 describe the SEAQ system, from which the data are taken, and the statistical features of the data, respectively. Section 4 outlines testing methodology employed in this study, while Section 5 presents our results and Section 6 offers a summary and suggests some implications for future research in this area.

2. A Description of the SEAQ System

SEAQ is a continuous quote driven screen based competitive market making system for domestic equity market securities. A SEAQ security is a domestic[1] equity market security for which a minimum of two market makers are obliged to display firm two way prices during the mandatory quote period (MQP) and for which it is possible to calculate a normal market size (NMS). The MQP is from 8:30 - 16:30 and all quotes are firm during this period. However, a normal SEAQ business day is as follows: 7:15 - 8:00 is known as the pre-market period when SEAQ is open for member firms to log on and report overnight trades; 8:00 - 8:30 is the period in which quotes may be opened, but prices are regarded as being indicative only; 8:30 - 16:30 is the MQP; 16:30 - 16:45 is when quotes become indicative again, must not be changed, and must close by 16:45

The NMS is a figure calculated for each stock and is based on a percentage of the stock's average daily customer turnover in the preceding year. This percentage, which is set at 2.5%, is intended to represent the normal institutional bargain. For practical purposes the NMS is used to define a number of market features: the minimum quote size in which the market makers are obliged to make their firm two-way quotes; trades in liquid (all stocks with an NMS of 2,000 shares or more are now called "liquid" while all others are called "less liquid") stocks are subject to immediate publication on the screen if they are below three times NMS; and any larger trades are delayed for 90 minutes.

The SEAQ screen displays each competing market maker's bid and ask quotes for an issue and the size in which these quotes are firm. The screen also shows the cumulative volume and the size and price of the most recent transaction. The best bid and ask price displayed on the screens is known as the "touch." In fact, inclusion of a market maker on the yellow strip is dependent on the price quotes, the size of the market being made, and the time the price was first submitted to SEAQ. Market makers, a total of thirty one[2] in 1994 and 1995,[3] are required not to display on another system prices that are better than the SEAQ prices. Moreover, market makers must also offer "best execution" to their clients which sometimes means that the dealers must execute trades at less than the minimum quote size or inside the prevailing touch, and dealers away from the touch have to execute orders at the touch or transfer the order to another market maker.

In practice,[4] market makers quote a number of different prices to different sets of clients in addition to the quoted SEAQ prices. For example, some market makers will fax a list of prices to selected clients at the beginning of the day. Market makers also quote prices on the IDB screens that are different from their IDB quotes. And most market makers will negotiate prices over the phone for orders which are not at the quoted size.

In addition, market makers on SEAQ are not obliged to maintain an orderly market (as are the specialists on the NYSE) to smooth prices or to maintain tight spreads. However, most market makers in a stock will maintain the same spread, and not vary it over lengthy periods of time. This is due, not only to the potential loss of business from having a wider spread, but also to a potential loss of reputation in the market.

The SEAQ system is largely modelled on NASDAQ but differs from it in certain key respects: 60% of trading on SEAQ is institutional compared to the retail base of the NASDAQ market; 35% of trades are done within the spread (58.3% take place at the touch and 5.8% outside the touch); and large orders (or block trades) are brought directly to the market maker as opposed to an upstairs market as in the U.S. Furthermore, market makers are willing to deal at the quoted price in much larger sizes than shown on the screen. One consequence of the institutional dominance of the market is that, due to their demand for immediacy, institutions typically trade on one side of the market only. As a result, the order flow is characterized by large, infrequent trades on one side of the market.

Finally, to help market makers unwind positions they have taken on or to set up intended positions, they may deal through Inter Dealer Brokers (IDBs). There are four principal IDBs in domestic securities and they may not trade with nor make available trading prices and trading information to persons other than the SEAQ market makers. It is estimated[5] that 17.1% of all trades are conducted via an IDB and a further 30% is market maker to market maker trade.

3. The Data

The data used in this study comprise a set of returns on ten companies selected at random from shares making up the FTSE-100 for the months of November 1995 and February 1996. We analyze two separate months in an attempt to ensure that our results are not time-specific and consequent upon a set of circumstances that occurred during one time period. The raw data are a set of all trades and their corresponding transactions volumes, and also the best bid and ask quotes (the "touch") posted by market makers on the SEAQ system. The last best bid and ask in every five minute interval are taken and their average calculated. A set of continuously compounded percentage returns are then constructed by taking 100 times the log price relatives:

$$r_t = 100 \cdot \ln(P_t / P_{t-1}) \tag{1}$$

where r_t is the return for the five minute interval and P_t is the average of the last best bid and ask quotes displayed on the SEAQ screen at time t. The use of mid-point quote returns should exclude any possibility of bid-ask bounce associated with the use of transactions data, which is important for the present study since bid-ask effects could plausibly lead to spurious nonlinear relationships. The volume measure used is simply the total number of shares in the firm under consideration that have been traded within the five minute interval.

The total number of daily five minute observations is ninety seven, and the number of trading days is twenty two in November 1995 and twenty one in February 1996, giving a total number of observations in each sample month of 2,134 and 2,037 respectively. The data are then divided up into non-overlapping windows of length one trading day for the purpose of analysis. The names of companies randomly selected for analysis are listed in Table 1.

4. Methodology

Hinich (1996) and Brooks and Hinich (1997) have recently proposed portmanteau bicorrelation and cross-bicorrelation tests for the analysis of time series data. The two tests are presented in Sections 4.1 and 4.2, respectively. The notation used for both is as follows. Let the data be a sample of length N from two stationary time series $\{x(t_k)\}$ and $\{y(t_k)\}$

Table 1. Names of Companies Randomly Selected for Analysis in Each Sample Month.

November 1995	February 1996
Dixon Goup	British American Tobacco
3i Group	Smithkline Beecham
General Accident	Tesco
Thames Water	GEC
Enterprise Oil	Prudential
Blue Circle Industries	British Aerospace
Pilkington	National Power
SIEBE	Zeneca
Shell	Cadbury Schweppes
Hanson	Bass

which have been standardized to have a sample mean of zero and a sample variance of one. In this application, $\{x(t_k)\}$ and $\{y(t_k)\}$ represent the return and volume series respectively. The bicorrelation test is applied to the residuals of an AR(2) model and the cross-bicorrelation test is applied to the residuals of a VAR(2,2) model. These filter models are of sufficient length to remove virtually all the evidence of linear dependence in the series, which is necessary to validly apply the tests for nonlinearity.

A window is defined as significant if the bicorrelation test statistic is significant at the 1% level, or if one of the cross-bicorrelation test statistics is significant at the 1% level (depending upon whether the null hypothesis is one of zero bicorrelations of zero cross-bicorrelations). We use a strict threshold so that only the most extreme deviations from these null hypotheses are flagged.

4.1 The Hinich Portmanteau Bicorrelation Test

A third order portmanteau statistic, which can in a sense be viewed as a time domain analogue of the bispectrum test, has been proposed by Hinich (1996). A full theoretical derivation of the test statistics and a number of Monte Carlo simulations to assess their size and power are given in Hinich (1996), and Hinich and Patterson (1995). We thus state without derivation that the test statistic, denoted H, is defined as:

$$H = \sum_{s=2}^{L} \sum_{r=1}^{s-1} [G^2(r,s)/(T-s)] \sim \chi^2((L-1)L/2), \tag{2}$$

$$\text{where } G(r,s) = \sum_{k=1}^{T-s} [Z(t_k)Z(t_k+r)Z(t_k+s)],$$

$Z(t_k)$ are the standardized observations at time $t=k$, and $L = T^c$ with $0<c<0.5$. c is a parameter under the choice of the user. Based on the results of Monte Carlo simulations, the use of $c = 0.4$ is recommended in order to maximize the power of the test while ensuring a valid approximation to the asymptotic theory. The test is of a null of pure white noise against an alternative that the process has M non-zero bicorrelations in the set $0 < r < s \leq L$, i.e., that there exists third order dependence in the data generating process. The test relies on the property of pure noise that it has zero bicovariances. The test is particularly useful in detecting nonlinear dependencies, since it has much better small-sample properties, and does not have such stiff data requirements as many of its competitors, such as the BDS test (Brock *et al.* (1996); see Brock *et al.* (1991) for simulations to assess the finite sample evidence on the BDS test.)

4.2 A Cross-Bicorrelation Test

A cross-bicorrelation test , which can be considered as a multivariate generalization to the bicorrelation test, is outlined in Brooks and Hinich (1997). The generalization is similar in spirit to the multivariate version of the BDS test (Brock *et al.* (1996)) proposed by Baek and Brock (1992). The null hypothesis for this test is that the two series are independent pure white noise processes, against an alternative that some cross-bicovariances $C_{xxy} = E[x(t_k)x(t_k+r)y(t_k+s)]$ are non-zero. As a consequence of the invariance of $E[x(t_1)x(t_2)y(t_3)]$ to permutations of (t_1,t_2), stationarity implies that the expected value is a function of two lags and $C_{xxy}(-r,s) = C_{xxy}(r,s)$. Thus, if the maximum lag used is $L < N$, then the principal domain for the bicovariances is the rectangle $\{1 \leq r \leq L, -L \leq s \leq L\}$.

Under the null hypothesis that $\{x(t_k)\}$ and $\{y(t_k)\}$ are pure white noise, then $C_{xxy} = 0 \ \forall \ r,s$ except when $r = s = 0$. This is also true for the less restrictive case that the two processes are merely uncorrelated, but demonstration of asymptotic normality of the test statistic requires independence between the series and pure white noise. If there is third order lagged dependence between the two series, then $C_{xxy}(r,s) \neq 0$ for at least one pair of r and s values. The following statistic gives the r,s sample xxy cross-bicorrelation:

$$C_{xxy}(r,s) = (N-m)^{(-1)} \sum_{t=1}^{N-m} x(t_k)x(t_k+r)y(t_k+s), \text{ where } m = \max(r,s). \tag{4}$$

It can be shown that:

$$E[C_{xxy}(r,s)] = 0 \tag{5}$$

$$E[C^2_{xxy}(r,s)] = (N-m)^{-1/2} \tag{6}$$

under the null hypothesis. Let $L = N^c$, where $0<c<0.5$. The test statistic for non-zero cross-bicorrelations is given by:

$$H_{xxy}(N) = \sum_{s=-L}^{L} \sum_{r=1}^{L} (N-m)^{(1/2)} C_{xxy}^{2}(r,s), \quad (s \neq 0). \tag{7}$$

Theorem 1 of Brooks and Hinich (1997) proves that H_{xxy} is asymptotically standard normal as $N \to \infty$.

5. Results

Tables 2 and 3 give results for separate applications of the univariate bicorrelations test to the returns and volume data for November 1995 and February 1996, respectively. The most salient feature of these results is that the proportion of days with significant bicorrelation statistics is much greater than the nominal 1% threshold. Specifically, we would expect at most one significant window per month purely by chance, but what we find is that there are typically at least three or four days per month with significant bicorrelation statistics, although the actual number varies greatly from one stock to another, as do the dates on which these large bicorrelations occur. The dates on which most rejections occur are November 2 and 17, 1995, and February 4 and 16, 1996, for the returns, and November 2 and 10, 1995, and February 6, 1996, for the volume data. Comparing the numbers and dates of significant bicorrelations between the returns and volume, there appears to be slightly more in the former than the latter, although the precise dates when these nonlinearities occur shows no relationship between the return and volume series. This indicates that although both sets of series show strong evidence of nonlinearities, the histories appear to contain different sets of information. A similar picture is presented by a plot of the H statistic transformed to a uniform distribution as shown in Figure 1.

In terms of the numbers of rejections of the null hypothesis, a similar result is observed for the cross-bicorrelations tests. The results for November 1995 and February 1996 are given in Tables 4 and 5. Again, a much greater proportion of days show evidence of significant cross-bicorrelations than one would have expected if the data were drawn from two independent white noise processes, although there are slightly fewer restrictions than for the individual series. Once again, there is considerable variation between the series. The days when rejections of the null of no significant cross-bicorrelations occur most frequently are November 2, 1995 and February 12, 1996.

Panel A of Tables 4 and 5 indicate whether the rejection was caused by the *xxy* statistic or the *yyx* statistic (or both). For the stocks during November 1995, there is no clear pattern as to which statistic causes most rejections, although it appears to be *xxy* that causes most rejections during February 1996. Finally, Panel B of Tables 4 and 5 give the simple correlations between the cross-bicorrelation and pure bicorrelation test statistics, between the cross-bicorrelation and (linear) cross-correlation, and between the two types of cross-bicorrelation test statistics. These correlations demonstrate, along with the dates on which the significant bicorrelations occur, that the univariate and multivariate nonlinearities need not occur at the same time. The results also show that the values of the *xxy* and *yyx* cross-bicorrelation test statistics are not closely related, an observation made more transparent by Figure 2, which plots the *xxy* and *yyx* statistics transformed to

Table 2. Bicorrelation Tests for Returns and Volume Data, November 1995

Series	Number (%) Sig. *xxy* Windows	Dates of Month for Sig. Windows	Number (%) Sig. *xxy* Windows	Dates of Month for Sig. Windows
	Returns		Volume	
Dixons Group	4 (18.2%)	2, 3, 15, 20	4 (18.8%)	2, 10, 21, 23
3i Group	8 (36.4%)	6, 13, 17, 22, 24, 28, 30	6 (27.3%)	3, 10, 17, 20, 21, 28
General Acident	9 (40.9%)	1, 2, 3, 7, 9, 14, 21, 22, 28	3 (13.6%)	2, 13, 22
Thames Water	5 (22.7%)	8, 13, 20, 22, 29	2 (9.1%)	9, 13
Enterprise Oil	0 (0%)	-	2 (9.1%)	1, 20
Blue Circle Industries	3 (13.6%)	2, 3, 20	3 (13.6%)	1, 17, 18
Pilkington	3 (13.6%)	2, 17, 23	4 (18.2%)	2, 8, 28, 29
SIBE	3 (13.6%)	6, 8, 24	6 (27.3%)	2, 8, 9, 13, 15, 29
Shell	5 (22.7%)	2, 7, 10, 14, 17	3 (13.6%)	10, 17, 30
Hanson	6 (27.3%)	3, 18, 15, 17, 21, 27	6 (27.3%)	1, 10, 14, 17, 21, 30

Table 3. Bicorrelation Tests for Returns and Volume Data, February 1996.

Series	Number (%) Sig. *xxy* Windows	Dates of Month for Sig. Windows	Number (%) Sig. *xxy* Windows	Dates of Month for Sig. Windows
	Returns		Volume	
British American Tobacco	4 (1915%)	5, 6, 16, 26	4 (19.1%)	5, 14, 20,21
SmithKline Beecham	4 (19.1%)	14, 19, 27, 28	1 (4.8%)	20
Tesco	6 (28.6%)	1, 5, 6, 12, 13, 28	2 (9.5%)	6, 29
GEC	3 (14.3%)	19, 22, 23	1 (4.8%)	7
Prudential	3 (14.3%)	2, 9, 21	1 (4.8%)	15
British Aerospace	5 (23.1%)	2, 7, 16, 19, 27	6 (28.6%)	6, 9, 13, 15, 28, 29
National Power	6 (28.6%)	5, 7, 12, 16, 19, 27	0 (0%)	-
Zeneca	4 (19.1%)	8, 14, 22, 26	4 (19.1%)	2, 12, 14, 16
Cadbury Schweppes	6 (28.6%)	5, 6, 7, 13, 20, 23	2 (9.5%)	6, 13
Bass	3 (14.3%)	7, 14, 16	2 (9.5%)	1, 17

Table 4. Cross-Bicorrelation Tests for Returns and Volume Data, November 1995.

Panel A

Series	No. (%) Sig. *xxy* or *yyx* Windows	Dates for Sig. Windows	No. Sig. *xxy* Windows	No. Sig. *yyx* Windows
Dixons Group	4 (18.2%)	7, 8, 21, 28	3	2
3i Group	4 (18.2%)	1, 21, 24, 30	2	3
General Acident	1 (4.5%)	3	1	-
Thames Water	6 (27.3%)	7, 13, 14, 15, 24, 28	2	4
Enterprise Oil	3 (13.6%)	1, 23, 27	2	2
Blue Circle Industries	4 (18.2%)	2, 10, 17, 27	2	2
Pilkington	4 (18.2%)	2, 6, 15, 29	2	2
SIBE	2 (9.1%)	6, 15	1	1
Shell	2 (9.1%)	6, 8	2	-
Hanson	3 (13.6%)	2, 9, 21	1	2

Panel B

Series	Corr (*xxy, xxx*)	Corr (*xxy,xy*)	Corr (*xxy,yyx*)
Dixons Group	0.25	0.26	0.28
3i Group	0.49	0.30	0.58
General Acident	0.48	0.72	-0.05
Thames Water	0.29	0.24	0.60
Enterprise Oil	-0.07	0.99	-0.04
Blue Circle Industries	-0.03	-0.10	-0.07
Pilkington	0.35	-0.13	0.26
SIBE	0.46	0.69	-0.10
Shell	0.17	0.26	-0.15
Hanson	0.30	0.50	0.30

Notes. *x* denotes the returns while *y* denotes volume. No. (%) Sig. Windows denotes the number (percentage) of significant windows during the month.

Table 5. Cross-Bicorrelation Tests for Returns and Volume Data, February 1996.

Panel A

Series	No. (%) Sig. *xxy* or *yyx* Windows	Dates for Sig. Windows	No. Sig. *xxy* Windows	No. Sig. *yyx* Windows
British American Tobacco	2 (9.1%)	6, 26	1	1
SmithKline Beecham	0 (0%)	-	-	-
Tesco	3 (14.3%)	12, 15, 22	3	1
GEC	1 (4.8%)	26	1	-
Prudential	1 (4.8%)	2, 7, 15	-	1
British Aerospace	3 (14.3%)	1, 8, 22	3	-
National Power	3 (14.3%)	2, 6	3	-
Zeneca	2 (9.1%)	2, 6	1	1
Cadbury Schweppes	2 (9.1%)	12, 20	2	2
Bass	5 (23.8%)	1, 2, 7, 12, 14	3	2

Panel B

Series	Corr (*xxy, xxx*)	Corr (*xxy,xy*)	Corr (*xxy,yyx*)
British American Tobacco	0.92	0.44	0.83
SmithKline Beecham	-	-	-
Tesco	0.66	0.68	0.63
GEC	0.20	0.25	-0.08
Prudential	-0.04	0.61	0.29
British Aerospace	0.21	0.54	-0.10
National Power	0.42	0.18	0.16
Zeneca	0.45	0.82	-0.15
Cadbury Schweppes	-0.01	0.69	0.69
Bass	0.40	0.29	-0.14

Notes. *x* denotes the returns while *y* denotes volume. No. (%) Sig. Windows denotes the number (percentage) of significant windows during the month.

Figure 1. The Values of the Uniform Transformed *H*-Statistic for the Volume and Returns Data (With Values Less Than 0.95 Set to Zero).

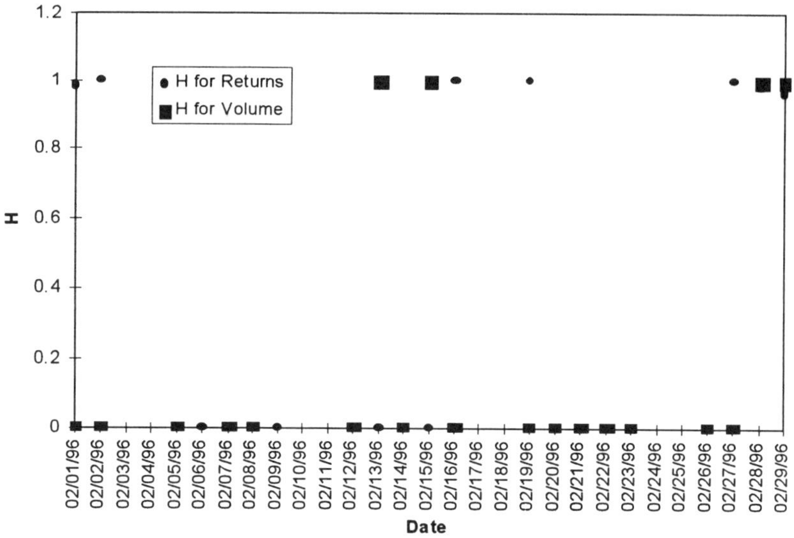

Figure 2. The Values of the Uniform Transformed *xxy* and *yyx* Statistics (With Values Less Than 0.95 Set to Zero).

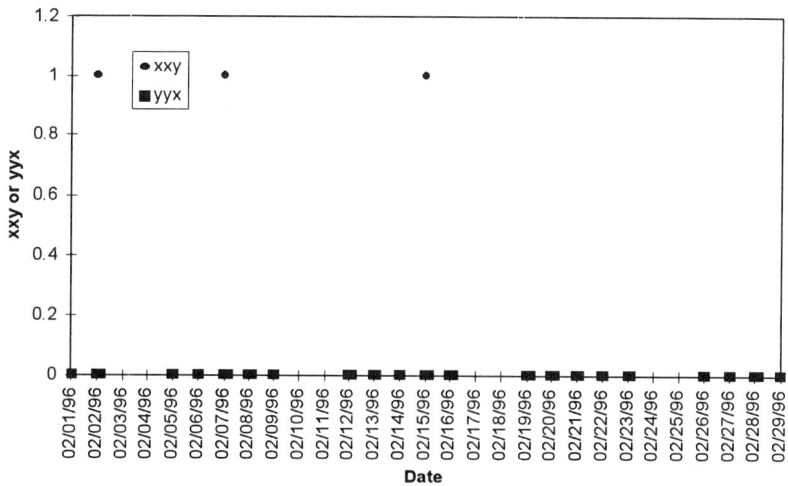

a uniform distribution. The periods of significant windows seem to last for two or three days and then disappear. There is, however, a reasonably strong link between the value of the (linear) cross-correlations and the (nonlinear) cross-bicorrelations for the majority of stocks.

6. Conclusions

We have presented tests for univariate and multivariate nonlinearity and applied them to two sets of stock returns and volume data. The most important feature of the results is that they show very strong evidence of a nonlinear relationship between the current value of a stock return or volume, and its own previous realizations. Furthermore, we observe strong evidence of multivariate nonlinearities between the returns and volume of an individual stock. Monte Carlo studies in Brooks and Hinich (1997) show that this cannot be attributed to the type of relationship captured by the GARCH family of models.

An obvious next step in this analysis would be an attempt to use the observed bicorrelation results to forecast the future values of the series. This is not a simple problem to solve, for the tests presented here, in common with all true portmanteau tests, do not directly imply a functional form for a parametric model to estimate. One possibility, however, would be to extend autoregressive models in the univariate case to incorporate terms in the squares of the lags (i.e., products of the own lags, such as $x_{t-i}x_{t-j}$), and to extend the VAR models to incorporate lagged cross-product terms (e.g., $x_{t-i}y_{t-j}$). Both extensions could be seen as types of Volterra expansions, and might provide better explanations of the data than standard AR and VAR models.

An important message stemming from our analysis is that researchers should be cautious about estimating models over long time series for financial data, since the implicit assumption of parameter constancy is most unlikely to be valid. Hsieh (1991) uses a Monte Carlo study to demonstrate that simple models which are *iid*, but with different means or different variances for part of the sample, lead to virtually 100% rejections of *iid* using the BDS test. Hence, we conjecture that many recent papers which have rejected this null hypothesis may simply be doing so as a consequence of a few large structural changes or regime shifts in the (extremely long) series under consideration, rather than some inherent nonlinearity in the data generating process.

Addendum

On October 20, 1997, the London quote-driven SEAQ market was replaced by an order-driven Stock Exchange Electronic Trading Service (SETS). This even occurred after the completion of the chapter, although it is considered likely that many of the features of the data we document here will still apply under the new trading mechanisms.

Notes

1. There is an international equity market, SEAQI, which is also a competing market maker system. Securities are eligible for trading in London's SEAQ if they are listed on any stock exchange recognized by the London Stock Exchange.
2. The number of market makers in individual stocks varies from eighteen+ down to zero, and the majority of shares have three market makers or less trading in them.
3. Source: *London Stock Exchange Fact Book*, 1995 and 1996.
4. Office of Fair Trading, *Rules of the London Stock Exchange Relating to Market Makers*, Report to the Chancellor of the Exchequer, March 1995.
5. London Stock Exchange, *Quality of Markets Review*, 1994.

References

Baek, E. and W. A. Brock (1992). "A Nonparametric Test for Independence of a Multivariate Time Series," *Statistica Sinica*, 2, 137-156.

Bollerslev, T., R. Y. Chou, and K. F. Kroner (1992). "ARCH Modelling in Finance: A Review of the Theory and Empirical Evidence," *Journal of Econometrics*, 52, 5-59.

Brock, W. A., W. D. Dechert, J. A. Scheinkman, and B. LeBaron (1996). "A Test for Independence Based on the Correlation Dimension," *Econometric Reviews*, 15, 197-235.

Brooks, C. (1996). "Testing for Nonlinearities in Daily Sterling Exchange Rates," *Applied Financial Economics*, 6, 307-317.

Brooks, C. and M. J. Hinich (1997). "Cross-Correlations and Cross-Bicorrelations in Sterling Exchange Rates," working paper, IMSA Centre, University of Reading.

Campbell, J. Y., S. J. Grossman, and J. Wang (1993). "Trading Volume and Serial Correlation in Stock Returns," *Quarterly Journal of Economics*, 108, 905-939.

Copeland, T. E. (1976). "A Model of Asset Trading under the Assumption of Sequential Information Arrival," *Journal of Finance*, 21, 1149-1168.

Epps, T. W. and M. L. Epps (1975). "The Stochastic Dependence of Security Price Changes and Transaction Volumes: Implications for the Mixture of Distributions Hypothesis," *Econometrica*, 44, 305-321.

Gallant, A. R., D. A. Hsieh, and G. Tauchen (1992). "Stock Prices and Volume," *Review of Financial Studies*, 5, 199-242.

Hiemstra, C. and J. D. Jones (1994). "Testing for Linear and Nonlinear Granger Causality in the Stock Price-Volume Relation," *Journal of Finance*, 49, 1939-1664.

Hinich, M. J. (1996). "Testing for Dependence in the Input to a Linear Time Series Model," *Journal of Nonparametric Statistics*, 6, 205-221.

Hinich, M. J. and D. M Patterson (1995). "Detecting Epochs of Transient Dependence in White Noise," working paper, Department of Government, University of Texas at Austin.

Hsieh, D. A. (1991). "Chaos and Nonlinear Dynamics: Application to Financial Markets," *Journal of Finance*, 46, 1839-1877.

Hsieh, D. A. (1989). "Testing for Nonlinear Dependence in Daily Foreign Exchange Rates," *Journal of Business*, 62, 339-368.

Karpoff, J. M. (1987). "The Relation Between Price Changes and Trading Volume: A Survey," *Journal of Financial and Quantitative Analysis*, 22, 109-126.

Ramsey, J. B. (1996). "If Nonlinear Models Cannot Forecast, What Use are They?" *Studies in Nonlinear Dynamics and Econometrics*, 1, 65-86.

Scheinkman, J. A. and B. LeBaron (1989). "Nonlinear Dynamics and Stock Returns," *Journal of Business*, 62, 311-337.

9 NONLINEAR ADJUSTMENT TOWARDS LONG-RUN MONEY DEMAND

Panos Michael
University of Liverpool
A.Robert Nobay
Financial Markets Group, LSE
David A. Peel
Cardiff University

1. Introduction

A stylized finding of the extensive empirical literature on money demand is the implausibly slow adjustment of real balances towards equilibrium. A possible explanation of this feature is provided by Goodfriend (1985) who suggests that the high persistence may be the result of measurement errors in the explanatory variables. An empirical test of this hypothesis has been undertaken recently by Taylor (1994); using the Goldfeld (1973) data, he finds that the observed persistence cannot be explained by serially correlated measurement errors.

The purpose of this paper is to consider the persistence in aggregate real balances from the perspective of the recent literature on adjustment processes implied when agents face non-convex rather than quadratic adjustment costs - see e.g. Caballero and Engel (1992, 1993), Bertola and Caballero (1990), Bar-Ilan and Blinder (1992) and Eberly (1992). These models generally imply a nonlinear process where the speed of adjustment varies with the extent of the deviation from equilibrium. The essential features of this process can be described by the exponential autoregressive (EAR) model originally proposed by Haggan and Ozaki (1981) and recently reconsidered by Teräsvirta (1994) and Granger and Teräsvirta (1993). We apply this model to two data sets for the US

(Goldfeld, 1973; Hoffman, Raasche and Tieslau, 1995) and find that the hypothesis of linear adjustment is strongly rejected in each case.

The rest of the paper is structured as follows. In the next section we discuss nonlinear adjustment to equilibrium money demand in the context of nonstationary integrated variables. Section 3 describes the procedure for testing linearity, while Section 4 reports our empirical findings. A summary and concluding comments are offered in the final section.

2. Long-Run Money Demand and Nonlinear Adjustment

The long-run money demand relationship may be written as:

$$M_t = \alpha_0 + \alpha_1 X_t + \alpha_2 R_t + y_t \tag{1}$$

where M is the logarithm of real money balances, X is the logarithm of real income, R is the nominal interest rate[1] and y is a stationary error term. Given the nonstationarity of the variables M, X and R, the cointegration framework of Engle and Granger (1987) provides a natural way of testing the validity of (1) as a long-run equilibrium relationship. If (1) is cointegrated then y can be interpreted as the short-run deviation of M from its equilibrium value.

There is by now a large number of empirical studies which test whether (1) is cointegrated.[2] However, the associated linear error correction models (like earlier models with only limited dynamics) imply a very slow speed of adjustment. These partial-adjustment models can be justified on the basis of quadratic costs of adjustment (Chow (1966), Cuthbertson and Taylor (1987)) or through buffer-stock arguments (Carr and Darby (1981), Laidler (1984)). However, quadratic costs are *a priori* hard to sustain (Goodfriend (1985)), while the evidence for buffer-stock models is weak (Cuthbertson and Taylor (1986)).

An alternative approach is to assume that individual agents face non-convex rather than quadratic adjustment costs. These (S,s) models show why the optimizing agent will not adjust to a shock if there exist lumpy costs of adjustment which are large relative to the shock. In particular, no adjustment will occur until the deviation from the desired level reaches given trigger points. This type of nonlinear adjustment in the deviations y_t may be modelled as a threshold autoregressive (TAR) process (Tong (1990)); for small deviations, y may exhibit unit root behavior, while beyond the trigger points the process for y is mean-reverting.

For the aggregate adjustment process, the above characterization would be appropriate only in the case of the representative agent. If agents are heterogenous, Bertola and Caballero (1990) demonstrate that the aggregate adjustment process will reflect the relative magnitude of idiosyncratic to common macroeconomic shocks. While the implied aggregate process is not straightforward, the linear partial-adjustment model will generally provide a poor approximation irrespective of the assumptions concerning heterogeneity.

A more general characterization of the nonlinear adjustment process for y is in terms of an exponential smooth transition autoregressive (ESTAR) model (Teräsvirta, 1994;

Granger and Teräsvirta (1993)). Here adjustment takes place in every period but the speed of adjustment varies with the extent of the deviation from equilibrium. In contrast with the TAR model, regime changes occur gradually rather than abruptly. As noted by Teräsvirta (1994), for aggregated processes the change in regime is more likely to be smooth rather than discrete since, even if agents make only dichotomous decisions, they are unlikely to change their behavior simultaneously.

In this paper we evaluate the null hypothesis of linear adjustment against the alternative that the deviations from equilibrium money demand follow the ESTAR process:

$$y_t = k + \sum_{j=1}^{p} \pi_j y_{t-j} + (k^* + \sum_{j=1}^{p} \pi_j^* y_{t-j})[1 - \exp\{-\gamma(y_{t-d} - c^*)^2\}] + u_t \tag{2}$$

where $\{y_t\}$ is assumed stationary and ergodic, $u_t \sim iid\ (0,\sigma^2)$, and $\gamma>0$. The regular EAR model of Haggan and Ozaki (1981) is the special case of (2) with $k^* = c^* = 0$. The transition function

$$F(y_{t-d}) = 1 - \exp\{-\gamma(y_{t-d} - c^*)^2\}$$

is U-shaped and the parameter γ determines the speed of the transition process between two extreme regimes. The middle regime corresponds to $y_{t-d} = c^*$ when $F=0$ and (2) becomes a linear AR(p) model:

$$y_t = k + \Sigma \pi_j y_{t-j} + u_t.$$

The outer regime corresponds to $y_{t-d} = \pm\infty$ when $F=1$ and (2) becomes a different AR(p) model:

$$y_t = k + k^* + \Sigma(\pi_j + \pi_j^*) y_{t-j} + u_t.$$

The ESTAR model may be viewed as a generalization of a symmetric two-threshold TAR model in which the autoregressive parameters in the outer regimes (i.e. $y_{t-d} > \delta$ and $y_{t-d} < -\delta$, with δ representing the threshold deviation) are the same;[3] such a model is the limiting case of (2) when $\gamma \to \infty$.

For our purposes, it is convenient to reparameterize the ESTAR model in (2) as follows:

$$\Delta y_t = k + \lambda y_{t-1} + \sum_{j=1}^{p-1} \phi_j \Delta y_{t-j} + (k^* + \lambda^* y_{t-1} + \sum_{j=1}^{p-1} \phi_j^* \Delta y_{t-j}) \cdot [1 - \exp\{-\gamma(y_{t-d} - c^*)^2\}] + u_t \tag{3}$$

In this form the crucial parameters are γ and λ^*. The general proposition that the larger the deviation the stronger is the tendency to move back to equilibrium implies that while $\lambda \geq 0$ is possible, we must have $\lambda^* < 0$ and $\lambda + \lambda^* < 0$; that is, for small deviations y_t may exhibit unit root or even explosive behavior, but for large deviations the process is mean-reverting.[4]

The analysis above has implications for the conventional cointegration test which is based on a linear AR(p) model, written below as an augmented Dickey-Fuller regression:

$$\Delta y_t = k' + \lambda' y_{t-1} + \sum_{j=1}^{p-1} \phi_j' \Delta y_{t-j} + v_t \tag{4}$$

Assuming that the true process for y is given by the nonlinear model (3), then the parameter λ' will lie between λ and $\lambda + \lambda^*$. Hence, the null hypothesis $H_0: \lambda' = 0$ (no linear cointegration) may not be rejected against the stationary alternative $H_1: ' < 0$, even though the true nonlinear process is globally stable $(\lambda + \lambda^* < 0)$. Thus, functional misspecification of (4), rather than simply low power due to a small sample, may explain the frequent failure of empirical studies to find cointegration. Moreover, this framework helps explain the frequent finding of a negative but small estimate of λ' which implies slow adjustment.

In the next section we describe the procedures for the testing of linearity, the determination of the delay parameter *d*, and the estimation of the ESTAR model.

3. Methodology

Given that the *(S,s)* model implies a nonlinear adjustment process, the first hypothesis we need to test is that of linearity. As the transition function in (2) implies that $F = 0$ when $\gamma = 0$, the linearity hypothesis may be expressed as $H_0: \gamma = 0$. It can be seen that (2) is only identified under the alternative $H_1: \gamma > 0$ since, if $\gamma = 0$ is true, the parameters $c^*, k^*, \pi_1^*, \ldots, \pi_p^*$ can take any value. To overcome this problem, Saikkonen and Luukkonen (1988) and Teräsvirta (1994) follow the approach of Davies (1977) and derive Lagrange Multiplier (*LM*)-type test statistics which follow a standard χ^2 distribution under H_0. If the delay parameter *d* is fixed, the linearity test against ESTAR consists of estimating by OLS the artificial regression

$$y_t = \beta_{00} + \sum_{j=1}^{p} (\beta_{0j} y_{t-j} + \beta_{1j} y_{t-j} y_{t-d}^2) + \epsilon_t \tag{5}$$

and testing the null hypothesis

$$H_{OL}: \beta_{1j} = \beta_{2j} = 0 \quad (j = 1, \ldots, p) \tag{6a}$$

against the general alternative that H_{OL} is not valid. In practice the ordinary *F*-test is used as an approximation to the *LM*-type test in order to improve size and power properties. This is essential when the order *p* of the linear AR model is large while the number of observations is small (see, for example, Harvey (1990, pp. 174-175)).

To determine the delay parameter, we follow the approach suggested by Tsay (1989) in his specification of TAR models. This involves repeating the linearity test for a set of plausible values of *d*, and choosing the delay parameter as the value which minimizes the *p*-value of the linearity test.

Since in our empirical analysis the series y_t represents the mean-corrected deviations, we may reasonably expect the ESTAR model to satisfy $k^* = c^* = 0$. In that case the artificial regression simplifies since $\beta_{1j} = 0$ for all j. Hence, in order to assist in the specification of the ESTAR model, we carry out another F-test based on (5) by testing

$$H_{ok}: \beta_{1j} = 0 \quad (j = 1,..,p). \tag{6b}$$

The test statistics associated with (6a) and (6b) are denoted by F_L and F_k respectively.[5]

Assuming that linearity is rejected, we can proceed to estimate and evaluate the nonlinear model. Estimation of (3) is by nonlinear least squares, which provides estimators that are consistent and asymptotically normal. The necessary and sufficient conditions for this are stated in Klimko and Nelson (1978). Tong (1990, pp. 299-302) shows that these conditions hold for STAR models provided $u_t \sim iid(0, \sigma^2)$ and $\{y_t\}$ is stationary and ergodic. To make it easier to select a good starting value for γ, Teräsvirta (1994) suggests standardizing the exponent of F by dividing it by the sample variance of y; $\gamma = 1$ is then an appropriate starting value. To check whether a global minimum has been achieved, estimation may be carried out with γ fixed at different values. Model evaluation includes a check of whether the estimates seem reasonable, examining the long-run properties of the model and, of course, checking the residuals for autocorrelation, ARCH, and normality.

4. Empirical Analysis

4.1 Long-Run Relationships

We analyze two U.S. data sets: the Goldfeld (1973) data covering the period 1952.2 - 1972.4 and the extended data considered recently in Hoffman, *et al.* (1995) covering the period 1955.2 - 1990.4. The Goldfeld data, having been subjected to extensive scrutiny, serves as a convenient benchmark for evaluating the nonlinear framework discussed above. The extended data set, which covers the "Missing Money" and "New Operating Procedures" periods, provides a useful check of the robustness of our findings.

Application of the multivariate maximum likelihood technique of Johansen (1988) confirms the existence of a cointegrating vector for each data set.[6] For the Goldfeld data, Taylor (1994) obtains the following cointegrating relationship:

$$M_t = 0.687X_t - 0.137R1_t - 0.103R2_t + \hat{\alpha}_0 + y_t \tag{G-T}$$

while for the extended data, Hoffman, *et al.* (1995) obtain:

$$M_t - X_t = -0.569R_t + \hat{\alpha}_0 + y_t \tag{HRT}$$

Below we will refer to the two data sets as G-T and HRT. In both cases M is the logarithm of narrow money ($M1$) deflated by the GNP deflator, and X is the logarithm of real GNP; the three interest rates are all in logarithms, with R representing the 90-day Treasury bill rate, $R1$ the rate on commercial paper $R2$ the rate on time deposits. Since the mean ($\hat{\alpha}_0$) of the residuals reflects the units of measurement, below we analyze the

Table 1. Linear AR Models: Equation (4).

G-T (n = 79)		
$\Delta y_t = 0.66\Delta y_{t-1} - 0.41\Delta y_{t-2} + 0.36\Delta y_{t-3} - 0.26 y_{t-1} + \hat{v}_t$ (6.01) (3.79) (3.21) (3.78)		
$R^2 = 0.78$	$JB = 3.32$	$RE(2) = 29.60^{***}$
$s = 0.014$	$Q(4) = 0.07$	$A(1) = 17.67^{***}$
$DW = 1.99$	$Q(8) = 0.45$	$A(4) = 25.19^{***}$
HRT (n = 139)		
$\Delta y_t = 0.54\Delta y_{t-1} - 0.38\Delta y_{t-2} + 0.30\Delta y_{t-3} - 0.23 y_{t-1} + \hat{v}_t$ (6.52) (4.74) (3.75) (4.50)		
$R^2 = 0.76$	$JB = 82.76^{***}$	$RE(2) = 7.41^{**}$
$s = 0.067$	$Q(4) = 1.23$	$A(1) = 10.52^{***}$
$DW = 1.99$	$Q(8) = 4.86$	$A(4) = 13.94^{***}$

Note: The deviations y_t are the residuals from the cointegrating relationships reported in Section 4.1. The numbers in brackets are t-ratios and n is the number of observations. R^2 is the proportion of the variation in y (not Δy) explained by the model; s is the standard error of regression; DW is the Durbin-Watson statistic; JB is the Jarque-Bera (1980) normality test statistic; $Q(k)$ is the Ljung-Box (1978) statistic for residual autocorrelation up to order k; $A(k)$ is the Engle (1982) test for ARCH up to order k; and $RE(k)$ is the RESET test of Ramsey (1969). The JB statistic is distributed as $\chi^2(2)$ under the null hypothesis, while $Q(k)$, $A(k)$ and $RE(k)$ are distributed as $\chi^2(k)$. ***,**,* indicate significance at the 1, 5, 10 percent levels respectively.

mean-corrected residuals y_t.

4.2 Linearity Tests

The linear models which form the basis of our linearity tests are shown in Table 1. In specifying the autoregressive order p, we rely on the partial autocorrelation function rather than an information criterion since the latter penalizes high order terms. Overspecification of the linear AR(p) model is preferable to underspecification since autocorrelated errors may affect the linearity test (it may have power against omitted serial correlation). The reported Ljung-Box statistics confirm that the AR(4) model for each series is not underspecified. However, the strong significance of the ARCH and RESET test statistics

Table 2. Linearity Tests.

	p-Values of F_L				p-Values of F_k
	$d=1$	$d=2$	$d=3$	$d=4$	$d=1$
G-T	0.0000	0.0143	0.0446	0.4154	0.0248
HRT	0.0048	0.0858	0.1293	0.1314	0.0859

Notes: The statistics F_L and F_k correspond to the tests of (6a) and (6b), with $p=4$ for both series.

provides clear evidence of functional misspecification in both models.

Table 2 reports the *LM* tests of linearity. We consider $d=1,..,4$ as plausible values for the delay parameter, and find that linearity is rejected most strongly when $d=1$ in both cases. For this value of *d*, the test of (6b) suggests that the restrictions $k^*=c^*=0$ are likely to hold for the HRT residuals but not for the G-T series; while the test of (6a) shows that in both cases linearity is decisively rejected against the ESTAR alternative.

4.3 ESTAR Models

We report the ESTAR models, in the form of equation (3), in Table 3. The residual diagnostic statistics do not reveal any model inadequacy apart from the rejection of normality for the HRT series - most likely due to an outlier in 1958.2. Particularly noteworthy is the absence of ARCH effects, in marked contrast with the linear models. Also, the variance ratio V indicates a substantial reduction in the residual variances compared to the linear models (39 and 9 percent respectively, for the G-T and HRT series).

The standardized transition parameter γ^* shows that the speed of transition between regimes is faster for the Goldfeld data, as indicated by Figure 1 which plots the two transition functions. Interestingly, the limiting case of $F=1$ is actually attained over the range of observed deviations in the case of the G-T series. As for the long-run properties, our simulation results (using various sets of starting values and with the noise set to zero) show that both models converge to a unique equilibrium point ($y_\infty=-0.024$ for the G-T series and $y_\infty=0$ for HRT), thus confirming the global stability of the ESTAR models.

Finally, the parameter estimates are consistent with the implications of the (*S,s*) model. In particular, the likelihood ratio test statistic *LR(2)* provides strong support for

Table 3. Restricted ESTAR Models: Equation (3)

G-T (n = 79)			
$$\underset{(8.60)}{\Delta y_t = 1.27\,\Delta y_{t-1}} \underset{(3.75)}{- 0.51} \underset{(5.72)}{+ 0.65} + (-1.27\,\Delta y_{t-1} \underset{(4.92)}{+ 0.51\,\Delta y_{t-2}} \underset{(6.44)}{- 1.28\Delta y_{t-3}} - y_{t-1} - 0.024)\hat{F} + \hat{u}_t$$ $$\hat{F} = 1 - \exp\{\underset{(5.64)}{-385.6}(y_{t-1} \underset{(7.70)}{- 0.019})^2\}$$			
$R^2 = 0.87$	$JB = 3.26$	$A(1) = 0.13$	$\hat{\gamma}^* = 0.31$
$s = 0.011$	$Q(4) = 1.28$	$A(4) = 0.57$	$LR(2) = 2.31$
$DW = 2.09$	$Q(8) = 2.53$	$V = 0.61$	$LR(5) = 7.73$
HRT (n = 139)			
$$\underset{(8.35)}{\Delta y_t = 0.51\,\Delta y_{t-1}} \underset{(3.69)}{- 0.29\Delta y_{t-2}} \underset{(3.80}{+ 0.29 y_{t-3}} + (0.51\,\Delta y_{t-1} - y_{t-1})\hat{F} + \hat{u}_t$$ $$\hat{F} = (1 - \exp\{\underset{(3.97)}{-6.39} y_{t-1}^2\}$$			
$R^2 = 0.78$	$JB = 178.90^{***}$	$A(1) = 3.35^{*}$	$\hat{\gamma}^* = 0.12$
$s = 0.064$	$Q(4) = 0.82$	$A(4) = 3.46$	$LR(2) = 0.46$
$DW = 2.00$	$Q(8) = 5.24$	$V = 0.91$	$LR(8) = 10.52$

Note: $\hat{\gamma}^* = s_y^2$ and $V = S_E^2 / S_L^2$, where S_y^2 is the variance of y, while S_E^2 and S_L^2 are the residual variances from the ESTAR and linear AR models, respectively. The statistic *LR(2)* is a likelihood ratio test of the restrictions $\lambda = 0$ and $\lambda^* = -1$, while *LR(5)* and *LR(8)* provide tests of all the restrictions implicit in the reported models against the general unrestricted ESTAR model (3) with p = 4 (coefficient estimates without a *t*-ratio are restricted; e.g. $\phi_j^* = \phi_j ; j = 1,2$ in the case of the G-T series). All other diagnostic tests are explained in the notes to Table 1.

the restrictions $\lambda = 0$ and $\lambda^* = -1$ in the ESTAR model (3). This implies random walk behavior for small deviations (from y = 0.019 in the case of the Goldfeld data) but fast adjustment for large deviations from equilibrium.

Figure 1. Transition Functions: $F(y_{t-1})$ Against y_{t-1}

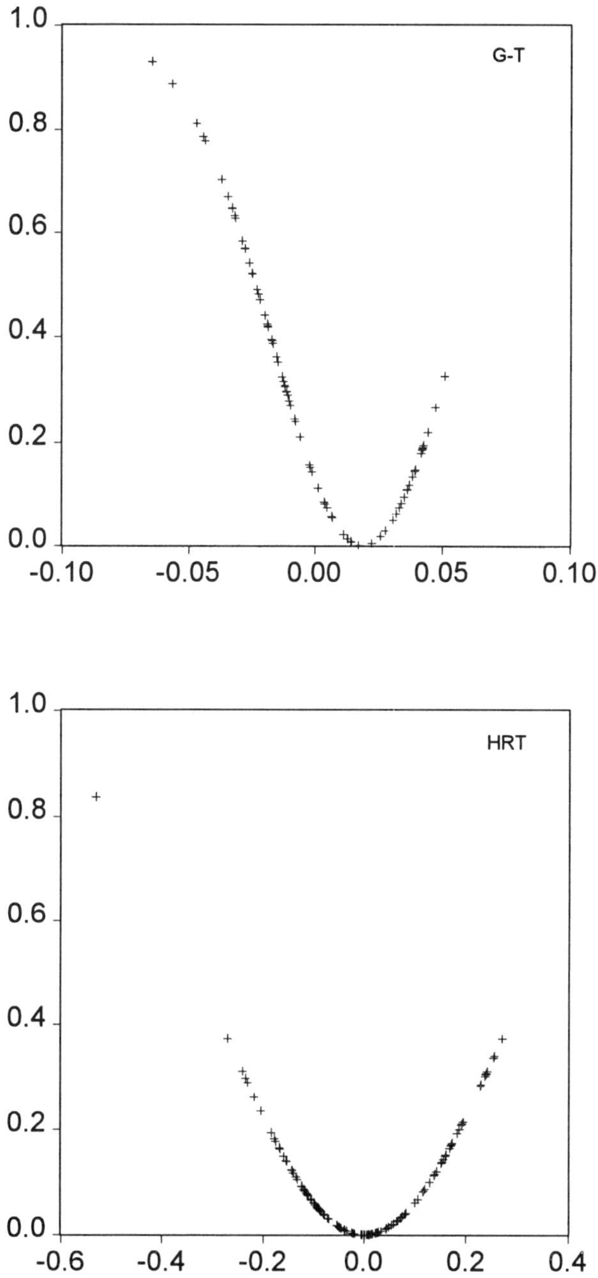

5. Conclusion

A stylized empirical finding is the slow adjustment towards equilibrium money demand. This paper provides an explanation of this feature by considering the implications of non-convex costs of adjustment. A key prediction of this literature is that a linear adjustment model will be inappropriate. Although the precise nonlinear form of the aggregate process implied by explicit micro underpinnings may be intractable, its distinct features can be parsimoniously represented in terms of an ESTAR model subject to appropriate parameter restrictions.

We evaluate the hypothesis of linear adjustment against the alternative of an ESTAR process using two sets of quarterly data for the US, and find that linearity is strongly rejected in each case. The estimated ESTAR models imply no adjustment for small deviations, but fast adjustment for large deviations from equilibrium. The strong evidence in favor of a nonlinear but mean-reverting process for the deviations y_t suggests that linear error correction models are misspecified.

Acknowledgements

We acknowledge, with the usual disclaimers, the helpful comments of Michael Wickens and seminar participants at the National University of Singapore, and thank Mark Taylor and Robert Raasche for generously making available the two data sets utilized in our study. A. R. Nobay acknowledges the support of the H .G. B. Alexander Research Foundation at the University of Chicago.

Notes

1. In empirical analysis both a short-term and a long-term rate may be used (in either levels or logarithms).
2. Studies for the U.S. include Miller (1991), Hafer and Jansen (1991), and Baba, *et al.* (1992).
3. If different intercepts are desired for the outer regimes, then the intercept k^* in (2) could be replaced by $k^+D^+ + k^-D^-$ where $D^+ = \{1 \text{ if } y_{t-d} > 0; 0 \text{ otherwise}\}$ and $D^- = \{1 \text{ if } y_{t-d} < 0; 0 \text{ otherwise}\}$.
4. In Michael, *et al.* (1997) we apply the ESTAR model to deviations from purchasing power parity and find that the restrictions $\lambda = 0$ and $\lambda^* = -1$ hold for every exchange rate considered.
5. Provided (6b) is valid, a more powerful test of linearity is obtained by testing the null hypothesis $H_{0L}^*: \beta_{2j} = 0 | \beta_{1j} = 0 \ (j = 1, \ldots, p)$..
6. Our argument at the end of the previous section notwithstanding, the simulation results of Balke and Fomby (1997) show that the Johansen procedure, although based on a linear VAR process, has reasonable power against TAR alternatives; as already noted, the TAR model can be a special case of the ESTAR model.

References

Baba, Y., D.F. and R.M. Starr (1992). "The Demand for M1 in the USA, 1960-1988," *Review of Economic Studies*, 59, 25-61.

Balke, N.S. and T.B. Fomby (1997). "Threshold Cointegration," *International Economic Review*, 38, 627-646.

Bar-Ilan, A. and A.S. Blinder (1992). "Consumer Durables: Evidence on the Optimality of Usually Doing Nothing," *Journal of Money, Credit, and Banking*, 24, 258-272.

Bertola, G. and R.J. Caballero (1990). "Kinked Adjustment Costs and Aggregate Dynamics," in

Blanchard, O.J. and S. Fischer (eds.) *NBER Macroeconomic Annual*. Cambridge: MIT Press.

Caballero, R.J. and E.M.R.A. Engel (1993). "Microeconomic Adjustment Hazards and Aggregate Dynamics," *Quarterly Journal of Economics*, 108, 359-383.

Caballero, R.J. and E.M.R.A. Engel (1992). "Beyond the Partial Adjustment Model," *American Economic Review, Papers and Proceedings*, 82, 360-364.

Carr, J. and M.R. Darby (1981). "The Role of Money Supply Shocks in the Short-Run Demand for Money," *Journal of Monetary Economics*, 8, 183-200.

Chow, G. (1966). "On the Long-Run and Short-Run Demand for Money," *Journal of Political Economy*, 74, 111-131.

Cuthbertson, K. and M.P. Taylor (1986). "Monetary Anticipations and the Demand for Money: Testing the Rationality of Buffer-Stock Money," *Journal of Applied Econometrics*, 1, 355-365.

Cuthbertson, K. and M.P. Taylor (1987). "The Demand for Money: A Dynamic Rational Expectations Model," *Economic Journal*, 97, 65-76.

Davies, R.B. (1977). "Hypothesis Testing When a Nuisance Parameter is Present Only Under the Alternative," *Biometrika*, 64, 247-254.

Eberly, J.C. (1994). "Adjustment of Consumers' Durables Stocks: Evidence from Automobile Purchases," *Journal of Political Economy*, 102, 403-436.

Engle, R.F. (1982). "Autoregressive Conditional Heteroskedasticity with Estimates of the Variance of United Kingdom Inflation," *Econometrica*, 50, 987-1007.

Engle, R.F. and C.W.J. Granger (1987). "Cointegration and Error Correction: Representation, Estimation and Testing," *Econometrica*, 55, 251-276.

Goldfeld, S.M. (1973). "The Demand for Money Revisited," *Brookings Papers on Economic Activity*, 3, 577-638.

Goodfriend, M.S. (1985). "Reinterpreting Money Demand Regressions," *Carnegie-Rochester Conference Series on Public Policy*, 22, 207-242.

Granger, C.W.J. and T. Teräsvirta (1993). "Modelling Nonlinear Economic Relationships." Oxford: Oxford University Press.

Hafer, R.W. and D.W. Jansen (1991). "The Demand for Money in the United States: Evidence from Cointegration Tests," *Journal of Money, Credit, and Banking*, 23, 155-168.

Haggan, V. and T. Ozaki (1981). "Modelling Nonlinear Random Vibrations Using an Amplitude - Dependent Autoregressive Time Series Model," *Biometrika* 68, 189-196.

Harvey, A.C. (1990). "The Econometric Analysis of Time Series," (2nd ed.), New York: Philip Allan.

Hoffman, D.L., R.H. Rasche, and M.A. Tieslau (1995). "The Stability of Long-Run Money Demand in Five Industrial Countries," *Journal of Monetary Economics*, 35, 317-339.

Jarque, C.M. and A.K. Bera (1980). "Efficient Tests for Normality, Homoscedasticity and Serial Independence of Regression Residuals," *Economics Letters*, 6, 255-259.

Johansen, S. (1988). "Statistical Analysis of Cointegration Vectors," *Journal of Economic Dynamics and Control*, 12, 231-254.

Klimko, L.A. and P.I. Nelson (1978). "On Conditional Least Squares Estimation for Stochastic Processes," *The Annals of Statistics*, 6, 629-642.

Laidler, D.E.W. (1984). "The Buffer Stock Notion in Monetary Economics," *Economic Journal, Conference Supplement*, 94, 17-34.

Ljung, G.M. and G.E.P. Box (1978). "On a Measure of Lack of Fit in Time Series Models," *Biometrika*, 65, 297-303.

Michael, P., A.R. Nobay, and D.A. Peel (1997). "Transactions Costs and Nonlinear Adjustment in Real Exchange Rates: An Empirical Investigation," *Journal of Political Economy*, 105, 862-879.

Miller, S.M. (1991). "Monetary Dynamics: An Application of Cointegration and Error - Correction Modelling," *Journal of Money, Credit, and Banking*, 23, 139-154.

Saikkonen, P. and R. Luukkonen (1988). "Lagrange Multiplier Tests for Testing Non-Linearities in Time Series Models," *Scandinavian Journal of Statistics*, 15, 55-68.

Taylor, M.P. (1994). "On the Reinterpretation of Money Demand Regressions," *Journal of Money, Credit, and Banking*, 26, 851-866.

Teräsvirta, T. (1994). "Specification, Estimation and Evaluation of Smooth Transition Autoregressive Models," *Journal of the American Statistical Association*, 89, 208-218.

Tong, H. (1990). "Non-Linear Time Series: A Dynamical System Approach." Oxford: Clarendon Press.

Tsay, R. (1989). "Testing and Modelling Threshold Autoregressive Processes," *Journal of the American Statistical Association,* 84, 231-240.

10 ASYMMETRIC NONLINEAR SMOOTH TRANSITION GARCH MODELS

Heather M. Anderson
Texas A&M University
Kiseok Nam
Texas A&M University
Farshid Vahid
Texas A&M University

1. Introduction

Many studies of time varying volatility in stock markets have focused on an asymmetry in the volatility process known as "the leverage effect." This asymmetry, first noted by Black in 1976, is characterized by the market's asymmetric reaction to news, in the sense that stock market returns become more volatile after a negative price shock, than they do after a positive shock of the same absolute size. The first empirical models which explicitly attempted to capture the leverage effect include Engle's (1990) A-GARCH model, Nelson's (1991) E-GARCH model, and then the switching GARCH model proposed by Glosten, *et al.* (1993). Nowadays, there are many models which attempt to incorporate this asymmetry,[1] and they are often conveniently summarized and compared using Engle and Ng's (1993) "news impact curves," which plot the predicted volatility response implied by a model, against past price shocks of given sign and size.

It is now well known that empirical models which allow for "leverage" dominate the standard GARCH specifications, but it is also recognized that factors other than leverage are behind the observed asymmetries. Christie (1982) and Schwert (1989) have pointed out that leverage is too small to account for the large volatility swings which characterize the data, and they have experimented with models which allow volatility to depend on factors other than price shocks, such as interest rates and measures of macroeconomic activity. Generally, however, researchers have relied on nonlinear specifications of volatility to generate asymmetries, and most of these models have simply allowed volatility to respond differently to negative and positive price shocks. One feature of such models is that the volatility response to (positive or negative) price shocks does not depend on the current volatility state. This implies that a single asymmetric news impact curve can adequately characterize the volatility process, regardless of whether the market is calm or turbulent.

This chapter departs from most of the literature on asymmetric GARCH models, by considering models which allow the predicted volatility response to price shocks to depend, nontrivially, on volatility itself. Our model is nonlinear in both past price shocks and past volatility, allowing the GARCH parameters as well as the ARCH parameters to vary with news. This feature gives rise to a news impact curve which changes *shape* as volatility varies, with volatility responding more strongly to news which strikes when the market is nervous and volatility is already high, than when such news reaches a calm and quiet market. A further implication of the time-varying GARCH parameters in our models is that they can model the possibility that current volatility might affect the market's interpretation of different types of news. If we define the best news to be news which minimizes next period's volatility, then in a quiet market in which volatility is low, the best news is no news. However, once the market is upset and volatility is high, then the best news need not be zero news. We find that in high volatility situations, the best news is generally positive.

Time-varying GARCH parameters have been employed before, by Fornari and Mele (1997), who were specifically interested in the question of whether past volatility *surprises* could explain a reversion anomaly they associated with leverage. The anomaly was that small positive price shocks might have a larger effect on volatility than similar sized negative shocks, even though large negative shocks have the usual leverage effects and increase volatility by more than similar sized positive shocks. Fornari and Mele (1997) argued that past volatility shocks were potentially useful in explaining this reversion effect, because a small negative price shocks might not be bad news if it increased volatility by less than the market expected, and further, a small positive price shock might not be good news if it increased volatility by more than the market expected. The two regime switching GARCH model which these authors employed supported their hypothesis that past volatility surprises could affect current volatility, but they did not discuss how this might affect the implied news impact curves.

We find the two regime switching set-up used by Fornari and Mele (1997) to be restrictive, and to overcome this shortcoming, we use a smooth transition function which ensures that the volatility parameters are smooth functions of the past news. Smooth transition autoregressive models have been used extensively in models of conditional mean,[2] but their use in modeling conditional variance has been limited to a single recent study undertaken by González-Rivera (1996). Our model is more general than the González-Rivera (1996) model because it incorporates an additional assumption that the volatility response to price shocks depends on the current volatility state. It is this particular assumption which is responsible for the significant improvements in likelihoods which we observe. In compliance with the tradition observed in the ARCH literature, we call our asymmetric nonlinear smooth transition GARCH model, an ANST-GARCH model.

Section 2 introduces our model, and discusses its theoretical properties which distinguish it from others which have been used in the literature. Here, we use the principles used by Bollerslev, *et al.* (1994), to extend the concept of a news impact curve to incorporate the past conditional variance. We take our model to the data in Section 3, and show that the response of the volatility process to price shocks depends in a non-trivial way on past volatility as well as the news. This section also presents the associated news impact curves for these volatility processes, and discusses the various behavioral patterns we observe. Section 4 summarizes and concludes.

2. Alternative Models for Volatility

Let r_t denote the rate of return of a stock from time $t-1$ to t and let Ψ_{t-1} denote the investors' information set which contains relevant information up to and including time $t-1$. When investors make decisions at time $t-1$, the expected return is given by $E(r_t) = f(\Psi_{t-1})$, and the unexpected part of the return is given by $\epsilon_t = r_t - E(r_t)$. The unexpected return ϵ_t, which measures a price shock, is frequently interpreted as "the news." The conditional variance of the return, $h_t = Var(\epsilon_t | \Psi_{t-1})$, provides a natural measure of predictable volatility, and since Engle (1982) first suggested his ARCH specification in which he modeled h_t for inflation as a function of lagged ϵ_t^2, ARCH specifications have been widely employed in the literature on stock returns. The quantity denoted by $\xi_t = \epsilon_t^2 - h_t$ measures volatility surprises. For the purposes of this paper, we assume that $\epsilon_t = u_t\sqrt{h_t}$, where $u_t \sim i.i.d.$ with $E(u_t) = 0$ and $Var(u_t) = 1$, and we also assume that $\epsilon_t | \Psi_{t-1} \sim N(0, h_t)$.

The most popular specification for the conditional variance is Bollerslev's (1986) GARCH(p,q) specification, which is given by

$$h_t = \omega_0 + \sum_{i=1}^{q} \alpha_i \epsilon_{t-i}^2 + \sum_{j=1}^{p} \beta_j h_{t-j}$$

$$\text{with } \omega_0, \alpha_i, \text{ and } \beta_j > 0, \ \sum_{i=1}^{q} \alpha_i + \sum_{j=1}^{p} \beta_j < 1,$$

and GARCH(1,1) models appear to fit financial data well, in practice. However, GARCH(p,q) models do not allow for asymmetries such as the leverage effect, and it has been this shortcoming which has led to nonlinear GARCH specifications which allow positive and negative news to have different effects on volatility. One of the best known of these nonlinear GARCH models is due to Glosten, *et al.* (1993), who specify the following GJR(1,1) model for the conditional variance:

$$h_t = \omega + \alpha^+ I(\epsilon_{t-1} > 0)\epsilon_{t-1}^2 + \alpha^- I(\epsilon_{t-1} \leq 0)\epsilon_{t-1}^2 + \beta h_{t-1}.$$

This model appears to fit stock market returns data much better than a standard GARCH(1,1) model, but it is restrictive in the following two senses. The first is that the time varying alpha coefficients can take only two possible values, and this implies only two distinct behavioral regimes, with an abrupt switch from one to the other according to the sign of the news. Secondly, it assumes that the impacts of current news and current volatility are additively separable, so that the impact of current news on future volatility does not depend on the current volatility state. These features imply that the "best news" is always no news, regardless of the volatility state. We show below that this restriction is rejected by the data.

González-Rivera (1996) introduces a smooth transition GARCH specification which addresses the first of these problems. In her model,[3]

$$h_t = \omega_0 + \alpha_0 \epsilon_{t-1}^2 + \beta_0 h_{t-1} + F(\epsilon_{t-1}) \times \alpha_1 \epsilon_{t-1}^2, \text{ with}$$

$$F(\epsilon_{t-1}) = \frac{1}{1 + e^{\gamma \epsilon_{t-1}}},$$

and with the $\alpha_1 > 0$, this model implies that negative shocks increase volatility by more than positive shocks of the same absolute size. The interesting feature of this model is that the effective alpha coefficient at any given time is given by $\alpha_t = \alpha_0 + \alpha_1 F(\epsilon_{t-1})$, which decreases monotonically with the price shock last period. If γ is small, then α_t changes quite smoothly with ϵ_{t-1}, but for larger γ, such changes become more abrupt. The GJR (1993) model is a limiting case of the González-Rivera (1996) model when $\gamma \to \infty$.

Fornari and Mele (1997) address the other restriction noted above, i.e., that the effects of ϵ_{t-1} and h_{t-1} on current volatility are additively separable. In their specification

$$h_t = \omega_0 + \alpha_0 \epsilon_{t-1}^2 + \beta_0 h_{t-1} + S(\epsilon_{t-1}) \times (\omega_1 + \alpha_1 \epsilon_{t-1}^2 + \beta_1 h_{t-1}) \quad \text{with}$$

$$S(\epsilon_{t-1}) = 1 \text{ for } \epsilon_{t-1} > 0 \text{ and } S(\epsilon_{t-1}) = -1 \text{ for } \epsilon_{t-1} < 0,$$

and now the volatility predicted last period also contributes towards the realized volatility regime. Noting that the effect of past volatility now depends on the sum of β_0 and β_1, it is evident that this specification allows for wide changes in how past volatility affects current volatility. This is potentially important, because in practice, the β term is usually found to be the larger and more significant parameter in GARCH and related specifications. Fornari and Mele (1997) do not focus on this point, and instead, they reparameterize the $(\omega_1 + \alpha_1 \epsilon_{t-1}^2 + \beta h_{t-1})$ term to obtain ϕv_{t-1}, with

$$v_{t-1} = \delta_0 \epsilon_{t-1}^2 - \delta_1 h_{t-1} - \delta_2,$$

and they interpret the v_{t-1} term as a measure of past unexpected volatility. It is this reparameterization which leads them to conclude that past volatility shocks contribute towards predictable volatility.

The model proposed in this chapter combines the features of González-Rivera (1996) and Fornari and Mele (1997) to obtain a generalized asymmetric nonlinear smooth transition (ANST)-GARCH model given by

$$h_t = \omega_0 + \alpha_0 \epsilon_{t-1}^2 + \beta_0 h_{t-1} + F(\epsilon_{t-1}) \times (\omega_1 + \alpha_1 \epsilon_{t-1}^2 + \beta_1 h_{t-1}),$$

where F is the logistic function described above. We impose the restrictions that the three coefficients ω_0, α_0, and β_0 are positive, and also that $\omega_0 + \omega_1 > 0$, $\alpha_0 + \alpha_1 > 0$, and $\beta_0 + \beta_1 > 0$, to ensure that the conditional variance is always positive. For simplicity of notation we have shown the (1,1) variant only, but extensions to incorporate longer lag-lengths are straightforward. Our interpretation of this model emphasizes the intuition that the market's interpretation of news shocks and the future volatility response to these news shocks, depend, in part, on the current volatility state.

It is interesting to reparameterize this volatility process into an ARMA process for ϵ_t^2. Our ANST-GARCH model implies that

$$\epsilon_t^2 = \omega_0 + (\alpha_0 + \beta_0)\epsilon_{t-1}^2 - \beta_0 \upsilon_{t-1} + F(\epsilon_{t-1}) \times \left(\omega_1 + (\alpha_1 + \beta_1)\epsilon_{t-1}^2 - \beta_1 \upsilon_{t-1}\right) + \upsilon_t,$$

which is a smooth transition ARMA(1,1) process for ϵ_t^2.

Our model nests many previously proposed models of asymmetric volatility.[4] For instance, if we impose the restrictions that $\omega_1 = 0$ and $\beta_1 = 0$, then our model becomes the González-Rivera (1996) model, and if we impose the further restriction that γ in the transition function is infinitely large, then we obtain the GJR (1993) specification. Relaxing the first two of these restrictions, but maintaining the third leads to the Fornari Mele (1997) model. One advantage of studying our proposed specification is that it allows us to examine questions such as whether the smoothness in the transition function is important, or whether regime shifts involve coefficient changes on the h_{t-1} terms rather than just on the ϵ_{t-1}^2 terms.

The unconditional variance of the returns can be derived from our specification by using the usual trick attributed to Kakwani (1967), which uses the fact that

$$(F(\epsilon_{t-1}) - 1/2) \times (\omega_1 + \alpha_1 \epsilon_{t-1}^2 + \beta_1 h_{t-1})$$

is an odd function of ϵ_{t-1}, and will have an expected value of zero, when ϵ_{t-1} is symmetrically distributed around zero. For specifications which involve only one lag, the second unconditional moment of ϵ_t in an ANST-GARCH(1,1) model is given by

$$E(\epsilon_t^2) = (\omega_0 + 0.5\omega_1)(1 - \alpha_0 - 0.5\alpha_1 - \beta_0 - 0.5\beta_1)^{-1}$$

provided that the transition variable is symmetrically distributed around zero (this is true for transition variables such as ϵ_{t-1} or $(\epsilon_{t-1}/\sqrt{h_{t-1}})$, although we only consider ϵ_{t-1} here)[5]. The model will be covariance stationary if $0<(\alpha_0+0.5\alpha_1+\beta_0+0.5\beta_1)<1$, but this condition is sufficient, rather than necessary for stationarity. Persistence in the extreme volatility regimes is measures by $(\alpha_0+\beta_0)$ and $(\alpha_0+\beta_0+\alpha_1+\beta_1)$ respectively, and it is useful to note that the second one, which corresponds to extremely large negative shocks, can be greater than one without violating wide sense stationarity.[6]

It is of interest to think about the news impact curves associated with these models. Usually such curves hold the information dates *t-2* and earlier constant, and then plot the volatility response (h_t) predicted by the model, against the preceding news shock (ϵ_{t-1}). Here, however, this procedure is inappropriate, because the effect of the more distant past (i.e., $h_{t-1}|\epsilon_{t-2}, h_{t-2}$) on current volatility, changes with the news shock (ϵ_{t-1}). We follow Bollerslev, *et al.* (1994) and draw a set of conditional news impacts curves, where each curve in the set traces out the volatility response (h_t) to the past price shock (ϵ_{t-1}), given a particular value of h_{t-1}. These sets of curves show that news shocks of the same magnitude and sign may have different effects on predictable volatility, where the effect depends on how volatile the market is, when the news (ϵ_{t-1}) arrives.

The distinctive feature of our specification is its flexibility in defining the "best news." We define this to be the value of the news (ϵ_{t-1}) which minimizes the volatility (h_t) given the volatility state (h_{t-1}). This corresponds to what Engle and Ng (1993) call the "center" of the news impact curve. In the GJR and smooth transition GARCH models that we described above, the best news is always no (i.e., zero) news. Other nonlinear GARCH models proposed in the literature, such as the asymmetric GARCH model of Engle (1990) or the Q-GARCH model of Sentana (1991) allow the best news to be a non-zero constant, but force the news impact curve to be symmetric around the center. In contrast, our news impact curves show that the best news is a function of the volatility state, and that the news impact curve can be asymmetric around this center. We show in the Appendix that the sign of β_1 determines how the "best news" varies with h_{t-1}.

While the sets of conditional news impact curves show how the volatility response (h_t) to the news shock ϵ_{t-1} depends on the volatility sate (h_{t-1}), such curves cannot provide a clear visual picture of the relationship between (h_t), (h_{t-1}), and (ϵ_{t-1}). We therefore also draw news impact surfaces, which graph (h_t) as a function of (h_{t-1}) and the standardized news $(\epsilon_{t-1}/\sqrt{h_{t-1}})$. We choose standardized news rather than the news itself, in order to be able to show the volatility response to a range of shocks that might occur with fixed probability.

3. Empirical Models of Asymmetric Volatility

The data used to study the models suggested in this paper consist of six time series of ex-post returns calculated from value weighted stock market price indices for the United States, Japan, Germany, United Kingdom, France, and Italy. The data were used by Fornari and Mele, who provide a data description as well as some summary statistics in their 1997 paper. We study this particular data set because it covers a broad spectrum of different stocks being sold in different environments, and it is readily available from the *Journal of Applied Econometrics Data Archives.* The sample consists of 1493 daily observations, dating from January 2, 1990 to October 16, 1995.

Table 1. Smooth Transition GARCH Models

	USA	Japan	Germany	UK	France	Italy
μ_0	0.0275 (0.018)	-0.0651 (0.027)	0.0051 (0.021)	0.0129 (*NA*)	0.0000 (*NA*)	0.0360 (*NA*)
μ_1	0.0322 (0.029)	0.1371 (0.027)	0.0626 (0.026)	0.0609 (*NA*)	0.0531 (*NA*)	0.2345 (*NA*)
ω_0	0.0087 (0.011)	0.0002 (0.001)	0.0280 (0.020)	0.0203 (0.013)	0.1542 (0.045)	4.6×10^{-5} (8×10^{-4})
α_0	0.0520 (0.039)	0.0734 (0.047)	0.0201 (0.084)	0.0224 (0.022)	0.0000 (*NA*)	0.0584 (0.016)
β_0	0.4848 (0.345)	0.0026 (0.014)	0.0009 (0.011)	0.8706 (0.045)	0.7427 (0.052)	0.8590 (0.037)
ω_1	-0.0086 (0.011)	0.0755 (0.048)	-0.0281 (0.020)	-0.0203 (0.013)	-0.1542 (0.045)	0.0468 (0.021)
α_1	-0.0518 (0.039)	0.0771 (0.118)	0.0548 (0.134)	0.0422 (0.021)	0.0630 (0.029)	0.0104 (0.027)
β_1	0.9641 (0.646)	1.7655 (0.090)	1.8744 (0.062)	0.1390 (0.035)	0.3373 (0.214)	0.1251 (0.066)
γ	0.4340 (0.255)	0.3187 (0.080)	0.1903 (0.088)	∞	∞	∞
$\ln(L)$	-1575.81	-2287.80	-1916.23	-1591.18	-2253.04	-2296.44
λ	0.9929	0.9973	0.9855	0.9836	0.9743	0.9851

Notes: Estimated robust standard errors appear in parentheses. We have presented threshold models rather than smooth transition models for the UK, French, and Italian cases, because in these cases our estimation algorithms led to likelihoods which monotonically increased with γ as $\gamma\to\infty$. The mean of the return process has been specified by $r_t=\mu_0+\mu_1 r_{t-1}+\epsilon_t$, and standard errors for the mean parameters of the threshold models are not presented because the likelihood is not differentiable with respect to these parameters at $\epsilon_{t-1}=0$. For France, the estimated α_0 was very small, and was set to zero to avoid numerical precision problems. The parameter λ is defined by $\lambda=\alpha_0+\beta_0+0.5(\alpha_1+\beta_1)$, and as discussed in the text, the condition that $\lambda<1$ ensures covariance stationarity.

Table 2. *P*-Values of Likelihood Ratio Tests of Restrictions on the Table 1 Models.

Restriction	US	Japan	Germany	UK	France	Italy
Threshold ($\gamma \to \infty$)	0.0183	0.0043	0.0010	-	-	-
González-Rivera ($\omega_1 = 0, \beta_1 = 0$)	0.0041	0.0000	0.0014	0.0322	0.0000	0.0000
GJR ($\gamma \to \infty, \omega_1 = 0, \beta_1 = 0$)	0.0019	0.0000	0.0042	0.0123	0.0000	0.0000
GARCH (1,1) $\omega_1 = 0, \alpha_1 = 0, \beta_1 = 0$	0.0000	0.0000	0.0000	0.0000	0.0000	0.0000
$\omega_1 = -\omega_0$	1.0000	0.5529	1.0000	0.9840	0.9508	0.0112
$\alpha_1 = 0$	0.1841	0.5135	0.6825	0.0444	0.0298	0.7001
$\beta_1 = 0$	0.1356	0.0000	0.0000	0.0000	0.1164	0.0580

Notes: The parameter γ is not identified under the restriction that $\omega_1 = \alpha_1 = \beta_1 = 0$, and this implies that the likelihood ratio statistic has a non-standard asymptotic distribution. The *p*-values for the GARCH(1,1) restrictions in this table are based on a $\chi^2(3)$ distribution, and although they are not strictly correct, they indicate large improvements in the likelihood function as the restriction that $\omega_1 = \alpha_1 = \beta_1 = 0$ is relaxed.

Table 1 contains estimated coefficients and likelihoods for our model, in which both the sign and magnitude of the last news shock ϵ_{t-1} and the volatility state h_{t-1} determine the current volatility regime, and changes between such regimes can be smooth. Two broad patterns seem to emerge from this table. Firstly, it is apparent that for several countries, there *is* a smooth transition between volatility regimes. The γ parameter is estimated to be under one for the United States, Japan, and Germany. Formal tests (see the first line in Table 2), find evidence against the hypothesis that threshold models are appropriate for modelling the stock market returns for these countries. This suggests that researchers should, as a general rule, allow for less restrictive specifications in the first instance. The threshold specification does, however, appear to be appropriate for data relating to the United Kingdom, France, and Italy. When we tried to estimate smooth specifications for these latter countries, our algorithms always led to increasingly larger values of γ, which eventually caused numerical overflows. In these cases, we then imposed the threshold restriction, and reported the corresponding threshold model.

Figure 1. News Impact Curves

1a: For the US

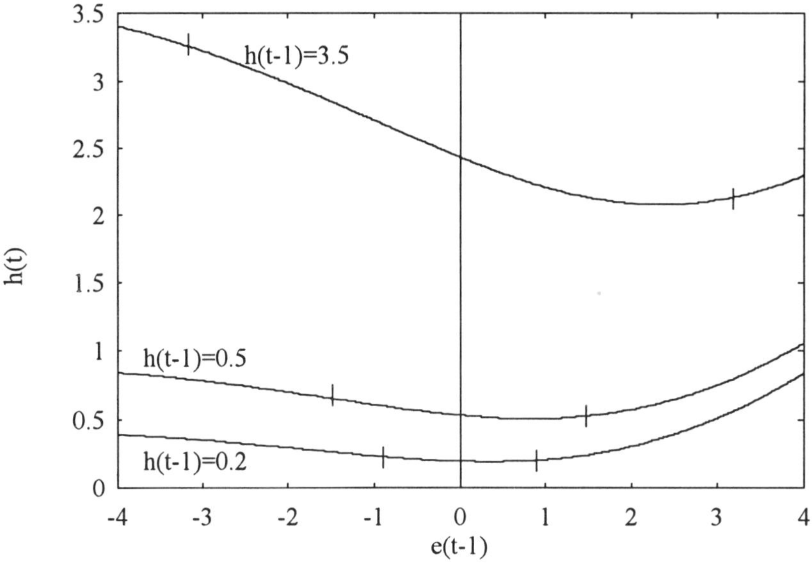

1b: For Japan

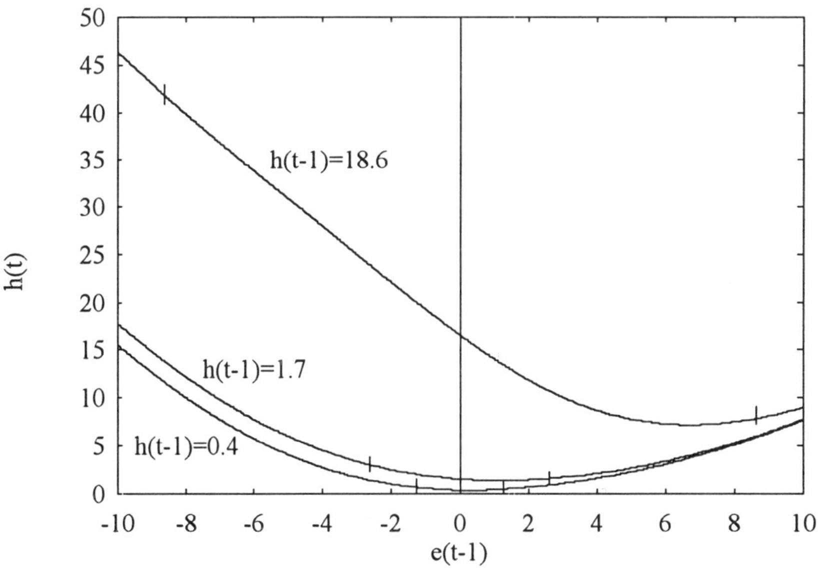

Figure 1, Continued

1c: For Germany

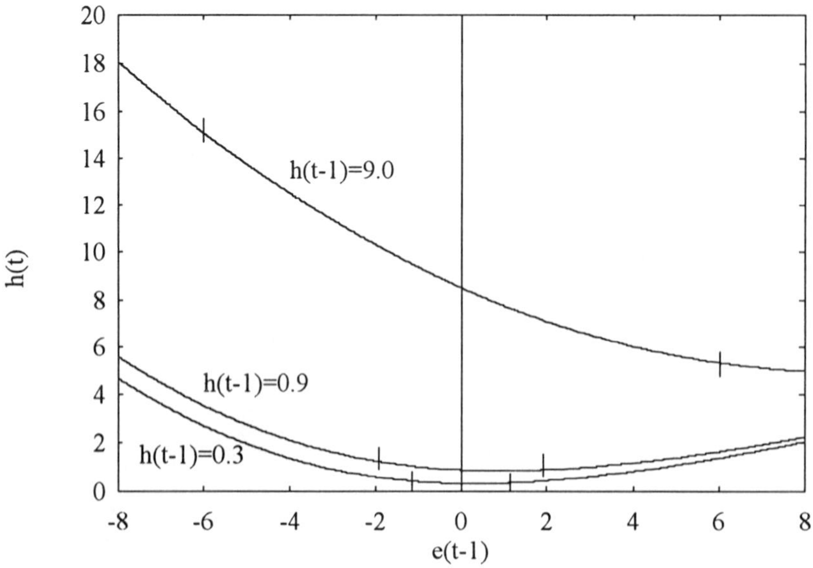

Notes: The bars on each news impact curve indicate the range of h_t in response to shocks within minus/plus two standard deviations. The three lines correspond to h_{t-1} being the minimum, the unconditional, and the maximum value of estimated volatility within the sample period.

The other broad feature of Table 1 which merits discussion is that the volatility process undergoes substantial changes as news shocks vary, and in particular, the GARCH (β) coefficient undergoes a considerable change as volatility moves between different regimes. Tests that $\beta_1 = 0$ reject the null for all but two countries. Consistent with these test results are further test results, which show that our general model fits the data much better than the GJR and González-Rivera models, which allow only the ARCH, and not the GARCH coefficients to vary. Note (from line 6 in Table 2) that after allowing for time variation in β, the null hypothesis that the α parameter does not vary with the news cannot be rejected for four out of six countries, which explains why those models which allow only the ARCH coefficients to vary with news do not fit the data as well. A final interesting point is that in five out of the six cases considered, the constant terms in the volatility process approaches zero as $\epsilon_{t-1} \to -\infty$, and the null hypothesis that $\omega_1 = -\omega_0$ is rejected for only one country. This suggests that the non-negativity constraint in the constant term is binding, which should be taken into account in any specification tests done on these models.

Even though for each of the estimated models, the sum of the alpha and beta parameters becomes greater than 1 for large negative shocks, these processes are nevertheless covariance stationary. The last row of Table 1 shows the magnitude of $(\alpha_0 + 0.5\alpha_1 + \beta_0 + 0.5\beta_1)$ for the estimated models which, as discussed in Section 2, should be less than 1 to ensure covariance stationarity. It is interesting to note that this condition is satisfied for all of the estimated models, even though it was not imposed in the estimation stage.

Figure 1 graphs the news impact curves for the US, Japan, and Germany. Three news impact curves are plotted for each country, and these respectively plot predictable volatility as a function of news for three different values of h_{t-1}. The middle curve in each picture corresponds to $h_{t-1} = E(\epsilon_t^2)$, while the lowest and highest curves respectively correspond to the minimum and maximum values of conditional variance, estimated for the relevant country. We have marked the ± 2 standard deviation range on each curve, in order to highlight the range of volatility which is likely to occur with 95% probability. Obviously, this range is different for each curve.[7]

An important implication of our models is that the effect of a news shock ϵ_{t-1} on h_t depends in a non-trivial manner on h_{t-1}, which measures agents' assessment of volatility at the time of the shock. This can be seen quite clearly in Figure 1. In standard GARCH models with a constant β, higher predictable volatility h_{t-1} when the news shock ϵ_{t-1} arrives, simply moves the news impact curve vertically upwards, with no change in shape. However, in these ANST-GARCH models (in which $\beta_1 > 0$), higher predictable volatility again moves the news impact curve upwards, but now the distance moved depends on $F(\epsilon_{t-1})\,h_{t-1}$ rather than on just h_{t-1}. Those portions of the news impact curve relating to negative news shocks will move up by more than the corresponding portions relating to positive shocks. Thus the *shape* of the impled news impact curves changes with h_{t-1}. Noting that the bars on each curve enclose those portions of the curves which are likely to observed in practice, it is clear that news impact curves are relatively flat for low h_{t-1}, but become much more curved as h_{t-1} increases. As these news impact curves change shape, their centers, which correspond to the values of news ϵ_{t-1} which will minimize volatility h_t next period, shift to the right. As discussed in the Appendix, this rightward shift is not imposed on our specification, but is implied by the fact that our estimated β_1 coefficients are positive. A corollary of the observation that these centers move to the right as volatility h_{t-1} increases, is that when volatility h_{t-1} is high, positive news can be better than no news, because a positive shock corresponding to the minimum of the news impact curve, ensures the lowest possible volatility in the subsequent period.

News impact surfaces for each country are presented in Figure 2. These surfaces show the volatility response to standardized shocks, for ± 4 standard deviations, at each level h_{t-1} between 0 and 10. The probability of shocks being outside ± 4 standard deviations is less than 10^{-4}, which indicates that these surfaces cover the empirically relevant range of possible shocks. The surfaces for the US, Japan, and Germany show the two distinguishing features of our model very clearly; the asymmetry of the news impact curve for each h_{t-1}, and the positive relationship between the "best" news and h_{t-1}. They also show that functional forms which do not allow for asymmetry, or which restrict the center of the news impact curves to be zero, cannot fit the data well. The news impact surfaces for the UK, France, and Italy show an abrupt switch at zero, which is due to the threshold specification chosen by the data.

Figure 2. News Impact Surfaces

2a: For the US

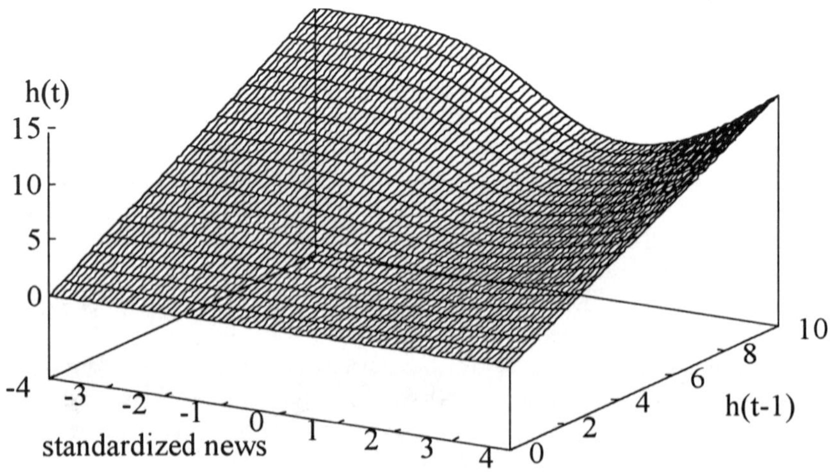

2b: For Japan

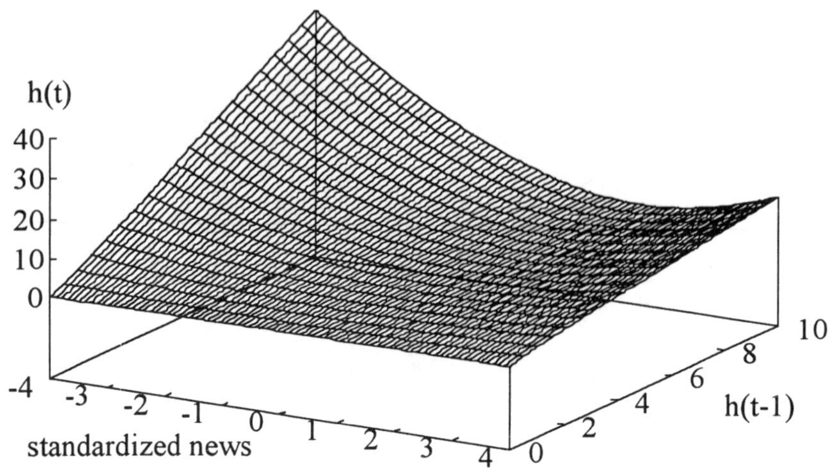

Figure 2, Continued

2c: For Germany

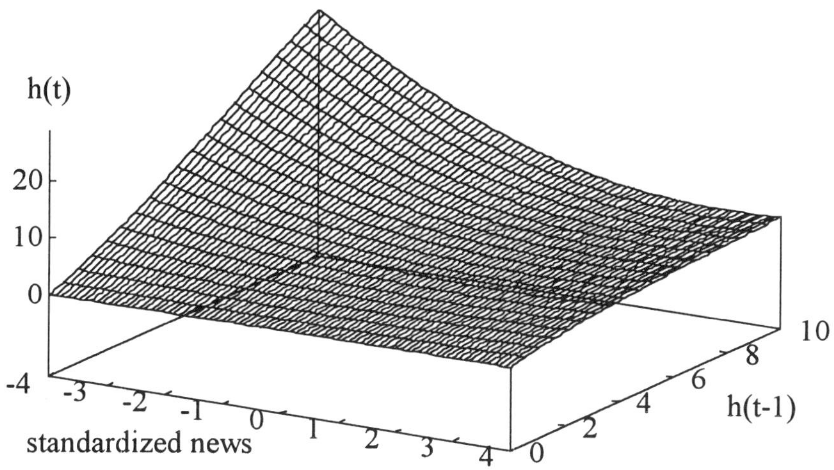

2d: For the UK

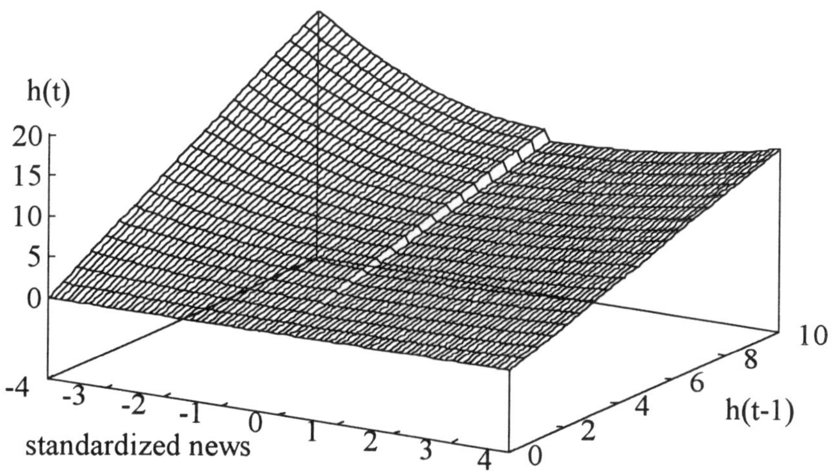

Figure 2, Continued

2e: For France

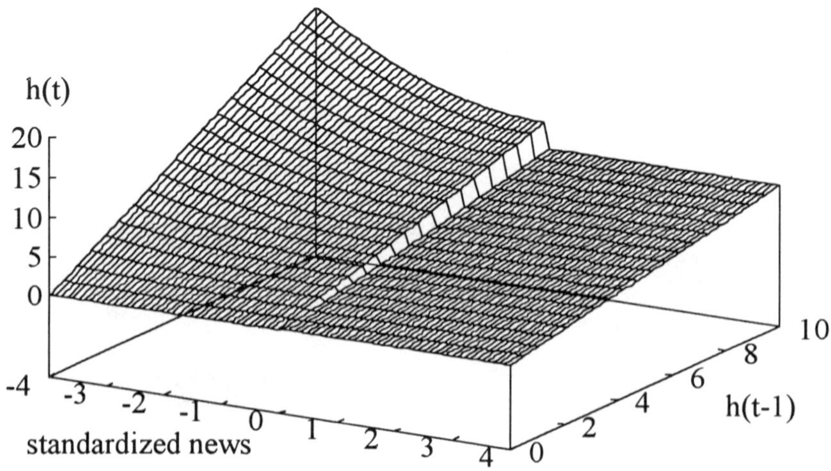

2f: For Italy

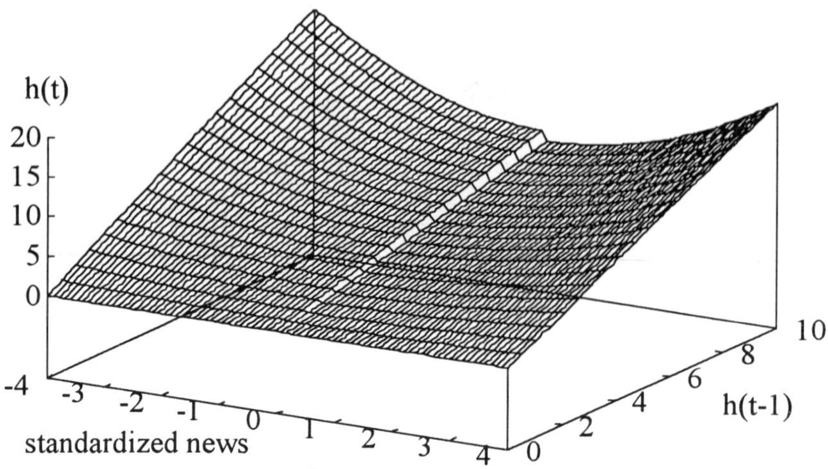

4. Concluding Remarks

In this paper, we have shown that the effect of news on future volatility depends, in a nonlinear fashion, on current volatility as well as characteristics of the news itself. We do this in the context of an ANST-GARCH model, in which the conditional variance evolves according to a process that is the analogue of smooth transition ARMA models for the conditional mean. An interesting implication of our estimated models is that the best news is not necessarily no news. As volatility increases, some positive news will be better than no news at all.

We have shown that ANST-GARCH models fit the international stock market data better than alternative well-known simple models of asymmetry in conditional variance, and that the reason for this better fit is due to our assumption that the volatility response to price shocks depends on the interaction between the current volatility state and the news. Our graphical news impact analysis highlights the distinguishing features of ANST-GARCH specification.

We have relied on test statistics and likelihood based measures for evaluating our models, and future research should compare our ANST-GARCH models with other nonlinear models of volatility using utility based measures or other economic criteria. It would also be interesting to determine the existence of higher moments, and to study the persistence of volatility in this nonlinear framework, but these topics are left for future research.

Appendix

In this appendix we analyze how the "best" news, i.e., the news that minimizes volatility given the past volatility, depends on past volatility in the asymmetric nonlinear smooth transition GARCH model presented in this paper. Let's denote the best news by ϵ^*, i.e.,

$$\epsilon^*(h_{t-1}) = \begin{cases} \arg\min\limits_{\epsilon_{t-1}} (h_t) \\ \arg\min\limits_{\epsilon_{t-1}} \left(\omega_0 + \alpha_0\epsilon_{t-1}^2 + \beta_0 h_{t-1} + F(\gamma\epsilon_{t-1})(\omega_1 + \alpha_1\epsilon_{t-1}^2 + \beta_1 h_{t-1})\right). \end{cases}$$

The first order condition for this minimization implicitly defines ϵ^*,

$$2\alpha_0\epsilon^* - \gamma F(\gamma\epsilon^*)(1 - F(\gamma\epsilon^*))\left(\omega_1 + \alpha_1\epsilon^{*2} + \beta_1 h_{t-1}\right) + 2\alpha_1\epsilon^* F(\gamma\epsilon^*) = 0.$$

Even though it is not possible to obtain an explicit solution for ϵ^*, we can study how it changes with h_{t-1}. Using the implicit function theorem, and after some substitution and algebraic manipulations we obtain:

$$\frac{d\epsilon^*}{dh_{t-1}} = \frac{\beta_1 \gamma F(\gamma\epsilon^*)(1 - F(\gamma\epsilon^*))}{2[\alpha_0 + \alpha_0\gamma\epsilon^* - (2\alpha_0\gamma\epsilon^* + \alpha_1\gamma\epsilon^* - \alpha_1)F(\gamma\epsilon^*)]}.$$

Observe that this formula tells us that when $\gamma = 0$ (the regular GARCH model) or when $\beta_1 = 0$ (as in the smooth transition model of González-Rivera), the best news does not depend on h_{t-1}. Defining $z = \gamma\epsilon^*$, we can show that the denominator

$$D = 2(\alpha_0 + \alpha_0 z - (2\alpha_0 z + \alpha_1 z - \alpha_1)F(z))$$

is positive, provided that $\alpha_0 > 0$ and $-0.95 < (\alpha_1/\alpha_0) < 20$. The easiest way we have been able to show this is by graphing the line $1 + z$ and the function $(2z + cz - c)/(1 + e^z)$, and showing that the former never crosses and always lies above the latter for $-0.95 < c < 20$. Given $D > 0$, the sign of the above derivative is the same as the sign of β_1 (recall that $\gamma > 0$). Note that the maintained assumption of $\beta_0 + \beta_1 > 0$ does not place any a-priori restrictions on the sign of β_1.

Acknowledgements

We thank Gloria González-Rivera for helpful comments. Heather Anderson and Farshid Vahid wish to thank the Department of Econometrics at Monash University, Australia, where part of this research was undertaken.

Notes

1. See, for instance, Ding, *et al.* (1993), Hamilton and Susmel (1994), Zakoian (1994), and González-Rivera (1996).
2. See, for instance, Teräsvirta and Anderson (1992), and Teräsvirta (1994).
3. This is a slightly reparameterized version of González-Rivera (1996), which uses

$$F(\epsilon_{t-1}) = \frac{1}{1 + e^{\gamma\epsilon_{t-1}}} - 1/2$$

as the transition function. Her parameterization, though theoretically more convenient, leads to less straightforward restrictions on parameters to ensure non-negativity.

4. One notable exception is Nelson's (1990) EGARCH model.
5. See Nam (1998) for some examples of models which use $\epsilon_{t-1}/\sqrt{h_{t-1}}$ and $(\epsilon_{t-1}^2 - h_{t-1})$ as transition variables.
6. As with nonlinear ARMA processes, necessary and sufficient conditions for stationarity of nonlinear GARCH processes are not known. In nonlinear autoregressive processes, it is customary to provide simulation evidence that the process stays bounded in all arbitrarily chosen replications.
7. An alternative way of representing news impact curves would be to plot h_t against the standardized ϵ_{t-1}, in which case the 95% range of news would be the same for all h_{t-1}. We have plotted our news impact surfaces (discussed later in the paper) in this way.

References

Black, F. (1976). "Studies in Stock Price Volatility Changes," *Proceedings of the 1976 Business Meeting of the Business and Economic Statistics Section, American Statistical Association*, 177-181.

Bollerslev, T. (1986). "Generalized Autoregressive Conditional Heteroskedasticity," *Journal of Econometrics*, 31, 307-327.

Bollerslev, T., R. F. Engle, and D. Nelson (1994). "ARCH Models," *Handbook of Econometrics*, Vol. 4, Chp. 49, 2959-3038.

Christie, A. A. (1982). "The Stochastic Behavior of Common Stock Variances: Value, Leverage and Interest Rate Effects," *Journal of Financial Economics*, 10, 407-432.

Ding, Z. and C. W. J. Granger (1996). "Modeling Volatility Persistence of Speculative Returns: A New Approach," *Journal of Econometrics*, 73, 185-216.

Ding, Z., C. W. J. Granger, and R. F. Engle (1993). "A Long Memory Property of Stock Market Returns and a New Model," *Journal of Empirical Finance*, 1, 83-106.

Engle, R. F. (1982). "Autoregressive Conditional Heteroskedasticity with Estimates of the Variance of United Kingdom Inflation," *Econometrica*, 50, 987-1007.

Engle, R. F. (1990). "Discussion on Schwert," *The Review of Financial Studies*, 3, 103-106.

Engle, R. F. and V. K. Ng (1993). "Measuring and Testing the Impact of News on Volatility," *Journal of Finance*, 48, 1749-1778.

Fornari, F. and A. Mele (1997). "Sign and Volatility-Switching ARCH Models: Theory and Applications to International Stock Markets," *Journal of Applied Econometrics*, 12, 49-66.

Glosten, L., R. Jagannathan, and D. Runkle (1993). "On the Relation Between the Expected Value and the Volatility of the Nominal Excess Return on Stocks," *Journal of Finance*, 48, 1779-1801.

González-Rivera, G. (1996). "Smooth Transition GARCH Models," working paper, Department of Economics, University of California, Riverside.

Hamilton, J. D. and R. Susmel (1994). "Autoregressive Conditional Heteroskedasticity and Changes in Regime," *Journal of Econometrics*, 64, 307-333.

Kakwani, N. C. (1967). "The Unbiasedness of Zellner's Seemingly Unrelated Regression Equations Estimator," *Journal of the American Statistical Association*, 62, 141-142.

Nam, K. (1998). "Essays on Stock Market Volatility," Ph.D. dissertation, Department of Economics, Texas A&M University.

Nelson, D. (1990). "Conditional Heteroskedasticity in Asset Returns: A New Approach," *Econometrica*, 59, 347-370.

Schwert, G. W. (1989). "Why Does Stock Market Volatility Change Over Time?" *Journal of Finance*, 44, 1115-1153.

Sentana, E. (1991). "Quadratic ARCH Models: A Potential Reinterpretation of ARCH Models as Second-Order Taylor Approximations," working paper, Department of Economics, London School of Economics.

Teräsvirta, T. and H. M. Anderson (1992). "Characterizing Nonlinearities in Business Cycles Using Smooth Transition Autoregressive Models," *Journal of Applied Econometrics*, 7, S119-S136.

Teräsvirta, T. (1994). "Specification, Estimation and Evaluation of Smooth Transition Autoregressive Models," *Journal of the American Statistical Association*, 89, 208-218.

Zakoian, J. (1994). "Threshold Heteroskedastic Models," *Journal of Economic Dynamics and Control*, 18, 931-955.

11 TESTING THE PRESENT VALUE HYPOTHESIS FROM A VECTOR AUTOREGRESSION WITH STOCHASTIC REGIME SWITCHING

John Driffill
University of Southampton
Martin Sola
University of London
Universidad Torcuato Di Tella
London Business School

1. Introduction

In this paper we propose a method for testing the present value hypothesis from a VAR which allows for regime switching, and we examine the simple efficient market relationship, which asserts that stock prices equal the present value of expected future dividends discounted by a constant real discount rate, allowing for the public's forecasts of future changes of regime. This procedure, suggested in Sola and Driffill (1994), modifies Hamilton's (1988) analysis in order to allow an unobserved state of the system to affect both variables included in a VAR in a general way.

Present value models of stock prices and dividends have been widely studied by many investigators, including LeRoy and Porter (1981), Shiller (1981) and Campbell and Shiller (1987), using a variety of different techniques and data sets. Most of them claim that stock markets are too volatile to accord with such models. One of the most controversial issues has been how to induce stationarity in or detrend the data (see Kleidon (1986) and Marsh and Merton (1986)). Shiller (1981) detrended prices and dividends by a simple exponential trend. Since the rate of growth was estimated over the entire sample, the detrended price for data t depended on information not available at time t. To overcome the detrending problem Campbell and Shiller (1987) (C&S

henceforth) estimated a VAR which included first differences of dividends and a variable created as the difference between the stock prices and a multiple of dividends. In this way they test the restrictions implied by the present value model from a stationary VAR. Surprisingly, using this different methodology their evidence was far from conclusive.

A possible reason for their inability to reject the present value model is that the validity of their test was undermined by structural breaks in the data. Our initial examination of it reported below suggests that there were indeed such structural changes. The innovations in the rate of change of real dividends and a linear combination of prices and lagged dividends show skewness and excess kurtosis. The parameters in simple reduced form models for these variables do not appear to be constant over the whole sample period. The variability of these variables appears to have changed over time. The procedure adopted in this paper is capable of accommodating these changes and giving a valid test of the present value model.

Given that the evidence provided in C&S (1987) is inconclusive (with respect to stock prices and dividends) and that they do not explicitly model expected future changes in fundamentals, we propose to extend the analysis of C&S in two main directions. First, we allow the public to learn about the dividend process and incorporate this in their projection of the future. Second, we allow the unconditional means and the variances and covariances of the variables included in the VAR to switch between two states.

In this paper, we follow C&S in modeling a VAR in two stationary variables: (i) the difference between real stock prices and a multiple of dividends and (ii) first differences of real dividends. But we allow the evolution of the system to be dependent on current and lagged values of the unobserved state of the system. The present value model can then be used to derive restrictions on the VAR and to test their validity. There are at least two reasons why regime shifts are relevant from an econometric point of view. One is the direct effect of the regime shift, which causes a break in the time-series properties of the data. Another is that even during periods when no shift actually occurs, the anticipation of possible future shifts changes the behavior of market participants. This is a manifestation of a kind of "Peso problem" in stock markets. We allow for these possibilities by allowing for stochastic changes of regime, following Hamilton (1988).

As pointed out by West (1987) and C&S, stocks and dividends are not measured contemporaneously. The literal application of the general restrictions derived in Section 3 of this paper (which are analogous to those derived by C&S (1987) for the present framework) might lead us to a spurious rejection of the model if dividends are known only at the beginning of the subsequent period.

C&S overcome this problem by constructing a variable $SL = P_t - \theta D_{t-1}$, where P_t denotes real stock prices and D_t real dividends. They then include this transformed variable in the VAR. To test the present value model they amend the cross equation restrictions appropriately. This transformation ensures that if $S = P_t - \theta D_t$ is integrated of order zero, then SL is also integrated of order zero.

Table 1. Unit Root Statistics

Variable	ADF	Lags	Z_α	Z_t
Stock Prices	-2.81	4	-14.61	-2.63
Dividends	-2.83	4	-21.77*	-3.38

Notes: Both regressions include a significant deterministic trend. The number of lagged first differences included in the ADF test regressions were determined using the sequential stop-down testing procedure discussed in Ng and Perron (1995). An asterisk indicates that the null hypothesis is rejected at the 5% significance level.

2. Data

The data used are as in C&S (1987). That is, real annual prices and dividends on a broad stock index from 1871 to 1985. P_t is the Standard and Poor's composite stock price index for January, divided by the January producer price index (base 1967). The nominal dividend series is, starting in 1926, dividends per share adjusted to index, four-quarter total, for Standard and Poor's composite index. The nominal dividends before 1926 were taken from Cowles (1939), who extended Standard and Poor's series backwards in time. Finally D_t is the nominal Dividend series, divided by the average producer price index (base 1967).

C&S pointed out that real prices and dividends appear to be integrated of order one. Table 1 reports the outcome of tests for the presence of a unit autoregressive root in the non-deterministic components of the data. These are based on the Said and Dickey (1984) Augmented Dickey-Fuller (ADF) statistic and the Z_a and Z_t statistics of Phillips (1987) and Phillips and Perron (1988).[1] We cannot reject the unit root null for any of the series. Also C&S pointed to the fact that these variables are cointegrated. Let us now turn our attention to the possibility of obtaining a cointegrating relationship between real stock prices and dividends. We are unable to reject the null hypothesis that the series are not cointegrated using the C&S data and sample size. However, the power of these tests is notoriously low, and therefore this does not provide overwhelming evidence that the series are indeed not cointegrated. We therefore analyze the sensitivity of the C&S result with respect to changes in the sample period. We found that when the last observation is excluded, there is evidence which favors cointegration and also makes our results comparable with those of C&S. Table 2 records the outcome of several cointegrating tests for the sample 1871-1985, including the Johansen (1991) rank test, the Augmented Dickey-Fuller (ADF) test recommended in Engle and Granger (1987), and the $\hat{Z}_\alpha$ and $\hat{Z}_t$ tests of Phillips and Ouliaris (1990). We reject the hypothesis that there is no cointegrating relationship.

Since these tests provide clear evidence in favor of cointegration for the 1871-1985 sample, we will proceed on the assumption that the series are cointegrated. Notice that cointegration between Stock Prices and Dividends is a necessary condition for the efficient market hypothesis to hold. Therefore despite the fact that the

Table 2. Cointegration Test Statistics and Estimated Cointegrating Vector.

LR	*ADF*	$\hat{Z}_\alpha$	$\hat{Z}_t$
20.347*	-3.512*	-2.677*	-3.80
	Estimated Cointegrating Vector		
	y	x	
	1.000	30.514	
	(1.000)	(31.984)	

Notes: An asterisk indicates that the null hypothesis is rejected at the 5 significance level. Numbers in parentheses are fully modified least-squares estimates (see Phillips and Hansen (1990)).

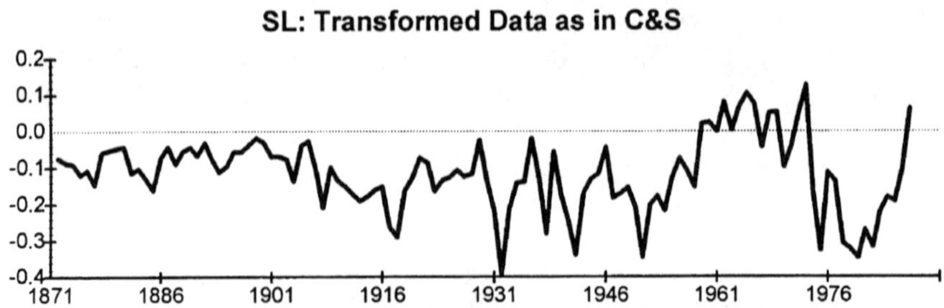

Figure 1

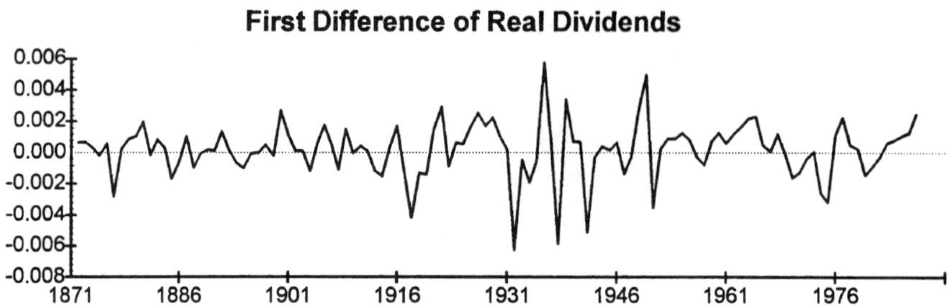

Figure 2

cointegrating results are affected by the choice of the sample size, we continue the investigation of the validity of that hypothesis under the scenario where we reject the null hypothesis of no-cointegration against the alternative of cointegration for prices and dividends. Then a VAR in *SL*, the cointegrating relationship, and the first difference of real dividends, ΔD, can validly be estimated, and the expectations model can be tested by imposing on the VAR the restrictions implied by the present value model. Therefore, we proceed on this basis and investigate the validity of the present value model allowing for the possibility of regime shifts. To simplify the notation we redefine $y = SL$ and $x = \Delta D$ (where Δ is the difference operator); time series plots of these two series are found in Figures 1 and 2.

In this study we use as θ the OLS estimate 30.5 from the static cointegrating relationship, (the fully modified least squares estimate is 31.981); the corresponding real discount rate is 3.39% rather than the sample mean return of 8.2% (these values may be compared with the values 31.092 and 3.2% which C&S obtained for a sample period which includes 1986). There are two main reasons for doing this. First, this rate is consistent with the theoretical long run relationship predicted by the Efficient Market Model. Secondly and more importantly, for this discount rate, the C&S results are inconclusive. In this paper we question whether their results change if we allow the public to forecast expected changes in regime.

Figures 3 and 4 plot rolling estimates of the constant in the cointegrating relation and the unconditional mean of differenced dividends. Both plots suggest that regime shifts may be present.[2]

3. A Bivariate Model with Unobserved States.

We consider a VAR process in two variables, with *m* lags, with the feature that the mean of each variable and the innovation variance-covariance matrix are allowed to switch between two possible states. The two equations that define the VAR are influenced by the same state variable. The state is not observed and has to be inferred from the observed data. Therefore, we consider the following centered bivariate autoregressive process:

$$\begin{aligned} y_t' &= \Phi_1 y_{t-1}' + \cdots + \Phi_m y_{t-m}' + \psi_1 x_{t-1}' + \cdots + \psi_m x_{t-m}' + (w_0 + w_1 s_t)\upsilon_t \\ x_t' &= \varphi_1 y_{t-1}' + \cdots + \varphi_m y_{t-m}' + \Omega_1 x_{t-1}' + \cdots + \Omega_m x_{t-m}' + (\tau_0 + \tau_1 s_t)\eta_t, \end{aligned} \tag{1}$$

where the centered variables are defined as follows:

$$\begin{aligned} y_t' &= y_t - \alpha_0 - \alpha_1 s_t \\ x_t' &= x_t - \beta_0 - \beta_1 s_t \,. \end{aligned} \tag{2}$$

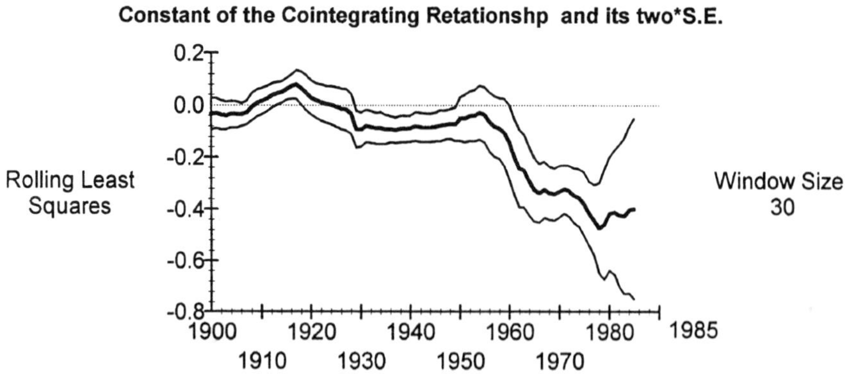

Figure 3

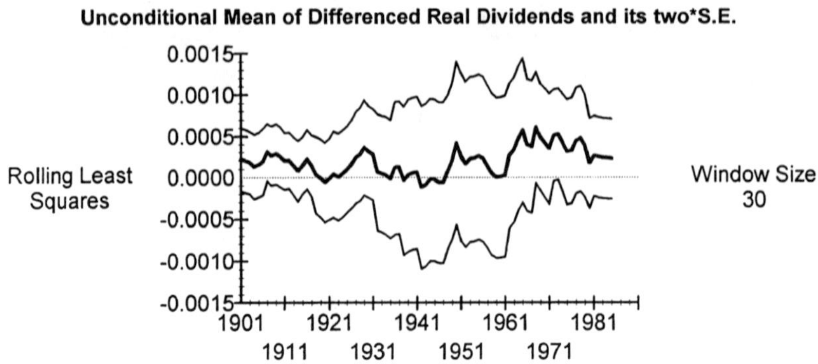

Figure 4

In (2), s_t denotes the unobserved state of the system and takes on the values 0 and 1. s_t is assumed to be governed by a Markov process, summarized by the probabilities: $\text{prob}(s_t = 0 \mid s_{t-1} = 0) = q$ and $\text{prob}(s_t = 1 \mid s_{t-1} = 1) = p$. Substituting the centered variables into the VAR and rearranging terms, we obtain the following expression for y_t and x_t.

$$y_t = \alpha_0(1-\Phi_1-\cdots-\Phi_m) + \beta_0(-\psi_1-\cdots\psi_m) + \Phi_1 y_{t-1} + \cdots + \Phi_m y_{t-m} +$$
$$\psi_1 x_{t-1} + \cdots + \psi_m x_{t-m} + \alpha_1(s_t - \Phi_1 s_{t-1} - \cdots - \Phi_m s_{t-m}) +$$
$$\beta_1(-\psi_1 s_{t-1} - \cdots - \psi_m s_{t-m}) + (w_0 + w_1 s_t)\upsilon_t$$

(3)

$$x_t = \alpha_0(-\varphi_1 - \cdots - \varphi_m) + \beta_0(1-\Omega_1-\cdots-\Omega_m) + \varphi_1 y_{t-1} + \cdots + \varphi_m y_{t-m} +$$
$$\Omega_1 x_{t-1} + \cdots \Omega_m x_{t-m} + \alpha_1(-\varphi_1 s_{t-1} - \cdots - \varphi_m s_{t-m}) +$$
$$\beta_1(s_t - \Omega_1 s_{t-1} - \cdots - \Omega_m s_{t-m}) + (\tau_0 + \tau_1 s_t)\eta_t.$$

The errors υ_t, η_t are assumed to be independent of all $s_{t-j}, j \geq 0$.

4. Results of the Bivariate Filter for SL and ΔD

The model in (3) was fitted to US data for the period 1871 - 1985. With no restrictions imposed, the results are as follows. The estimated unconditional means are:[3]

$$\hat{\text{mean}}_y(s_t) = \underset{(-5.512)}{-0.138926} + \underset{(5.155)}{0.058009}\, s_t$$

$$\hat{\text{mean}}_x(s_t) = \underset{(-0.173)}{-0.000032} + \underset{(1.137)}{0.000178}\, s_t$$

The estimated covariance matrices for state 0 and state 1 are:

$$\Sigma_0 = \begin{pmatrix} \underset{(7.397)}{0.005684} & \underset{(75.797)}{0.000096} \\ & \underset{(2.366)}{0.000004} \end{pmatrix} \qquad \Sigma_1 = \begin{pmatrix} \underset{(4.342)}{0.000287} & \underset{(52.898)}{0.000011} \\ & \underset{(2.518)}{0.000001} \end{pmatrix}$$

The probabilities of remaining in the same state as the previous period are:

$$\hat{p} = \underset{(34.127)}{0.936434}$$

$$\hat{q} = \underset{(79.570)}{0.973282}$$

with expected duration of the states of $(1-\hat{p})^{-1} = 15.73$ years and $(1-\hat{q})^{-1} = 37.43$ years.

The results of the centered vector autoregressive process are presented below.

$$\hat{y}'_t = \underset{(30.181)}{1.014565}y'_{t-1} - \underset{(-5.384)}{0.230104}y'_{t-2} + \underset{(9.112)}{0.342417}y'_{t-3} - \underset{(-2.454)}{0.227078}y'_{t-4}$$

$$-\underset{(-14.432)}{29.313852}x'_{t-1} - \underset{(-5.080)}{1.795371}x'_{t-2} - \underset{(-4.384)}{7.186795}x'_{t-3} + \underset{(5.073)}{1.120904}x'_{t-4}$$

$$\hat{x}'_t = \underset{(1.918)}{0.003740}y'_{t-1} + \underset{(0.320)}{0.000712}y'_{t-2} - \underset{(-1.015)}{0.002472}y'_{t-3} - \underset{(-0.137)}{0.00303}y'_{t-3}$$

$$-\underset{(-0.524)}{0.028481}x'_{t-1} - \underset{(-5.514)}{0.183454}x'_{t-2} + \underset{(0.766)}{0.041586}x'_{t-3} - \underset{(-2.270)}{0.132776}x'_{t-4}$$

The unrestricted estimates show that the mean in *SL* is significantly different from zero and negative, larger in absolute value in state 0 than in state 1. The variance covariance matrix differs between states. The variance of innovations in *SL* is bigger by a factor of 19 in state 0 than in state 1, and the variance of differences in the dividends by a factor of 4. The covariance between innovations is bigger by a factor of 9 in state 0 than in state 1.

On the basis of these estimates, we can calculate the probabilities of having been in each of the two states at any date in the sample, i.e., $\text{prob}(s_t = 1 \mid y_t, y_{t-1}, y_{t-2}, .. x_t, x_{t-2}, ..)$. These are shown in Figure 5. The unrestricted estimates show several regime changes. The economy appears to have been in the low variance regime (state 1) most of the time in the period up to 1914, when the NYSE closed due to the first world war, reopening on 12 December 1914. Within this period there was a shift to the high variance regime (state 0) in 1901, associated with the collapse of the Morgan and Harriman fight for North Pacific (see Schwert (1989)), and in 1908, at the time of a bank panic, when there was a suspension of convertibility of deposits. The period that starts in the first world war is clearly associated with high variances. The filter has clearly identified dramatic crises in 1917-22, 1929-1933, 1936, 1938-1942, 1947-1950, 1958-1960, 1974-1975 and again in 1978-1982. For all these periods the filter sets the probability of being in the high variance-covariance state to 1. These results are broadly the same in this respect as those obtained by Schwert (1989).

Probabilities of State 1 from the Unrestricted VAR

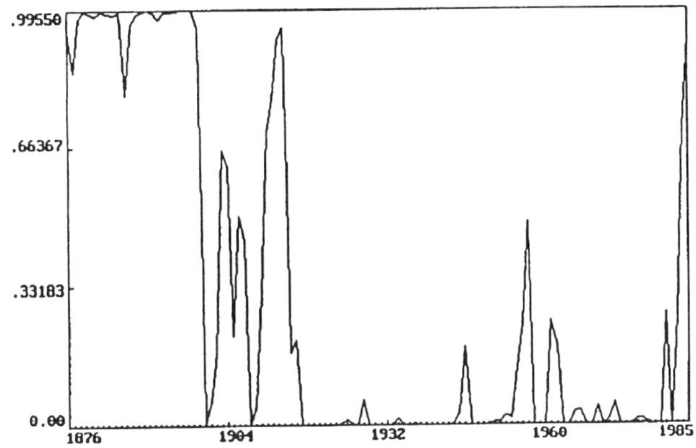

Figure 5

Another issue studied by the literature is the influence of margin requirements on stock volatility (see Hardouvelis (1990) and Schwert (1989)). Our results broadly show that increases in margin requirement follow periods of high volatility and increases in volatility usually took place after reductions in margin requirements.

To test the validity of the efficient market hypothesis we assume that the public forecasts future dividends using a Markov-switching VAR. Under this assumption, the efficient market hypothesis imposes one more cross equation restriction on the coefficients of the VAR than those derived in C&S. The derivation of these restrictions is presented in the Appendix.

The results for the restricted VAR are as follows. The estimated unconditional mean is:

$$\widehat{\text{mean}}_x(s_t) = \underset{(1.148)}{0.000352} - \underset{(-1.856)}{0.000594}\, s_t$$

The estimated covariance matrices for state 0 and state 1 are:

Probabilities of State 1 from the Restricted VAR

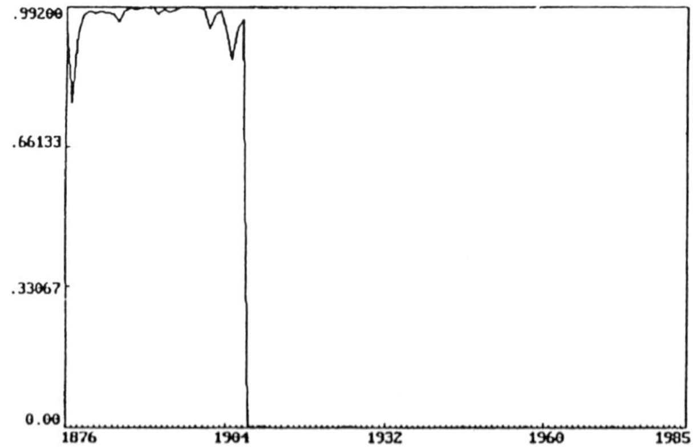

Figure 6

$$\Sigma_0 = \begin{pmatrix} 0.012613 & -0.000113 \\ (4.776) & (-49.553) \\ & 0.000010 \\ & (2.205) \end{pmatrix} \qquad \Sigma_1 = \begin{pmatrix} 0.000947 & -0.000033 \\ (2.706) & (-27.382) \\ & 0.000005 \\ & (2.296) \end{pmatrix}$$

The estimated transition probabilities are:

$$\hat{p} = \underset{(74.976)}{0.967280}$$

$$\hat{q} = \underset{(134.376)}{0.999894}$$

The results of the centered vector autoregressive process are presented below.

$$\hat{x}_t' = \underset{(114.598)}{0.673206}\, y_{t-1}' - \underset{(-5.508)}{0.253990}\, y_{t-2}' + \underset{(5.639)}{0.271771}\, y_{t-3}' - \underset{(-4.857)}{0.173311}\, y_{t-4}'$$

$$-\underset{(-6.783)}{20.026801}\, x_{t-1}' + \underset{(13.588)}{0.562743}\, x_{t-2}' - \underset{(-6.363)}{7.428130}\, x_{t-3}' - \underset{(-9.359)}{1.050802}\, x_{t-4}'$$

The likelihood ratio test statistic for the validity of the restrictions imposed by the efficient market hypothesis on the restricted VAR takes the value 266.335. Since the statistic has an asymptotic $\chi^2(9)$ distribution under the null, the hypothesis is decisively rejected.

Notice that the theoretical framework presented in Sections 3-4 assumes the existence of only two states. If this condition does not hold the results presented above are not valid because there could be structural breaks within each regime. We discuss this possibility and other possibility and other potential problems in the next section.

5. Discussion

In applying the C&S methodology to get a stationary VAR, we are implicitly assuming that there exists a single long-run relationship between prices and dividends. Other authors who have explored the possibility of change in the cointegrating relationship include Granger and Lee (1991) who uses a Kalman filter, effectively assuming that the cointegration vector can take a continuum of values. Hall, *et al.* (1997) consider the possibility of a discrete change in the cointegration vector. More importantly, Driffill and Sola (1998) found that when the public is assumed to observe the state, the theoretical model predicts two long-run solutions (one for each state). Using a shorter sample period (1900-1987) they found clear evidence of a difference in the long-run relation in the two different states. Incorporating this extra feature will make y a function of two possibly independent state variables. Under this assumption y_t would be a function of s_t and s_t', where s_t' would be a second unobserved state variable which would dictate the changes in the long run relationship. Allowing for this feature in the present framework would make the analysis intractable.[4]

To assess how much our results may be affected by the (untested) underlying assumptions of time-invariance of the cointegrating relationship we conduct several parameter constancy tests. To test for constancy in the cointegrating regression between stock prices and dividends, while treating the change point as unknown, we use the sequential *LM*-type tests developed by Hansen (1992a).[5] Table 3 reports the Phillips-Hansen fully modified OLS estimates of the long run parameters, along with tests for parameter constancy. The tests show some weak evidence in favor of structural breaks. We have to bear in mind however that these tests have low power when the breaks occur near the end of the sample, so we should interpret these results cautiously. Further information can be obtained by looking at Figure 7 which shows the rolling estimates of the coefficient of stock prices on real dividends. This coefficient takes values around 25 from the beginning of the sample to 1960, and then

Table 3. Temporal Stability Tests for the Cointegrating Relationship.

Regressor x_t	Deterministic Terms	*supLM*	*meanLM*	L_c
31.981 (3.404)	*INT*	6.96	3.38	0.29
48.991 (5.02)	*INT, TR*	9.14	4.61	0.37*

Notes: An asterisk indicates that the null hypothesis of constancy is rejected at the 10% significance level. Numbers in parentheses are standard errors. *INT* and *TR* denote an intercept and a linear trend, respectively.

Cointegrating Coefficient of Real Dividends and its two*S.E.

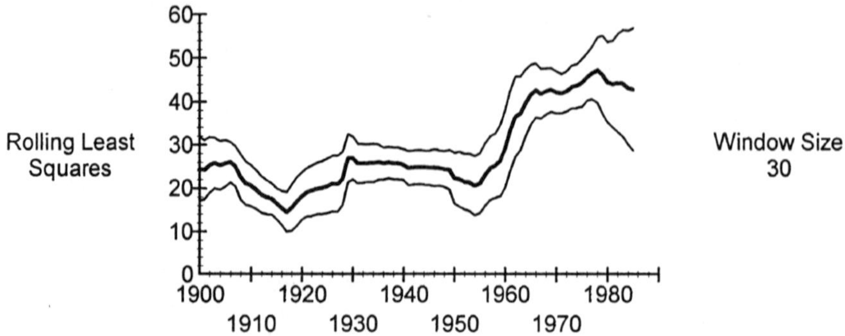

Figure 7

takes values of about 45, suggesting that the cointegrating relationship changed and that the value that we use for the transformation is just an average of these two values.

In a related paper, Blix (1997) extends the analysis by also allowing the slopes to change between states in the Markov-switching VAR. In this case the theoretical restrictions will be somewhat different than the ones derived here. This may lead to the efficient markets hypothesis being rejected in one model and not in another, and the problem this raises is that it is very difficult to determine which is the appropriate VAR. If all VAR parameters are allowed to change between states, then the model will typically be unidentified, as Hamilton (1994) notes. Nevertheless, it is possible that the switching parameters are slopes rather than the unconditional means. Hansen (1992b) has explored this issue for the univariate case. Thus, our view is that this methodology should be applied with caution, preferably in situations in which the theoretical long-run relation cannot switch between states, as in the case of the term structure of interest rates, and also examining the univariate series included in the VAR for switches at similar dates.

6. Conclusions

This paper has shown that when allowance is made for stochastic regime changes, the expectations present value model appears to be inconsistent with US data. We have allowed for non-stationarity in stock prices and dividends, following C&S (1987). In contrast to C&S (1987), however, we are able to reject the present value model. This is due to the fact that their inference comes from a misspecified VAR. We conclude, as does Schwert (1989), that the sample can be divided into periods of high variability and periods of low variability. The fact that we allow agents to predict changes in regime has increased the power of the test. The period from 1871 to 1912 appears to have been broadly characterized by low variabilitiy in both the rate of change of real dividends and real prices, a low mean change in real dividends, and on average low real prices relative to dividends. From time to time there have been periods of high variability associated with banking crises and other external events, which the regime-switching model is able to capture.

However, as we noted in the discussion in Section 5, while allowing for regime-switching strengthens the analysis, there remain some problems associated with Campbell and Shiller's methodology. In particular, the cointegrating relationship between stock prices and dividends has been assumed to remain constant across regimes, whereas it may in fact change, and thus an inappropriate restriction may have been imposed on the data. In principle it is possible to extend the analysis to allow for this, but at the cost of considerable additional complexity.

Appendix: Testing the Present Value Hypothesis From a VAR with Regime Switching

In this appendix we will first derive the restrictions for a present value model when prices and dividends are assumed to be integrated of order 0.[6] The transformations necessary to apply the Campbell and Shiller methodology using the switching VAR are also presented below.

The present value model predicts that the stock price of a share at the beginning of a period t is given by:

$$P_t = \sum_{i=0}^{\infty} \gamma^i \mathrm{E}_t D_{t+i} + \varepsilon_t \tag{A1}$$

where $E_t D_{t+i}$ is the public's forecast of the i-periods dividend that will prevail at time $t+i$ and $\gamma = 1/(1+r)$. The error term is assumed to be white noise and orthogonal to the discounted value of the expected future dividends. This error term can be due to measurement errors.

The econometric problem will be to test a rational expectations model from a Markov-switching regime VAR. To do this, it is convenient to express the regime-shifting VAR for the centered variables in the companion form:

$$Z_t' = \Delta Z_{t-1}' + \xi_t$$

where

$$\Delta = \begin{bmatrix} \Phi_1 & \Phi_2 & \Phi_3 & \Phi_4 & \psi_1 & \psi_2 & \psi_3 & \psi_4 \\ 1 & 0 & 0 & 0 & 0 & 0 & 0 & 0 \\ 0 & 1 & 0 & 0 & 0 & 0 & 0 & 0 \\ 0 & 0 & 1 & 0 & 0 & 0 & 0 & 0 \\ \varphi_1 & \varphi_2 & \varphi_3 & \varphi_4 & \Omega_1 & \Omega_2 & \Omega_3 & \Omega_4 \\ 0 & 0 & 0 & 0 & 1 & 0 & 0 & 0 \\ 0 & 0 & 0 & 0 & 0 & 1 & 0 & 0 \\ 0 & 0 & 0 & 0 & 0 & 0 & 1 & 0 \end{bmatrix} \qquad Z_t' = \begin{bmatrix} P_t - \alpha_0 - \alpha_1 s_t \\ P_{t-1} - \alpha_0 - \alpha_1 s_{t-1} \\ P_{t-2} - \alpha_0 - \alpha_1 s_{t-2} \\ P_{t-3} - \alpha_0 - \alpha_1 s_{t-3} \\ D_t - \beta_0 - \beta_1 s_t \\ D_{t-1} - \beta_0 - \beta_1 s_{t-1} \\ D_{t-2} - \beta_0 - \beta_1 s_{t-2} \\ D_{t-3} - \beta_0 - \beta_1 s_{t-3} \end{bmatrix}$$

$$\xi_t = \begin{bmatrix} (w_0 + w_1 s_t)\upsilon_t \\ 0 \\ 0 \\ 0 \\ (\tau_0 + \tau_1 s_t)\eta_t \\ 0 \\ 0 \\ 0 \end{bmatrix}$$

We proceed as in Hamilton (1988), first considering forecasts under the assumption that the information set includes observation of the regime directly. This information set can be written as:

$$I_t^* = \{P_t, P_{t-1}, \ldots, D_t, D_{t-1}, \ldots, s_t, s_{t-1}, \ldots\}.$$

This assumption will be relaxed below. Using this notation we can express the expected *j+1* periods ahead dividend conditional on information possessed by the agents at time *t-1* as:

$$\mathrm{E}(D'_{t+j}|I^*_{t-1}) = [0\,0\,0\,0\,1\,0\,0\,0]\Delta^{j+1}Z'_{t-1}.$$

The expected *j+1* periods ahead dividends can be written as:

$$\mathrm{E}(D_{t+j}|I_{t-1}) = \beta_0 + \beta_1\mathrm{E}(s_{t+j}|I^*_{t-1}) + \mathrm{E}(D'_{t+j}|I^*_{t-1}),$$

and by further substitution

$$\mathrm{E}(D_{t+j}|I^*_{t-1}) = \beta_0 + \beta_1\mathrm{E}(s_{t+j}|I^*_{t-1}) + [0\,0\,0\,0\,1\,0\,0\,0]\Delta^{j+1}Z'_{t-1}.$$

Substituting the expected value of a *j+1* periods ahead two state Markov process, the following expression is obtained:

$$\mathrm{E}(D_{t+J}|I^*_{t-1}) + \beta_0 + \beta_1[\rho + (s_{t-1} - \rho)\lambda^{j+1}] + [0\,0\,0\,0\,1\,0\,0\,0]\Delta^{j+1}Z'_{t-1}, \qquad \text{(A2)}$$

where $\rho = (1-q)/(2-p-q)$ and $\lambda = (p+q-1)$, where p and q are the parameters of the transition matrix as defined in Section 3.

A similar formula can be derived for stock prices:

$$\mathrm{E}(P_t|I^*_{t-1}) = \alpha_0 + \alpha_1\mathrm{E}(s_t|I^*_{t-1}) + \mathrm{E}(P'_t|I^*_{t-1}),$$

and by further substitution,

$$\mathrm{E}(P_t|I_{t-1}) = \alpha_0 + \alpha_1\mathrm{E}(s_t|I^*_{t-1}) + [1\,0\,0\,0\,0\,0\,0\,0]\Delta Z'_{t-1}.$$

Substituting the expected value of a 1 period ahead two-state Markov process, the following expression for the expected short rate is obtained:

$$E(P_t|I^*_{t-1}) = \alpha_0 + \alpha_1[\rho + (s_{t-1} - \rho)\lambda] + [1\,0\,0\,0\,0\,0\,0\,0]\Delta Z'_{t-1}. \tag{A3}$$

If we condition on information available at time *t-1* in both sides of equation (A1) we obtain[7]

$$E_{t-1}P_t = \sum_{i=0}^{\infty} \gamma^i E_{t-1}D_{t+i}. \tag{A4}$$

When substituting equations (A2) and (A3) into equation (A4) we obtain

$$\alpha_0 + \alpha_1[\rho + (s_{t-1} - \rho)\lambda] + [1\,0\,0\,0\,0\,0\,0\,0]\Delta Z'_{t-1}$$

$$= \left[\sum_{i=0}^{\infty} \gamma^i \left\{\beta_0 + \beta_1[\rho + (s_{t-1} - \rho)\lambda^{i+1}] + [0\,0\,0\,0\,1\,0\,0\,0]\Delta^{i+1}Z'_{t-1}\right\}\right],$$

which can be simplified to

$$\alpha_0 + \alpha_1[\rho + (s_{t-1} - \rho)\lambda] + [1\,0\,0\,0\,0\,0\,0\,0]\Delta Z'_{t-1}$$

$$= \frac{-\lambda\rho\beta_1}{1-\gamma\lambda} + \frac{(\beta_0 + \beta_1\rho)}{1-\gamma} + \frac{\lambda\beta_1}{1-\gamma\lambda}s_{t-1} + [0\,0\,0\,0\,1\,0\,0\,0]\Delta[I-\gamma\Delta]^{-1}Z'_{t-1}. \tag{A5}$$

Equation (A5) is linear in $s_{t-1}, s_{t-2}, s_{t-3}, s_{t-4}$. This means that to calculate the forecasts given information on I_{t-1}, we only have to replace $s_{t-1}, s_{t-2}, s_{t-3}, s_{t-4}$ in equation (A5) with their conditional expectation given I_{t-1}, where $I_t = \{P_t, P_{t-1}, ..., D_t, D_{t-1}, ...,\}$.

The restrictions on the VAR follow immediately from equating $[1\,0\,0\,0\,0\,0\,0\,0]\Delta = [0\,0\,0\,0\,1\,0\,0\,0]\Delta[I-\gamma\Delta]^{-1}$, and by equating the constants and coefficients of s_{t-1} of each side of equation (A5).

Notice that the restrictions can easily be derived from the equivalent equality:

$$[1\,0\,0\,0\,0\,0\,0\,0]\Delta[I-\gamma\Delta] = [0\,0\,0\,0\,1\,0\,0\,0]\Delta.$$

This restriction has the advantage that it only involves terms in Δ and Δ^2, where

$$\Delta^2 = \left[\begin{array}{c|c} A & B \\ \hline C & D \end{array}\right]$$

$$A = \begin{bmatrix} \Phi_1^2 + \Phi_2 + \varphi_1\psi_1 & \Phi_1\Phi_2 + \Phi_3 + \varphi_2\psi_1 & \Phi_1\Phi_3 + \Phi_4 + \varphi_3\psi_1 & \Phi_1\Phi_4 + \varphi_4\psi_1 \\ \Phi_1 & \Phi_2 & \Phi_3 & \Phi_4 \\ 1 & 0 & 0 & 0 \\ 0 & 1 & 0 & 0 \end{bmatrix}$$

$$B = \begin{bmatrix} \psi_1\Phi_1 + \psi_2 + \psi_1\Omega_1 & \psi_1\Omega_2 + \psi_3 + \psi_2\Phi_1 & \Phi_1\psi_3 + \psi_4 + \Omega_3\psi_1 & \Phi_1\psi_4 + \Omega_4\psi_1 \\ \psi_1 & \psi_2 & \psi_3 & \psi_4 \\ 0 & 0 & 0 & 0 \\ 0 & 0 & 0 & 0 \end{bmatrix}$$

$$C = \begin{bmatrix} \varphi_1\Phi_1 + \varphi_2 + \varphi_1\Omega_1 & \varphi_2\Omega_1 + \varphi_3 + \varphi_1\Phi_2 & \Phi_3\varphi_1 + \varphi_4 + \Omega_1\varphi_3 & \Phi_4\varphi_1 + \Omega_1\varphi_4 \\ \varphi_1 & \varphi_2 & \varphi_3 & \varphi_4 \\ 0 & 0 & 0 & 0 \\ 0 & 0 & 0 & 0 \end{bmatrix}$$

$$D = \begin{bmatrix} \Omega_1^2 + \Omega_2 + \Phi_1\psi_1 & \Omega_1\Omega_2 + \Omega_3 + \varphi_1\psi_2 & \Omega_1\Omega_3 + \Omega_4 + \varphi_1\psi_3 & \Omega_1\Omega_4 + \varphi_1\psi_4 \\ \Omega_1 & \Omega_2 & \Omega_3 & \Omega_4 \\ 1 & 0 & 0 & 0 \\ 0 & 1 & 0 & 0 \end{bmatrix}$$

The restrictions written in full are:

$$\varphi_1 = \frac{\Phi_1(1-\gamma\Phi_1)-\gamma\Phi_2}{(1+\gamma\psi_1)} \qquad \Omega = \frac{\psi_1(1-\gamma\Phi_1)-\gamma\psi_2}{1+\gamma\psi_1)}$$

$$\varphi_2 = \frac{\Phi_2(1-\gamma\Phi_1)-\gamma\Phi_3}{(1+\gamma\psi_1)} \qquad \Omega_2 = \frac{\psi_2(1-\gamma\Phi_1)-\gamma\psi_3}{(1+\gamma\psi_1)}$$

$$\varphi_3 = \frac{\Phi_3(1-\gamma\Phi_1)-\gamma\Phi_4}{(1+\gamma\psi_1)} \qquad \Omega_3 = \frac{\psi_3(1-\gamma\Phi_1)-\gamma\psi_4}{(1+\gamma\psi_1)} \tag{A6}$$

$$\varphi_4 = \frac{\Phi_4(1-\gamma\Phi_1)}{(1+\gamma\psi_1)} \qquad \Omega_4 = \frac{\psi_4(1-\gamma\Phi_1)}{(1+\gamma\psi_1)}$$

$$\alpha_1 = \frac{\beta_1}{1-\gamma\lambda}$$

When the series included in the VAR are integrated of order one, a simple transformation of this procedure is needed. We rewrite equation (A1) as:

$$P_t = \theta(1-\gamma)\sum_{i=0}^{\infty}\gamma^i \mathrm{E}_t D_{t+i} + \varepsilon_t,$$

where θ is a factor of proportionality and γ is the discounting factor.

Subtracting from both sides of the equation θD_t and rearranging terms we obtain:

$$S_t = P_t - \theta D_t = \theta \sum_{i=1}^{\infty}\gamma^i \mathrm{E}_t \Delta D_{t+i} + \varepsilon_t. \tag{A7}$$

This transformation provides a representation of the present value relationship in terms of integrated of order zero variables if the vector $(1, -\theta)$ cointegrates with the vector (P, D). Nevertheless, as was pointed out in the Introduction, the fact that stocks and dividends are not measured contemporaneously can lead to spurious rejection if we impose the restrictions implied by equation (A7) to a VAR of S and ΔD.

C&S propose the following transformation to equation (A7) to overcome the above mentioned problem:

$$SL_t = P_t - \theta D_{t-1} = \theta \Delta D_t + \theta \sum_{i=1}^{\infty} \gamma^i E_t \Delta D_{t+i} + \varepsilon_t, \quad \text{or}$$

$$y_t = \theta x_t + \theta \sum_{i=1}^{\infty} \gamma^i E_t x_{t+i} = \theta \sum_{i=0}^{\infty} \gamma^i E_t x_{t+i} + \varepsilon_t. \tag{A8}$$

The restrictions implied by equation (A8) can be derived in a similar fashion as in the present value model and an endogenous switching regime VAR. If we redefine Z as a vector which contains the current and three first lags of x' and y', x' and y' being centered x and y respectively, taking expectations conditional on information at time t-1 on both sides of equation (A8) we obtain:

$$\varphi_1 = \frac{\Phi_1(1-\gamma\Phi_1)-\gamma\Phi_2}{\gamma\psi_1+\theta} \qquad \Omega_1 = \frac{\psi_1(1-\gamma\Phi_1)-\gamma\psi_2}{\gamma\psi_1+\theta}$$

$$\varphi_2 = \frac{\Phi_2(1-\gamma\Phi_1)-\gamma\Phi_3}{\gamma\psi_1+\theta} \qquad \Omega_2 = \frac{\psi_2(1-\gamma\Phi_1)-\gamma\psi_3}{\gamma\psi_1+\theta}$$

$$\varphi_3 = \frac{\Phi_3(1-\gamma\Phi_1)-\gamma\Phi_4}{\gamma\psi_1+\theta} \qquad \Omega_3 = \frac{\psi_3(1-\gamma\Phi_1)-\gamma\psi_4}{\gamma\psi_1+\theta} \tag{A9}$$

$$\varphi_4 = \frac{\Phi_4(1-\gamma\Phi_1)}{\gamma\psi_1+\theta} \qquad \Omega_4 = \frac{\psi_4(1-\gamma\Phi_1)}{\gamma\psi_1+\theta}$$

$$\alpha_1 = \frac{\beta_1\theta}{1-\gamma\lambda}$$

Notes

1. Z_a and Z_t were computed using a Parzen kernel, an autoregressive prewhitening procedure, and automatic data-dependent bandwidth selection (see Andrews and Monahan (1992)).
2. By allowing this constant to switch in the VAR we can account for this feature of the data.
3. Numbers in parenthesis are t-values.
4. There is no intrinsic problem in allowing for four states (see Ravn and Sola (1995)). The main problem of this extension under our specification is that it is very likely that the model will not be identified.
5. This entails computing the *LM* test statistic for a subset of possible change points, and then basing a test on the maximum of these statistics (denoted here by *supLM*). The average *LM* (*meanLM*) and the L_c are alternative tests which are most powerful against alternatives involving parameter variation of a martingale form.

6. See for example Timmermann (1994), who first detrends prices and dividends by subtracting trend and then conducts his analysis using the detrended variables.
7. The presence of the error term in equation (A1) ensures that this procedure is valid. If the present value relationship is assumed to hold without this term, further restrictions should be imposed in the variance-covariance matrix.

References

Andrews, D. W. K. and J. C. Monahan (1992). "An Improved Heteroskedasticity and Autocorrelation Consistent Covariance Matrix Estimator," *Econometrica*, 60, 953-966.

Blix, M. (1997). "Rational Expectations in a VAR with Markov Switching," seminar paper No. 633, Institute of International Economic Studies, University of Stockholm.

Campbell, J. Y. and R. J. Shiller (1987). "Cointegration and Tests of Present Value Models," *Journal of Political Economy*, 95, 1062-1088.

Driffill, J. and M. Sola (1998). "Intrinsic Bubbles and Regime Switching," working paper, Department of Economics, University of Southampton. Forthcoming in *Journal of Monetary Economics.*

Engle, R. F. and C. W. J. Granger (1987). "Co-integration and Error Correction: Representation, Estimation and Testing," *Econometrica*, 55, 251-276. Reprinted in Engle, R. F. and C. W. J. Granger (eds.) (1991). *Long-Run Economic Relationships: Readings in Cointegration*. Oxford: Oxford Press.

Granger, C. W. J. and H. S. Lee (1991). "An Introduction to Time-Varying Parameter Cointegration, in Hack, P. and A. H. Westlund (eds.), *Economic Structural Change: Analysis and Forecasting*. Berlin: Springer-Verlag.

Hall, S., Z. Psaradakis, and M. Sola (1997). "Cointegration and Changes in Regime: The Japanese Consumption Function," *Journal of Applied Econometrics*, 12, 151-168.

Hamilton, J. D. (1988). "Rational Expectations econometric Analysis of Changes in Regime. An Investigation of the Term Structure of Interest Rates," *Journal of Economic Dynamics and Control*, 12, 385-423.

Hamilton, J. D. (1989). "A New Approach to the Economic Analysis of Nonstationary Time Series and the Business Cycle," *Econometrica*, 57, 357-384.

Hamilton, J. D. (1990). "Analysis of Time Series Subject to Changes in Regime," *Journal of Econometrics*, 45, 39-70.

Hamilton, J. D. (1994). *Time Series Analysis*. Princeton: Princeton University Press.

Hansen, B. E. (1992a). "Tests for Parameter Instability in Regressions with I(1) Processes," *Journal of Business & Economic Statistics*, 10, 321-335.

Hansen, B. E. (1992b). "The Likelihood Ratio Test Under Nonstandard Conditions: Testing the Markov Switching Model of GNP," *Journal of Applied Econometrics*, 7, S61-S82.

Hardouvelis, G. A. (1990). "Margin Requirements, Volatility, and the Transitory Component of Stock Prices," *American Economic Review* 80, 736-762.

Johansen, S. (1991). "Estimation and Hypothesis Testing of Cointegration Vectors in Gaussian Vector Autoregressive Models," *Econometrica*, 59, 1551-1580.

Kleidon, A. W. (1986). "Variance Bounds Tests and Stock Price Valuation Models," *Journal of Political Economy*, 94, 953-1001.

LeRoy, S. and R. Porter (1981). "The Present Value Relation: Tests Based on Variance Bounds," *Econometrica*, 49, 555-577.

Marsh, T. A. and R. C. Merton (1986). "Dividend Variability and Variance Bounds Tests for the Rationality of Stock Market Prices," *American Economic Review*, 76, 483-498.

Ng, S. and P. Perron (1995). "Unit Root Tests in ARMA Models with Data-Dependent Methods for the Selection of the Truncation Lag," *Journal of the American Statistical Association*, 90, 268-281.

Perron, P. (1989). "The Great Crash, the Oil Price Shock, and the Unit Root Hypothesis," *Econometrica*, 57, 1361-1401.

Perron, P. (1990). "Testing for a Unit Root in a Time Series with a Changing Mean," *Journal of Business and Economic Statistics*, 8, 153-162.

Phillips, P. C. B. (1987). "Time Series Regression with a Unit Root Hypothesis," *Econometrica*, 55, 277-301.

Phillips, P. C. B. (1991). "Optimal Inference in Cointegrated Systems," *Econometrica*, 59, 283-306.

Phillips, P. C. B. and B. E. Hansen (1990). "Statistical Inference in Instrumental Variables Regression with I(1) Variables," *Review of Economic Studies*, 57, 99-125.

Phillips, P. C. B. and S. Ouliaris (1990). "Asymptotic Properties of Residual Based Tests for Cointegration," *Econometrica*, 58, 165-193.

Phillips, P. C. B. and P. Perron (1988). "Testing for a Unit Root in Time Series Regression," *Biometrika*, 75, 335-346.

Ravn, M. O. and M. Sola (1995). "Stylized Facts and Regime Changes - Are Prices Procyclical?" *Journal of Monetary Economics*, 36, 497-526.

Said, S. E. and D. A. Dickey (1984). "Testing for Unit Roots in Autoregressive-Moving Average Models of Unknown Order," *Biometrika*, 71, 599-607.

Schwert, G. W. (1989). "Business Cycles, Financial Crises, and Stock Volatility," *Carnegie-Rochester Conference Series on Public Policy*, 31, 83-126.

Shiller, R. J. (1981). "Do Stock Prices Move Too Much to be Justified by Subsequent Changes in Dividends?" *American Economic Review*, 71, 421-436.

Shiller, R. (1989). *Market Volatility*. Cambridge: MIT Press.

Sola, M. and Driffill, J. (1994). "Testing the Term Structure of Interest Rates Using a Stationary Vector Autoregression with Regime Switching," *Journal of Economic Dynamics and Control*, 18, 601-628.

Timmermann, A. (1994). "Present Value Models with Feedback: Solutions, Stability, Bubbles, and Some Empirical Evidence," *Journal of Economic Dynamics and Control*, 18, 1093-1119.

West, K. D. (1988). "Dividend Innovations and Stock Price Volatility," *Econometrica*, 56, 37-61.

White, H. (1987). "Specification Testing in Dynamic Models," in Truman F. Bewley (ed.), *Advances in Econometrics, Fifth World Congress*, Vol. I. Cambridge: Cambridge University Press.

12 BUSINESS CYCLE DYNAMICS: PREDICTING TRANSITIONS WITH MACROVARIABLES

Morten O. Ravn
Universitat Pompeu Fabra
CEPR
University of Aarhus
Martin Sola
University of London
Universidad Torcuato Di Tella
London Business School

1. Introduction

This chapter investigates the dynamics of the business cycle and considers how various macroeconomic variables can predict transitions between periods of high and low activity in aggregate output. To perform this study, we model (the change in) aggregate activity as a two-state Markov process and investigate whether or not macroeconomic variables affect the stochastic process that generates output. Specifically, we study whether macroeconomic variables affect the probability at which the economy switches between "booms" and "recessions."

There is a substantial literature that has studied business cycle fluctuations using two-state Markov process frameworks. In the seminal contribution of Hamilton (1989), aggregate output was modelled as a stochastic process in which the unconditional mean is assumed to be governed by a discrete unobserved state variable generated by a homogeneous Markov process. Hamilton's main motivation for using this model was not only to capture correctly the nonlinear features of the data, but also to propose it as an alternative to the *NBER* method of classifying booms and recessions. Two main extensions to the basic Hamilton model have been proposed in the literature so as to develop better models for separating booms and recessions. One extension adds more structure to the models (e.g., Diebold and Rudebusch (1996)), while the other endogenizes the transition probabilities (e.g., Diebold, *et al.* (1994)). Subsequent research has obtained promising results by extending the Hamilton (1989) model to time-varying transition probabilities. Several authors have claimed that models with time-varying transition probabilities can successfully characterize the dynamics of the business cycle. This methodology has been used to study the relationship between seasonality and turning points (e.g., Ghysels (1994)), the duration dependence of the cycle (e.g., Durland and McCurdy (1994)), and the phases of the business cycle and the transitional dynamics (e.g., Filardo (1994)).

We apply the time-varying transition probability framework in order to study the following issues. First, we look at potential "asymmetric effects" that different macroeconomic policy variables may have on the probability of staying in a boom or a recession. An "asymmetric effect" occurs in this framework if a given variable affects the two transition probabilities differently. For example, it might be the case that a given variable significantly affects only one of the transition probabilities.

We also attempt to understand the extent to which the results can yield any causal interpretation. The issue at hand is whether finding that a variable has predictive power for the transition probabilities can be interpreted as implying that movements in this variable have led to transitions in aggregate activity between booms and recessions. This issue is clearly of great importance for economic policy. The variables that we, and others before us, include in the modelling of the transition probabilities can be broadly divided into "policy variables" and "lagged endogenous variables." The interpretation of the results differs depending on which of these categories the variable under scrutiny falls into. Finding that an endogenous variable such as the inflation rate is useful for predicting aggregate activity, because it enters significantly into the transition probability, should not be interpreted as if the variations in this variable have led to changes in regime. Only if a variable is exogenous could one make such a policy interpretation. Obviously, this possible interpretation might be problematic since we face the same interpretation problems as those in Granger causality testing using linear models. Nevertheless, in our opinion this analysis may yield important insights into the nature of transmission mechanisms.

A second issue that we address is the structural stability of the parameters of the time-varying transition probability. There are good reasons to believe that the type of nonlinear relationships between output dynamics and macroeconomic variables may well change over time due, for example, to changes in the conduct of monetary policy. This issue is important because estimating the model with time-varying probabilities for a sample period in which there have been structural changes may lead to misleading results.

In our application we look at quarterly postwar data for the US and use the changes in industrial production as our measure of aggregate activity. The starting point of the analysis is the discrete state changes-in-regime approach of Hamilton (1989), with a minor extension to allow for fatter tails in the distribution. Following Diebold, *et al.* (1994), the basic Hamilton (1989) model is extended to allow for time-varying state transition probabilities. Notice that the accuracy of the separation between regimes is a prerequisite to any analysis that searches for potential effects that macroeconomic variables may have on the business cycle. In this respect we obtain satisfactory results, because the separation of booms and recessions obtained by the filter are most close to the business cycle turning points identified by the *NBER*.

Nevertheless, we find that the use of time-varying transition probabilities is not exempt from potential problems in our application. We find that the relationship between output growth and many of the variables included in the transition probability is structurally unstable. In particular, the stochastic process which governs some of these variables seems to have changed in the late 1970s when the Federal Reserve adopted a new policy. In our model the states are associated with booms and recessions and we do not account for any independent break in the variables that enter in the transition probabilities. The relationship between output and variables such as interest rates or money expansion may change when there is a change in policy, and in such cases the parameters of the time-varying transition probabilities may also change. This issue is discussed in detail both in Section 2.1 and in the empirical application, where we find that not allowing for these changes in the parameters of the transition probabilities causes a misclassification of the states in the last part of the sample in some cases. To account for this we also consider a sub-sample which ends in the third quarter of 1978.

The remainder of the paper proceeds as follows. Section 2 describes the statistical models used to characterize the dynamics of industrial production and related inference methods. Section 3 discusses the macroeconomic variables considered in the analysis, and their relation to the theoretical literature. Section 4 presents and discusses the estimation results of models with both constant and time-varying transition mechanisms, focusing on the information content of the various macrovariables. Section 5 summarizes and concludes.

2. Econometric Models and Inference

As a first step in the analysis, the dynamics of the variable of interest (denoted below by y_t) are described using the following autoregressive specification:

$$y_t - \mu(s_t) = \sum_{j=1}^{m} \varphi_j [y_{t-j} - \mu(s_{t-j})] + \sigma\varepsilon_t, \quad (t = 1,...,T) \tag{1}$$

In (1), $s_t \in \{0,1\}$ is a random variable that indicates the unobserved 'state' or 'regime' operative at time t, $\{\varepsilon_t\}$ is a sequence of independently and identically distributed random variables with zero mean and unit variance, $\mu(\cdot)$ represents a mean shift function, and σ is the standard deviation. Following Hamilton (1989, 1990), the regimes $\{s_t\}$ are modelled as the outcome of a homogeneous, first-order Markov chain with transition probabilities

$$\begin{aligned} \Pr(s_t = 1 | s_{t-1} = 1) &= p, \\ \Pr(s_t = 0 | s_{t-1} = 1) &= 1 - p \\ \Pr(s_t = 0 | s_{t-1} = 0) &= q \\ \Pr(s_t = 1 | s_{t-1} = 0) &= 1 - q \end{aligned} \tag{2}$$

where p and q are unknown. The state-dependent mean function is parameterized linearly as

$$\mu(s_t) = \alpha_0 + \alpha_1 s_t.$$

We further specify that ε_t is independent of s_{t+j} for all j, and that all the roots of $1 - \sum_{j=1}^{m} \varphi_j z^j = 0$ lies outside the unit circle.

Model (1) extends the standard linear autoregressive specification by allowing the mean around which y_t clusters to be a function of the (stochastically chosen) regime that controls the process at time t. However, for the purpose of our analysis a potential limitation of such a formulation is that the transition probabilities are constant over time. Thus, by assumption, the persistence of each regime is independent of economic variables that one might think affect the state of the business cycle. It may, of course, be the case that the switches in the state variable, s_t, are related to macroeconomic variables, but such interdependencies are not modelled above.[1]

To assess the potential links between booms-recessions and macroeconomic variables, we generalize the model in (1)-(2) by allowing the transition probabilities to vary over time. In particular, following Diebold, *et al.* (1994), we specify the transition mechanism governing $\{s_t\}$ as

$$\begin{aligned} \Pr(s_t = 1 | s_{t-1} = 1, z_{t-1}) &= p_t = \exp(c_1 + \beta_1 z_{t-1})/[1 + \exp(c_1 + \beta_1 z_{t-1})], \\ \Pr(s_t = 0 | s_{t-1} = 0, z_{t-1}) &= q_t = \exp(c_0 + \beta_0 z_{t-1})/[1 + \exp(c_0 + \beta_0 z_{t-1})], \\ \Pr(s_t = 0 | s_{t-1} = 1, z_{t-1}) &= 1 - \exp(c_1 + \beta_1 z_{t-1})/[1 + \exp(c_1 + \beta_1 z_{t-1})], \\ \Pr(s_t = 1 | s_{t-1} = 0, z_{t-1}) &= 1 - \exp(c_0 + \beta_0 z_{t-1})/[1 + \exp(c_0 + \beta_0 z_{t-1})], \end{aligned} \tag{3}$$

where z_t is an economic variable that affects the state transition probabilities.

Under this formulation z_t influences the *probability* with which y_t switches from one state to the other. This allows for nonlinear relationships between y_t and z_t, since z_t affects y_t indirectly through the transition probabilities p_t or q_t. Notice that the effect of z_t on the transition probabilities does not need to be symmetric and it may only be present in one of them.[2]

It is, however, important to stress that one has to be somewhat careful about the causality interpretation of the results. Finding that some variable enters significantly into (3) does not mean that it is causal for y_t since z_t may be an endogenous variable and, hence, without further analysis one cannot interpret the results in terms of causality. Nevertheless, a variable that enters significantly into (3) helps to predict, i.e., it nonlinearly Granger causes y_t, and thus may help to obtain better estimates of the two regimes and their timing.

Notice that since β_1 and dp_t/dz_{t-1} have the same sign, $\beta_1>0$ ($\beta_1<0$) implies that an increase in z_{t-1} increases (decreases) the probability of remaining in state 1. Similarly, $\beta_0>0$ ($\beta_0<0$) implies that an increase in z_{t-1} increases (decreases) the probability of remaining in state 0. It is also worth noting that the model in (1)-(2) is nested within the more general formulation defined by (1) and (3); a model with a time-invariant transition mechanism can be obtained from (1)-(3) by imposing the restriction that $\beta_0=\beta_1=0$ (and these restrictions can be tested using likelihood ratio tests).

Statistical inference in the context of Markov-switching models like those described above can proceed by making use of a nonlinear algorithm similar to that described in Hamilton (1994, pp. 692-694). This gives as a by-product the sample log-likelihood function

$$\log \mathcal{L} = \sum_{t=m+1}^{T} \log f(y_t|\mathcal{F}_{t-1}),$$

where $\mathcal{F}_i=(y_1,z_1,...,y_i,z_i)$ $(i\geq 1)$, and $f(y_t|\mathcal{F}_{t-1})$ represents the conditional density of y_t given the set of information available at time t-1. The maximum likelihood estimates may then be found by numerically maximizing (4) with respect to the underlying unknown parameters. Furthermore, since another by-product of the filter is the joint conditional probability $\Pr(s_t,s_{t-1},...,s_{t-m}|\mathcal{F}_t)$, inferences about the unobserved regimes $\{s_t\}$ may be made on the basis of the filter probabilities $\Pr(s_t|\mathcal{F}_t)$ obtained as:

$$\Pr(s_t|\mathcal{F}_t) = \sum_{s_{t-1}=0}^{1} \sum_{s_{t-2}=0}^{1} \cdots \sum_{s_{t-m}=0}^{1} \Pr(s_t,s_{t-1},...,s_{t-m}|\mathcal{F}_t).$$

Finally, note that, since standardized residuals from Gaussian models are typically found to be leptokurtic, in the sequel we analyze models where ε_t in (1) is treated as drawn from a (standardized) Student-t distribution with $2+h$ $(h\geq 0)$ degrees of freedom. The parameter h is estimated by maximum likelihood along with the other parameters in (1), (2), and (3).

2.1 Breaks in the Parameters of the Time-Varying Transition Probabilities

The purpose of this section is to highlight the potential estimation and inference problems which may arise when the relationship between y_t and z_t is subject to changes. Since the model presented below is not intended to be an extension of the time-varying transition probabilities model, we present a simple example to illustrate the potential damaging effects that non-modeled breaks in the parameters of p_t and q_t may have in the separation of the states.

To illustrate our point we will assume the following parameterization of the breaks

$$\beta_0(x_t) = \beta_0^0(1-x_t) + \beta_0^1 x_t, \tag{5}$$

$$\beta_1(x_t) = \beta_1^0(1-x_t) + \beta_1^1 x_t, \tag{6}$$

where $x_t \in \{0,1\}$ follows a homogeneous first order Markov chain and x_t is independent of s_{t+j} for all j.

Then the transition probabilities p_t and q_t will be

$$\begin{aligned} p_t &= \frac{\exp(c_1 + (\beta_1^0(1-x_t) + \beta_1^1 x_t) z_{t-1})}{1 + \exp(c_1 + (\beta_1^0(1-x_t) + \beta_1^1 x_t) z_{t-1})} \\ q_t &= \frac{\exp(c_0 + (\beta_0^0(1-x_t) + \beta_0^1 x_t) z_{t-1})}{1 + \exp(c_0 + (\beta_0^0(1-x_t) + \beta_0^1 x_t) z_{t-1})} \end{aligned} \tag{7}$$

Assume that $\beta_0 \in \{\beta_0^0, \beta_0^1\}$ and $\beta_1 \in \{\beta_1^0, \beta_1^1\}$.

When we do not allow for switches in x_t the transitions probabilities are

$$\begin{aligned} p_t &= \exp(c_1 + \beta_1 z_{t-1})/[1 + \exp(c_1 + \beta_1 z_{t-1})], \\ q_t &= \exp(c_0 + \beta_0 z_{t-1})/[1 + \exp(c_0 + \beta_0 z_{t-1})], \end{aligned} \tag{8}$$

while the correct transition probabilities are

$$\left.\begin{aligned} p_t(0) &= \exp(c_1 + \beta_1^0 z_{t-1})/[1 + \exp(c_1 + \beta_1^0 z_{t-1})] \\ q_t(0) &= \exp(c_0 + \beta_0^0 z_{t-1})/[1 + \exp(c_0 + \beta_0^0 z_{t-1})] \end{aligned}\right. \quad \text{when } x_t = 0 \tag{9}$$

and

$$\left.\begin{aligned} p_t(1) &= \exp(c_1 + \beta_1^1 z_{t-1})/[1 + \exp(c_1 + \beta_1^1 z_{t-1})] \\ q_t(1) &= \exp(c_0 + \beta_0^1 z_{t-1})/[1 + \exp(c_0 + \beta_0^1 z_{t-1})]. \end{aligned}\right. \quad \text{when } x_t = 1 \tag{10}$$

Under this scenario p_t and q_t are biased and therefore may adversely affect the inferences of the state, s_t, of the business cycle.

Notice that allowing for time-varying transition probabilities seemed to be an innocuous extension to the constant probabilities model. As mentioned above the time-varying probability model embeds the model with constant probability and the filter should return estimated of β_0 and β_1 which are not significantly different from zero whenever z_t is not useful to predict the state of the business cycle.

It should be apparent at this stage that this is not the only possibility, since z_t may be useful to predict the states of the business cycle and indeed greatly improve the fit for a part of the sample, for example when $x_t = 0$. However, it might not be a useful predictor of the states for other parts of the sample, for example when $x_t = 1$. Then, not allowing for this may produce totally inadequate inferences of the states for the periods when there is a break in the relationship between y_t and z_t, but provide estimates of β_0 and β_1 significantly different from zero. In Section 5 we report several examples in which this happens.

3. The Data

In this chapter we investigate whether there are nonlinear relationships between macroeconomic activity and relevant macroeconomic variables. We study postwar quarterly data for the United States for the period 1959.4-1995.4, using seasonally adjusted industrial production as a measure of aggregate activity.

In order to assess whether there are nonlinear relationships between these variables and aggregate activity, we use a number of different macroeconomic variables that one might think of as representing policy variables. Our list of variables includes real government expenditure (as a measure of fiscal policy) and several monetary measures. We experiment with using the monetary base, *M1*, *M2*, non-borrowed reserves (*NBR*), and the federal funds rate (*FFR*). The last of these variables has lately become popular as a measure of the stance of monetary policy (see, e.g., McCallum (1983), Sims (1986), or Christiano, *et al.* (1996)) and changes in non-borrowed reserves has been put forward as a measure of the exogenous changes in monetary policy (see, e.g., Chari, *et al.* (1995)).

It is probably important to recognize that some of the movements in these variables may reflect endogenous policy responses rather than exogenous changes in economic policy. The Fed may, for example, react to changes in money demand by changing the money supply, meaning that changes in the money stock measures may reflect, as least partially, endogenous responses of the Fed. The same may be the case with the federal funds rate, and it has indeed been suggested that the "price puzzle," i.e., the fact that the price level tends to rise after a contraction in monetary policy (as traditionally measured), can be accounted for by recognizing that some of the movements in this variable reflect endogenous policy responses. In fact, Sims and Zha (1994) and Christiano, *et al.* (1996) show that the "price puzzle" can be resolved if the federal funds rate is allowed to react to changes in an index of sensitive commodity prices. Similarly, it is not uncontroversial to argue that non-borrowed reserves reflect endogenous changes in monetary policy. Coleman, *et al.* (1996) argue that non-borrowed reserves are endogenous because they react (through changes in borrowed reserves) to changes in the spread between the discount window and the federal funds rate. Neither is it, of course, hard to think of reasons why parts of government spending are endogenous.

In some studies the problems of interpretation associated with potential endogeneity of the variables under scrutiny are "resolved" by controlling for the set of variables which might cause responses by the Fed. Christiano, *et al.* (1996), for example, measure exogenous shocks to monetary policy by the (smoothed) innovations to either the federal funds rate or non-borrowed reserves from a vector autoregression including variables such as GDP, the GDP deflator, and an index of sensitive commodity prices. In fact, Christiano, *et al.* (1997) suggest a general procedure with which the exogenous parts of such policy variables can be recovered from VARs that allow for endogenous responses of policy variables.

Rather than proceeding along these lines, we use an alternative strategy of checking the plausibility of the assumption that movements in a certain variable might be interpreted as exogenous. To get an idea of the possible transmission mechanism between the variables under scrutiny, we look at bivariate causality tests, which are useful for interpreting the results of the switching model.

Apart from the above mentioned variables, we also look at another set of variables that can be characterized as lagged endogenous and thus should be given no policy interpretation. These variables are the inflation rate (*INF*), the real interest rate (*RFFR*), and the real money stock. These variables are analyzed in order to see whether they help in estimating the two regimes of output growth, and to investigate possible transmission mechanisms.

4. Basic Empirical Findings and Discussion

We start by discussing the Granger causality patterns between the change in industrial production and the chosen macrovariables and estimate the Markov switching model with a constant transition mechanism. This model constitutes the baseline model to which the results of the subsequent section should be compared.

Since we found that most of the variables included in our data set seem to change substantially with the changes in the operating procedures of the Federal Reserve, we also consider a sub-sample ending in the third quarter of 1978. This date precedes the Volcker regime, but our results indicate this to be the proper break point. This may be related to expectations or other changes in the economy. Another reason for breaking the sample at this date is based on the behavior of the monetary aggregates. The research on the appropriateness of the use of simple monetary aggregates as initiated by Barnett (1980) has shown that there might be significant differences between standard monetary aggregates and more sophisticated aggregates. Barnett, *et al.* (1995) provide a plot of *M1* and *M2* together with the Divisia measures of these two series. This plot illustrates that not only did these measures diverge in the 1980s, but that *M1* and Divisia *M1* already started diverging by the late 1970s.

4.1 Causality Testing

As a preliminary step, we examine the predictive content of each of the variables for all the remaining variables considered in the analysis. The purpose of this is both to analyze the predictive content of the fiscal and monetary variables discussed in Section 3 on the change in industrial production, and to inquire into the extent to which these variables can be labeled as exogenous or response policy variables.[3] To do so, we

Table 1a. Full Sample Granger Causality Tests: *F*-Statistics.

Var.	y	*FFR*	*M0*	*M1*	*M2*	*NBR*	*INF*	*GOV*
y		2.0325 (0.094)	0.5457 (0.702)	2.9766 (0.022)	4.5928 (0.002)	2.3265 (0.060)	3.1644 (0.016)	2.9230 (0.024)
FFR	7.7459 (0.000)		5.5894 (0.000)	13.253 (0.000)	4.3939 (0.002)	1.4517 (0.221)	3.1583 (0.008)	0.3285 (0.858)
M0	0.8669 (0.486)	1.1939 (0.317)		1.8793 (0.118)	4.1311 (0.004)	0.3306 (0.857)	1.4410 (0.224)	1.0127 (0.403)
M1	1.8788 (0.118)	8.2278 (0.000)	4.1240 (0.004)		3.7609 (0.006)	1.0512 (0.384)	0.1137 (0.978)	0.7314 (0.572)
M2	3.2166 (0.015)	2.5546 (0.042)	2.7418 (0.031)	1.7837 (0.136)		0.2861 (0.887)	0.2268 (0.923)	0.3441 (0.848)
NBR	1.7829 (0.136)	0.8044 (0.525)	2.2017 (0.072)	2.2222 (0.070)	1.5604 (0.189)		2.5161 (0.045)	1.3751 (0.246)
INFL	4.9141 (0.001)	1.3795 (0.245)	1.0142 (0.403)	1.9618 (0.104)	3.1719 (0.016)	1.0028 (0.409)		0.3678 (0.831)
GOV	0.3251 (0.861)	0.1787 (0.949)	0.3664 (0.832)	0.1262 (0.973)	0.1023 (0.982)	0.9492 (0.438)	0.3038 (0.875)	

Notes: Sample period is 1959.1-1995.4. All statistics are based on regressions with four lags of the column variable and of the row variable. The table reports tests of the hypothesis that each element of the rows does not cause each element of the column. Numbers in parentheses are *p*-values.

employ the conventional test for Granger causality in two-variable systems. The results of this analysis are shown in Tables 1a and 1b, where the columns report the results of the test that each of the variables in the rows do not Granger cause the column variable.

Column 1 of Table 1a indicates that the standard *F*-tests reject the hypothesis that *FFR*, *INF*, and the log-difference of *M2* do not Granger cause output. We also find that the log difference of *M0*, *M1*, *NBR*, and government expenditure (*GOV*) do not seem to be useful linear predictors of the change in industrial production. In addition, the first row reports the results for tests of reverse causality from output to each of the explanatory variables. The interesting information from this row is that one cannot reject the null hypothesis that output does not Granger cause *INF* and the log-difference of *M2*. Thus, the *FFR* is the only variable that seems to cause output and at the same time is not caused by output in the Granger sense. These results will be used as a benchmark for comparison with the results obtained using the nonlinear filter in order to establish whether there exists a richer nonlinear interaction between the above variables.

Table 1b. Sub-Sample Granger Causality Tests: *F*-Statistics.

Var.	y	*FFR*	*M0*	*M1*	*M2*	*NBR*	*INF*	*GOV*
y		1.9585 (0.111)	0.5925 (0.670)	0.8413 (0.504)	1.7594 (0.002)	3.8611 (0.007)	2.4729 (0.053)	2.3326 (0.065)
FFR	6.234 (0.000)		0.783 (0.540)	1.7698 (0.145)	7.6501 (0.000)	8.5629 (0.000)	9.4309 (0.000)	0.6256 (0.646)
M0	1.3884 (0.248)	3.7905 (0.008)		1.4814 (0.218)	1.8499 (0.130)	1.5427 (0.200)	4.5879 (0.003)	0.8342 (0.508)
M1	1.7174 (0.157)	6.9428 (0.000)	2.4820 (0.052)		6.8052 (0.006)	2.5318 (0.049)	1.7584 (0.148)	0.8326 (0.509)
M2	3.5392 (0.015)	2.6167 (0.043)	3.1905 (0.019)	5.4932 (0.001)		2.9196 (0.028)	0.5764 (0.681)	0.4837 (0.748)
NBR	6.3459 (0.000)	5.7573 (0.001)	0.7204 (0.581)	1.1585 (0.337)	2.0260 (0.101)		4.8583 (0.002)	0.2937 (0.881)
INFL	5.8801 (0.000)	1.3802 (0.250)	1.2153 (0.313)	1.2586 (0.295)	2.183 (0.081)	1.3253 (0.270)		0.853 (0.497)
GOV	0.1024 (0.981)	0.6682 (0.616)	0.8175 (0.519)	0.9555 (0.438)	0.8779 (0.482)	0.9950 (0.417)	0.3321 (0.855)	

Notes: Sample period is 1959.1-1978.3. All statistics are based on regressions with four lags of the column variable and of the row variable. The table reports tests of the hypothesis that each element of the rows does not cause each element of the column. Numbers in parentheses are *p*-values.

In order to look into these results in a bit more detail, it is also useful to investigate the causality among the explanatory variables. The results of these tests are quite interesting. First, the results indicate that *NBR* is the only monetary aggregate that is not caused by some of the other variables, except output. This provides some support for the hypothesis that movements in this variable reflect a good measure of exogenous changes in monetary policy. The remaining monetary aggregates are either caused by output or by other monetary aggregates, and may thus be viewed as partly endogenous. Second, the results for the *FFR* indicate that it is caused by *M1* and *M2*, and thus that some of its movements reflect responses to movements in other variables. We may therefore think of the *FFR* as a "response" variable.

Third, the Granger causality results show that government expenditure is only caused by output and it does not cause any other variable. Thus these tests do not reject Granger non-causality of *GOV*, and therefore indicate limited predictive power for this variable. Finally, the *F*-tests indicate that we reject the null that inflation is non-caused by changes in industrial production, non-borrowed reserves, and by the federal funds rate.

As discussed above, the change in the operative procedure of the Federal Reserve can substantially change our results. In particular, because Granger causality tests are not robust to these types of breaks, the results obtained for the whole sample may be misleading. Table 1b presents the results for the sub-sample 1959-1978.3, from which we can clearly observe that some of the results reported in Table 1a change dramatically. The two most important changes are: (a) in the sub-sample non-borrowed reserves is also a useful linear predictor of output, and (b) non-borrowed reserves are now found to be caused by output, the federal funds rate, *M1*, and *M2*.

The results presented in Table 1a should be interpreted cautiously since they may change if we perform a more compete analysis including all the variables in a VAR. We did not proceed in such a way since when we carried out some preliminary analysis, we only found a satisfactory specification for a high-order VAR. Therefore, we interpret these results as providing an indication of the causal relationships between these variables.

It must also be borne in mind that such tests are designed to detect linear relationships between variables. Given that the center of our inquiry is to infer which variables help to forecast the probability of staying either in a boom or recession, the econometric framework employed in the sequel is undoubtedly more appropriate.

4.2 Time-Invariant Transition Mechanism

Before turning to the model with time-varying transition probabilities, we estimate a standard univariate switching regime model for the change in industrial production. This serves as the baseline model to which the results with time-varying transition probabilities should be compared. This is the standard Hamilton (1989) model except for the fact that we allow for fatter tails in distribution of errors.

Table 2 presents estimation results for the change in industrial production, using model (1)-(2) with $m=4$ (i.e., with four lagged values of the change in industrial production).[4] We also report the outcome of tests for linear and nonlinear temporal dependence in the standardized residuals. These are based on the Ljung-Box portmanteau Q statistics for the first k autocorrelations of the standardized residuals and their squares, respectively denoted by $Q(k)$ and $Q_2(k)$. Under the null hypothesis of no serial correlation, $Q(k)$ and $Q_2(k)$ have an asymptotic χ^2-distribution with k degrees of freedom (see Ljung and Box (1978), McLeod and Li (1983)). To facilitate an easier presentation of the results, we label state 0 (high growth in industrial production) as the boom state and state 1 (low growth in industrial production) as the recession state.

Overall, the estimated coefficients reveal significant differences between the two regimes. The mean of output growth in "booms" is estimated to be 1.28 percent per quarter (equivalent to an annual growth rate of 5.2 percent) while the mean in "recessions" is estimated to be -0.15%. The change in the mean between the two states is significant at any conventional level of significance. The estimates imply that the boom state is more persistent than the recession state, i.e., the expected duration of a boom is approximately three and a half years, whereas the expected duration of a recession is between one and a half and two years. Notice that this implies that the

Table 2. Maximum Likelihood Estimation Results (Time-Invariant Transition Probabilities).

Parameter	Full Sample	Sub-Sample
α_0	1.276 (0.305)	1.713 (0.114)
α_1	-1.434 (0.908)	-2.754 (0.763)
σ_0	1.008 (0.200)	1.054 (0.377)
p	0.849 (0.113)	0.725 (0.126)
q	0.929 (0.056)	0.915 (0.043)
φ_1	0.450 (0.109)	0.307 (0.089)
φ_2	-0.096 (0.108)	-0.203 (0.106)
φ_3	0.025 (0.077)	-0.050 (0.082)
φ_4	-0.113 (0.068)	-0.157 (0.072)
d	3.808 (1.956)	3.088 (1.704)
$\log \mathcal{L}$	-232.103	-127.707
$Q(1)$	1.212	0.881
$Q(5)$	6.766	6.438
$Q_2(1)$	1.131	1.370
$Q_2(5)$	3.175	6.342

Notes: Numbers in parentheses are asymptotic standard errors.

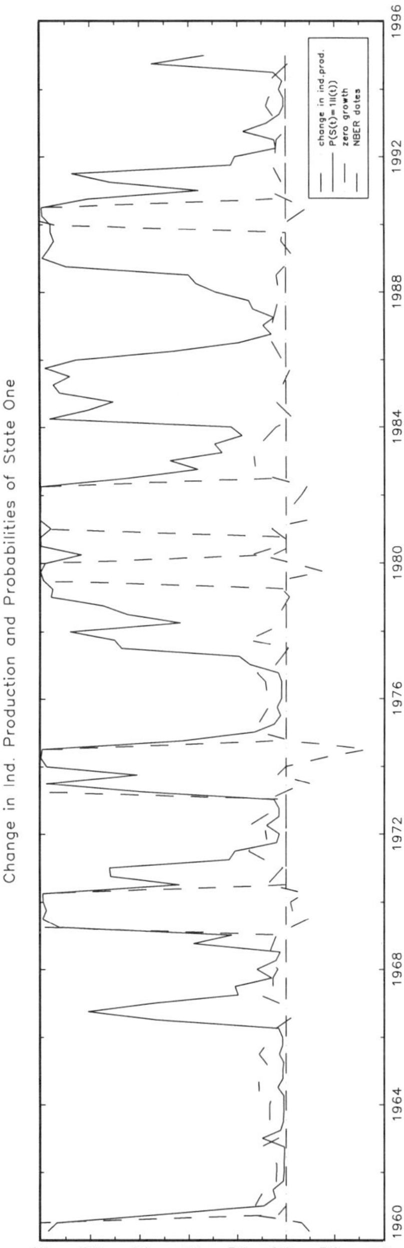

Figure 1a

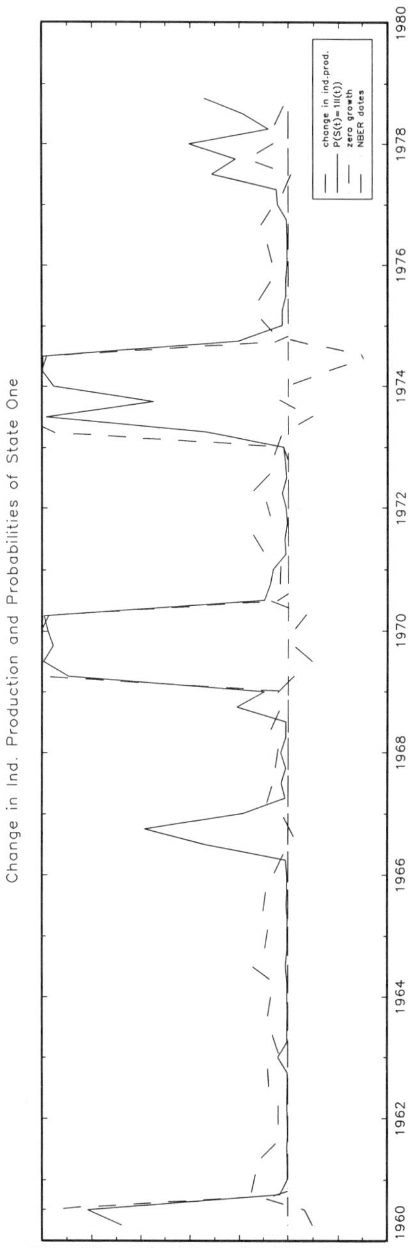

Figure 1b

expected duration of a complete cycle is approximately five years, which is in line with traditional measures of the business cycle (see, e.g., Zarnowitz (1992)),as is the result that recessions are less persistent than booms.

Figure 1a shows the estimated filter probability of being in the recession state at each point in the sample, along with the *NBER* dates. The estimates of the two states coincide in general with the *NBER* chronology. The most significant difference between the *NBER* dates and the estimated probabilities is that the latter select the period 1979.4-1982.4 to be one long recession (possibly with the exception of 1980.3) whereas the *NBER* divides this period into two separate recessions. Two other minor differences in the classification of the states are that the interpretation of the filter probabilities implies that the recession in the early seventies (1969.4-1970.4) had a slightly longer duration than that identified by the *NBER*, and that for some small periods the probabilities of being in a recession increase for one or two quarters, which also differs from the *NBER* dates. However, the estimates of the two regimes seem to be very closely related with the *NBER* turning points.[5]

Thus, the two-state Markov switching regime model provides a useful baseline model; it is well specified, and it provides a dating of regimes very similar to (though not identical to) the traditional business cycle chronology of the *NBER*.

The results for the sub-sample 1959-1978.3 presented in the second column of Table 2 show that the mean in the boom state in now 1.71 and in the recession state is -1.1. We also find much smaller persistence of the recessions, with an approximate duration of the recession state of one year. The plot presented in Figure 1b shows that when we only consider a sub-sample of the data we obtain a much better classification of the states in booms and recessions. This may indicate that there have been structural changes in the economy since the end of the 1970s such that estimating the filter for the entire period leads to a worse fit of the regime switching model.

5. Time-Varying Transition Mechanism

We now extend the analysis of the previous section to the model with time-varying transition probabilities discussed in Section 2. We will investigate whether the model in which the transition probabilities vary over time through dependence on some other variables provides a useful extension of the standard Hamilton with constant transition probabilities. A central part of our inquiry is to establish whether these variables can help to predict a change from a boom to a recession, and vice versa, and whether this effect is symmetric.

We divide the discussion into two main parts: analysis of "policy variables" and analysis of lagged "endogenous variables." For the interpretation of the "policy variables" we should bear in mind the results obtained in Section 3. With respect of the lagged endogenous variables we note from the preceding discussion that one should interpret the results in terms of predictive power for changes in regime in industrial production. These results may also help to understand the different transmission mechanisms which may affect output.

5.1 Policy Variables

5.1.1 Government Spending

As an indicator of fiscal policy we use government spending, which might be responsive to changes in output (see Table 1a). Therefore we may think that military spending could be a better indicator as it captures the exogenous component of fiscal policy. However, we were unable to obtain this variable for sufficiently long sample period. Another candidate for fiscal policy is the primary deficit, which we chose not to use because tax revenues must certainly depend on the state of the business cycle.

Column 1 in Table 3a records the results when transition probabilities are modelled as functions of the change in government spending. The estimates provide no evidence that real government spending affects the transition probabilities because both β_0 and β_1 are insignificantly different from zero (recall also that the Granger causality tests fail to reject Granger non-causality from the change in government spending to the change in industrial production.)

This result may reflect reverse causality due to countercyclical use of fiscal policy, or it might be that the simple specification used here fails to identify the effects that government spending has on the economy. For example, there may be constraints on the use of fiscal policy due to inflationary consequences, or to budgetary issues which are not controlled for in the estimations. A third possibility is that there are longer lags in the effects of government spending than those we have controlled for. Rather than going into these issues in more detail, we simply leave the matter here noting that we do not find evidence of government spending affecting industrial production in either a linear or a nonlinear sense, but at the same time recognizing that this result could be due to a number of potential shortcomings in our analysis.

Figure 2a shows very little variation in the transition probabilities for the whole sample period. The results obtained for the sub-sample presented in Table 3b and Figure 2b are similar to the full sample results in showing that *GOV* does not significantly affect the transition probabilities. Since p_t and q_t vary very little for the whole sample, the results are strongly similar to those obtained using the simple Hamilton filter.

Finally, the results for government spending are not sensitive to the sample period. For the shorter sub-sample, 1959-1978.3, we find results that are basically identical to the full sample results.

5.1.2 Monetary Variables

As discussed in Section 3, our list of indicators of monetary policy includes the monetary base, *M1*, *M2*, non-borrowed reserved, and the federal funds rate. Notice that some of these variables may not only be endogenous, but also have varying degrees of importance for aggregate activity. We first present the results and then turn to their interpretation.

Table 3a summarizes the maximum likelihood estimates from Markov switching models with monetary variables used as the forcing macrovariables. All models appear to be well identified: the switching parameters are highly significant in most cases, and

Table 3a. Full Sample Maximum Likelihood Estimation Results With Policy Variables (Time-Varying Transition Probabilities).

	Policy Variables					
	GOV	*FFR*	*M0*	*M1*	*M2*	*NBR*
α_0	1.643 (0.106)	1.555 (0.121)	1.633 (0.120)	1.610 (0.136)	1.745 (0.100)	1.643 (0.117)
α_1	-1.411 (0.204)	-1.590 (0.254)	-1.414 (0.217)	-1.500 (0.241)	-1.292 (0.182)	-1.439 (0.228)
σ_0	1.645 (0.888)	1.034 (0.395)	1.587 (0.181)	1.193 (0.942)	1.277 (1.149)	1.705 (0.998)
c_1	2.205 (0.606)	-0.149 (1.196)	2.226 (1.544)	3.624 (0.846)	6.075 (2.581)	2.218 (0.409)
β_1	-0.174 (0.266)	0.276 (0.186)	-0.293 (1.707)	-8.716 (3.499)	-1.741 (1.005)	-0.563 (0.760)
c_0	1.710 (0.513)	4.990 (1.697)	1.587 (1.336)	0.973 (0.741)	-2.408 (3.135)	1.947 (0.454)
β_0	0.1048 (0.284)	-0.470 (0.210)	0.339 (1.707)	8.204 (6.753)	2.346 (1.818)	2.751 (1.845)
φ_1	0.333 (0.049)	0.340 (0.056)	0.335 (0.050)	0.332 (0.050)	0.315 (0.057)	0.333 (0.036)
φ_2	-0.171 (0.065)	-0.182 (0.069)	-0.167 (0.065)	-0.149 (0.066)	-0.148 (0.052)	-0.172 (0.066)
φ_3	0.018 (0.057)	0.025 (0.061)	0.017 (0.054)	0.008 (0.045)	0.022 (0.047)	0.020 (0.048)
φ_4	-0.142 (0.039)	-0.151 (0.049)	-0.141 (0.039)	-0.140 (0.026)	-0.138 (0.037)	-0.142 (0.027)
h	2.153 (0.600)	2.712 (1.015)	2.223 (0.705)	1.631 (0.900)	2.279 (0.637)	2.144 (0.601)
$\log \mathcal{L}$	-230.413	-230.246	-230.726	-227.729	-227.135	-228.549
$Q(1)$	2.872	1.839	3.926	2.886	2.882	2.581
$Q(5)$	8.109	4.977	10.75	8.673	8.878	8.740
$Q_2(1)$	2.702	1.761	2.786	3.416	4.104	2.581
$Q_2(5)$	5.369	3.290	6.741	7.005	5.591	6.437

Notes: Numbers in parentheses are asymptotic standard errors.

Table 3b. Sub-Sample Maximum Likelihood Estimation Results With Policy Variables (Time-Varying Transition Probabilities).

	Policy Variables					
	GOV	*FFR*	*M0*	*M1*	*M2*	*NBR*
α_0	1.713 (0.113)	1.704 (0.135)	1.683 (0.134)	1.695 (0.115)	1.694 (0.114)	1.702 (0.117)
α_1	-2.767 (0.746)	-2.852 (0.725)	-2.9476 (0.737)	-2.482 (0.541)	-2.532 (0.586)	-3.204 (0.416)
σ_0	1.058 (0.383)	1.044 (0.319)	1.096 (0.230)	1.161 (0.534)	1.076 (0.313)	1.910 (0.988)
c_1	0.990 (0.670)	-2.061 (2.270)	-0.474 (2.173)	4.911 (3.351)	8.406 (6.403)	0.794 (0.612)
β_1	-0.081 (0.658)	2.254 (1.160)	1.101 (0.875)	-3.060 (1.500)	-3.252 (1.663)	-0.172 (0.246)
c_0	2.424 (0.648)	5.055 (1.923)	1.762 (1.391)	1.236 (1.256)	-3.252 (2.911)	12.458 (8.592)
β_0	-0.084 (0.549)	-2.329 (1.109)	0.611 (1.325)	1.035 (1.385)	3.459 (1.718)	-7.371 (3.334)
φ_1	0.308 (0.089)	0.293 (0.100)	0.337 (0.101)	0.210 (0.105)	0.212 (0.147)	0.335 (0.061)
φ_2	-0.204 (0.104)	-0.200 (0.112)	-0.116 (0.121)	-0.133 (0.078)	-0.134 (0.085)	-0.265 (0.063)
φ_3	-0.049 (0.082)	-0.051 (0.087)	-0.090 (0.091)	-0.065 (0.064)	-0.073 (0.069)	-0.072 (0.061)
φ_4	-0.158 (0.071)	-0.161 (0.076)	-0.092 (0.069)	-0.153 (0.072)	-0.145 (0.076)	-0.153 (0.051)
h	3.068 (1.664)	3.294 (1.868)	4.001 (2.301)	2.844 (1.418)	3.324 (0.637)	2.181 (0.862)
$\log \mathcal{L}$	-127.686	-123.903	-127.223	-125.955	-120.357	-120.817
$Q(1)$	0.862	0.253	0.240	3.257	0.222	0.002
$Q(5)$	5.886	4.339	4.085	8.067	6.139	1.221
$Q_2(1)$	1.270	1.963	0.467	2.026	1.112	0.003
$Q_2(5)$	6.532	3.128	5.864	6.429	3.482	0.364

Notes. Numbers in parentheses are asymptotic standard errors.

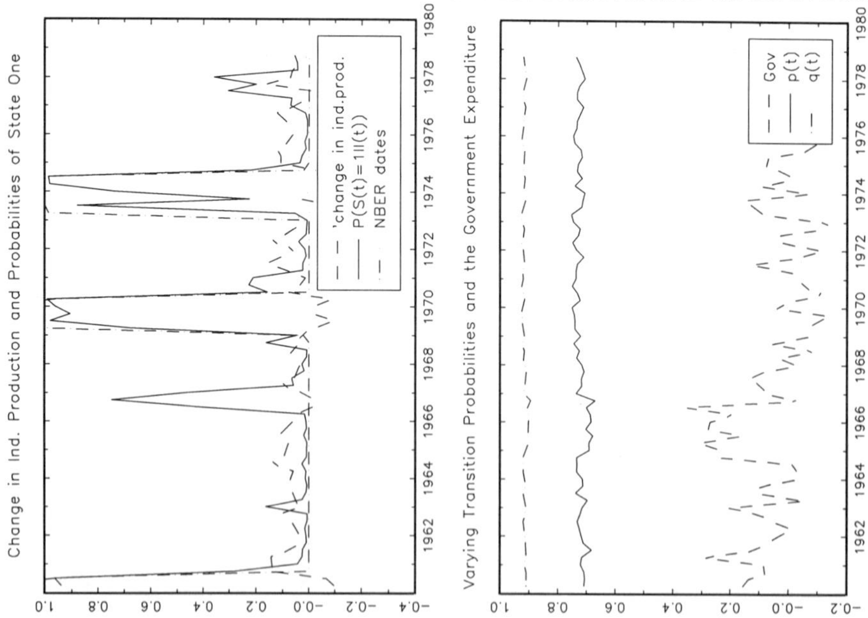

Figure 2b

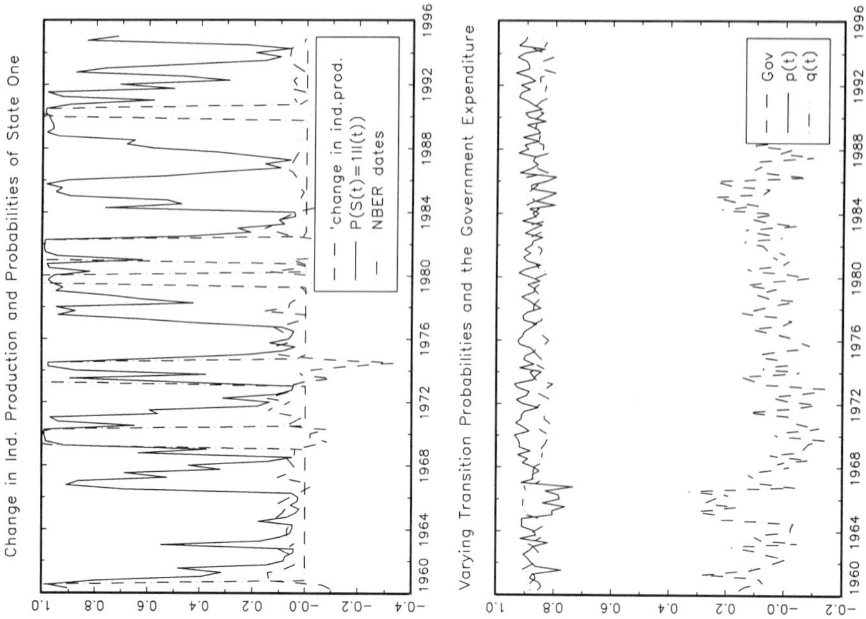

Figure 2a

the size of the coefficients is close to estimates from the Hamilton filter with constant transition probabilities (see Table 2). Moreover, the standardized residuals exhibit no signs of linear or nonlinear temporal dependence.

The Federal Funds Rate We start the discussion by considering the results for the federal funds rate, which was the only variable that Granger caused output without being caused by output itself. Column 2 in Table 3a shows that the federal funds rate is found to be significant in the boom state but insignificant in the recession state. Thus, the results imply that the federal funds rate apparently does not affect the probability that the economy remains in a recession (β_1 insignificantly different from zero), but does influence the probability that the economy goes out of a boom (β_0 negative and significant).

Figure 3a plots the filter probabilities associated with state 1, the recessionary state, as well as the transition probabilities p_t, the probability of remaining in a recession, and q_t, the probability of remaining in a boom, along with the federal funds rate. The lower panel reveals that the low federal funds rates of the 1960s were associated with very high values of the boom state probabilities, q_t, but that during the recession following the first oil-price shock (or the breakdown of the Bretton Woods agreement) the surge in the federal funds rate was associated with a substantial decrease in the probability of remaining in the boom state. After OPEC I the results indicate that the lower level of the federal funds rate in the period 1975-1977 led to a temporary increase in the transition probability associated with the boom state. Furthermore, it seems that the increase in the federal funds rate in the late seventies led to a decrease in the probability of remaining in a boom.

There are, of course, good reasons to believe that the results for the federal funds rate are affected by the changes in monetary policy that have taken place over the sample period. Because, in the period 1979-1982, the Federal Reserve changed its operating procedures towards a form of monetary base control, producing higher and more volatile rates than have been observed before or since, we compare the full-sample results to those for the sub-sample 1959-1978.3. As presented in Table 3b, the results for this period are qualitatively different in that, the federal funds rate not only influences the probability of remaining in a boom, but also significantly affects the probability of remaining in a recession (β_1 is significantly different from zero). The results indicate that an increase in the federal funds rate increases the time-varying transition probability of being in a recession, and decreases the one being in a boom.

The bottom panel of Figure 3b illustrates the estimated transition probabilities together with the federal funds rate. The plots are quite telling: the federal funds rate increased before both of the major recessions, and the higher interest rates were associated with a drop in the probability of staying in a boom and an increase in the persistence of the recession. This graph clearly shows that increases in interest rates preceded the two major recessions considered in this sample. Furthermore, the results also indicate that the lower interest rates towards the end of the recession around OPEC I increased the transition probability of the boom state and lowered the transition probability of the recession state and may thus have helped the economy out of the recession.

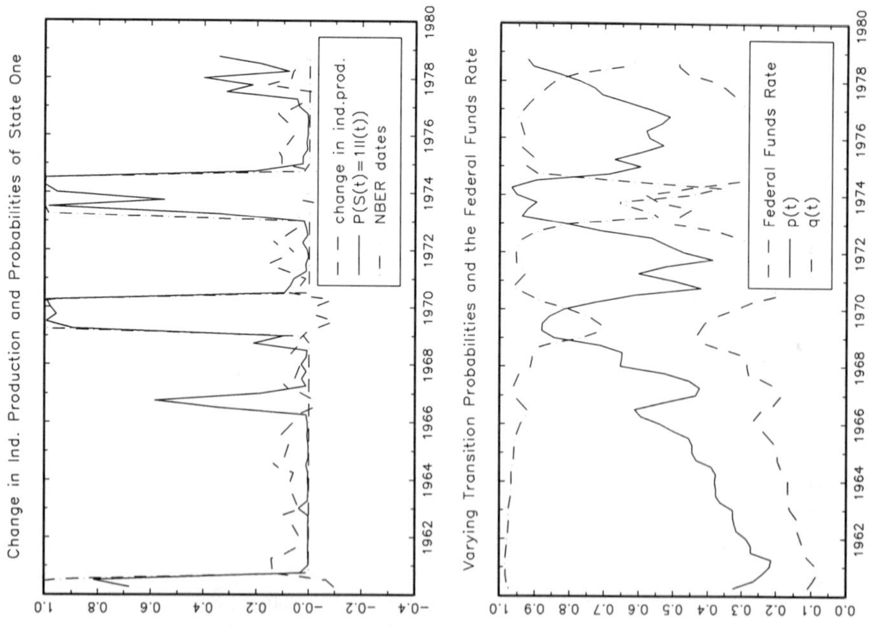

Figure 3b

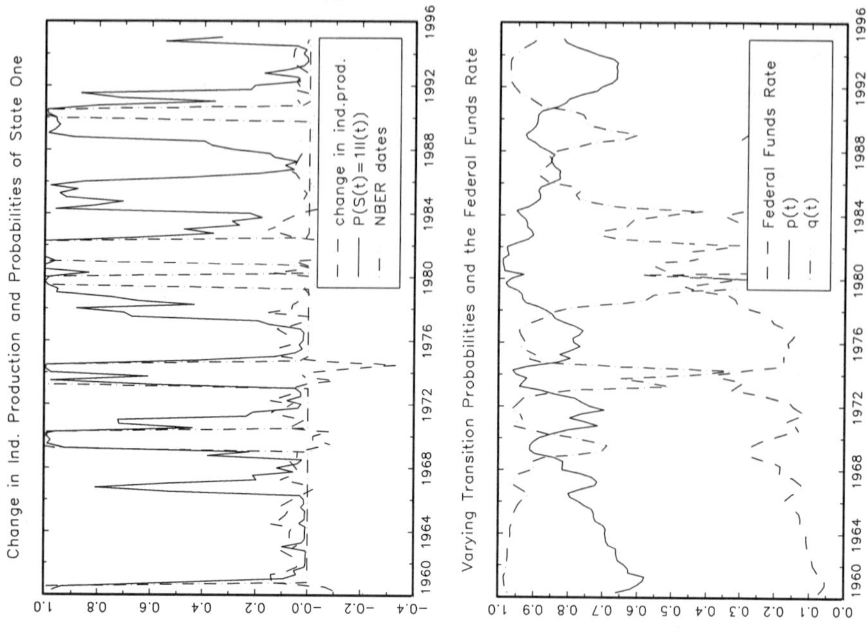

Figure 3a

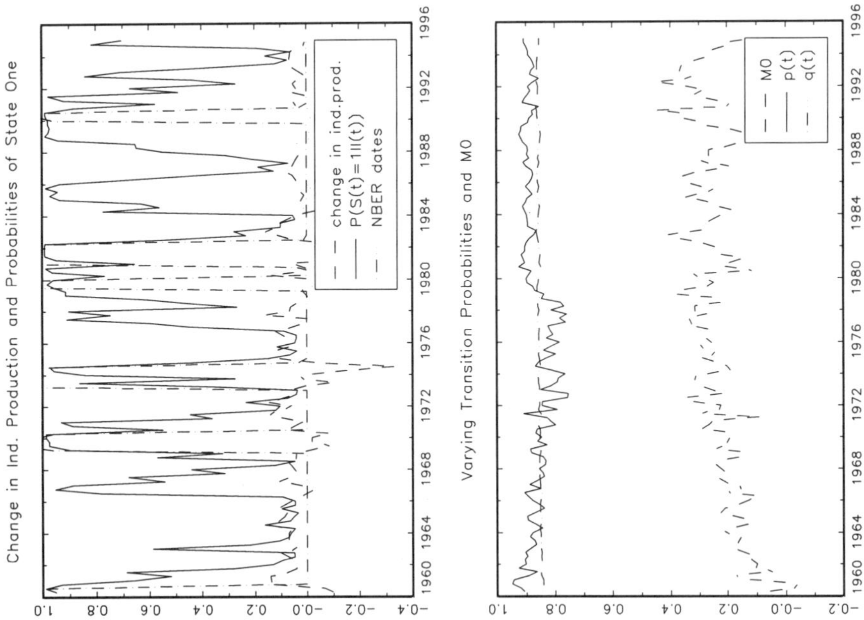

Figure 4a

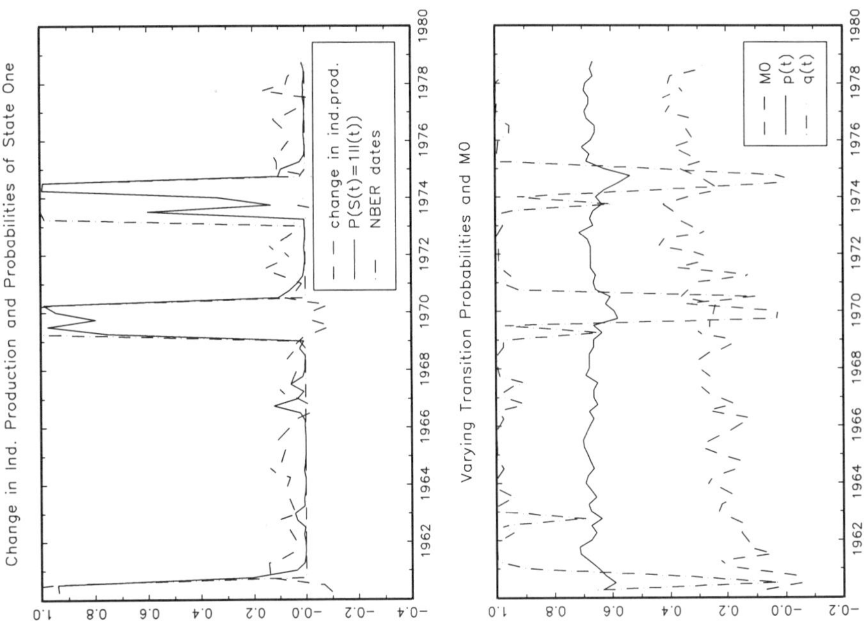

Figure 4b

Thus the results imply that the federal funds rate, the only variable that in a Granger sense causes output without itself being caused by output, is a useful predictor of transitions in output, but exactly how it helps depends on the sample period. For the early period, lower interest rates decrease (increase) the likelihood of remaining in a recession (boom). Furthermore, from our analysis one finds that the economy went into a recession following an increase in interest rates, and left it following a lowering of interest rates. If one ignores the possibility that there were changes in the operating procedures of the Fed, one would instead find that the federal funds rate only affects the likelihood with which the economy moves from a boom into a recession.

Monetary Aggregates Here we focus attention on four standard measures of monetary policy (initially measured in nominal values and in log first-differences): *M0*, *M1*, *M2*, and non-borrowed reserves.

The results for *M0* are particularly interesting since they highlight the potential problems that a break between output and the variables included in the transition probabilities may induce. On the one hand, we find that when the transition probabilities are functions of *M0* we obtain both the worst fit of all the models considered in this section, and a classification of the states into booms and recessions that is virtually unrelated to that of the *NBER* (see the top panel of Figure 4a). On the other hand the results recorded in column 3 of Table 3a show that changes in *M0* have a negligible effect on the probability of staying in a recession (β_1 insignificantly different from zero) and the affects the probability of staying in a boom ((β_0 positive and insignificant). Thus, the results imply that an increase in the money supply increases the probability of remaining in a boom.

The lower panel of Figure 4a shows that the probability of remaining in a boom, q_t, has abrupt drops in periods of very high contractions of the monetary base. On the other hand, for the sub-sample the results presented in Table 3b and Figure 4b are qualitatively the same, but the separation of the states is now very close to the *NBER* dates.

The results presented in column 4 of Table 3 show that when changes in *M1* are used as the variable affecting the transition probabilities, it has negligible effects on the probability of remaining in the boom state (β_0 insignificantly different from zero) but negative and significant effect on the probability of remaining in a recession ($\beta_1 < 0$ and significant at a 90% confidence level).

The top panel of Figure 5a shows that the identification of the two states is similar to that of the standard Hamilton filter, but somewhat different from the *NBER* dates. The plot of the transition probabilities using *M1* shows that both transition probabilities were relatively stable until around 1980, when the increased volatility of *M1* is associated with significant changes in the time-varying probabilities of remaining in the recession state. This is consistent with results obtained by Baba, *et al.* (1992), for example, which relate this period to structural instability of *M1* demand.

The results for the sub-sample are remarkably different. Even though *M1* enters qualitatively in the transition probabilities in the same way as in the full sample, the plots of the transition probabilities for the period 1960 to 1980 are very different (see Table 3b). For the short sample the probability of remaining in a recession, p_t, is very

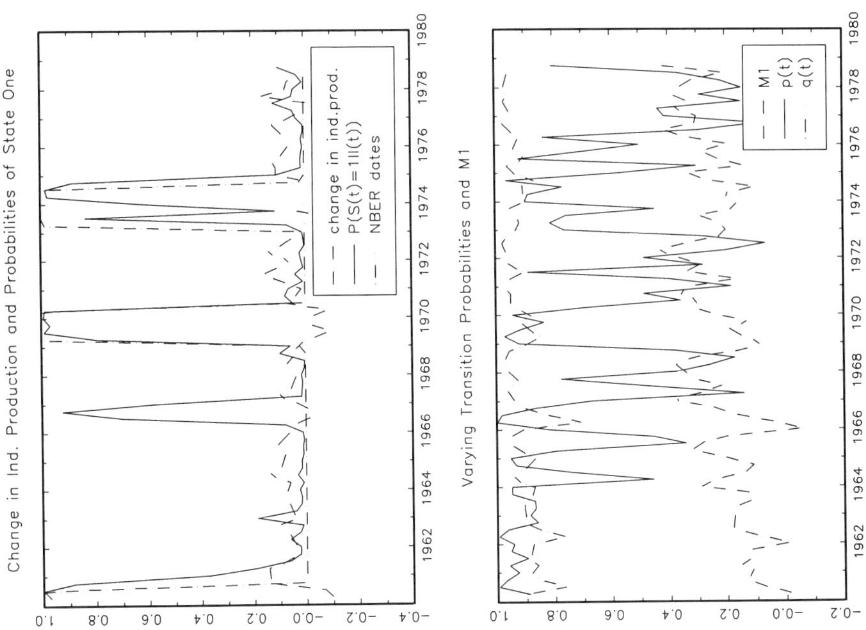

Figure 5b

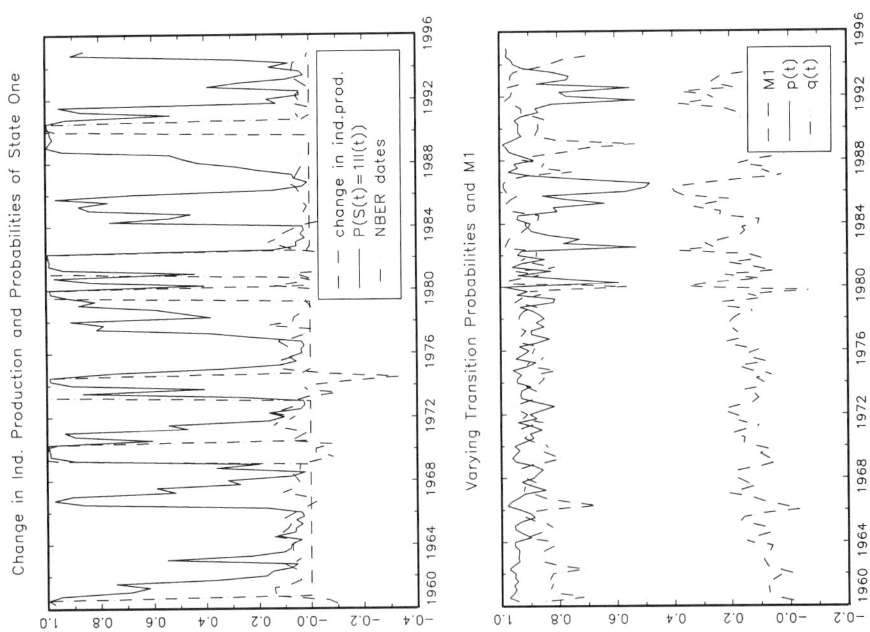

Figure 5a

volatile and takes values close to one in recessionary periods. The bottom panel of Figure 5b shows that (the change in) *M1* decreases before the economy enters into a recession, and increases before it goes out. Since *M1* enters with negative sign on p_t, these episodes are successfully forecasted. Notice also that the separation of the states is much better for the sub-sample than for the full sample.

The fact that the monetary measure seems to be significant only in the transition probability for the recession state and that an increase in *M1* is associated with a lower persistence of this state, might be interpreted as the money supply affecting aggregate activity only when the economy is recessionary. This means that there is a higher likelihood of switching from a recession to a boom as some theories of imperfect credit markets predict. According to this "credit view," the monetary transmission mechanism works through the credit available to economic agents, and when the economy is in a recession, credit is limited, and a loosening of monetary policy can therefore help to boost activity. Also, money would be neutral in booms because credit is readily available. Such a story seems plausible in terms of these results.

Column 5 in Table 3a presents the results when the change in *M2* is used as the variable affecting the transition probabilities. It has negligible effects in the probability of remaining in the boom state (β_0 insignificantly different from zero) but negative and significant effects on the probability of remaining in a recession ($\beta_1 < 0$ and significant at a 90% confidence level). The results are similar to those obtained with *M1* in that the monetary measure seems to be significant only in the transition probability for the recession state, and an increase in *M2* is associated with lower values of the probability of remaining in this state. Nevertheless, it should be noted that the change in the operating procedures of the Federal Reserve seems to have affected the stochastic properties of *M2* in such a way that the inference on the states obtained from the filter are misleading. The top panel of Figure 6a shows that the identification of the two states is only similar to those of the standard Hamilton filter and the *NBER* for the first part of the sample, while the filter always chooses the recessionary state for the second part of the sample. The plot of the transition probabilities when using *M2* as the monetary measure shows that the time-varying transition probabilities seem to be successfully capturing the dynamics of the data until 1980. For example, the probability q_t of staying in a boom decreases with the contraction of *M2* at the beginning of the 1970s, while the probability p_t of staying in a recession increases. Similar results are obtained for the mid-1970s. Very disappointing results are obtained for the second part of the sample, which seems to be a period of structural instability in the demand for *M2*.

On the other hand, the results for 1959-1978.3 reported in Table 3b and in Figure 6b are fare more encouraging. The change in *M2* enters significantly in both transition probabilities, the separation of the states mimics that of the *NBER*, and the change in *M2* seems to help forecast the changes from booms to recessions, and vice versa. The bottom panel of Figure 6b shows that a decrease in *M2* preceded the two major recessions, while increases in *M2* preceded the end of the recessions. Consequently, p_t increases in periods when *M2* decreases, typically before the recessions of the early and middle 1970s, decreases dramatically for the boom periods, which are concomitant with high money expansion, and is close to one in periods of very low money expansion. The pattern of q_t is similar, only that it typically decreases when p_t

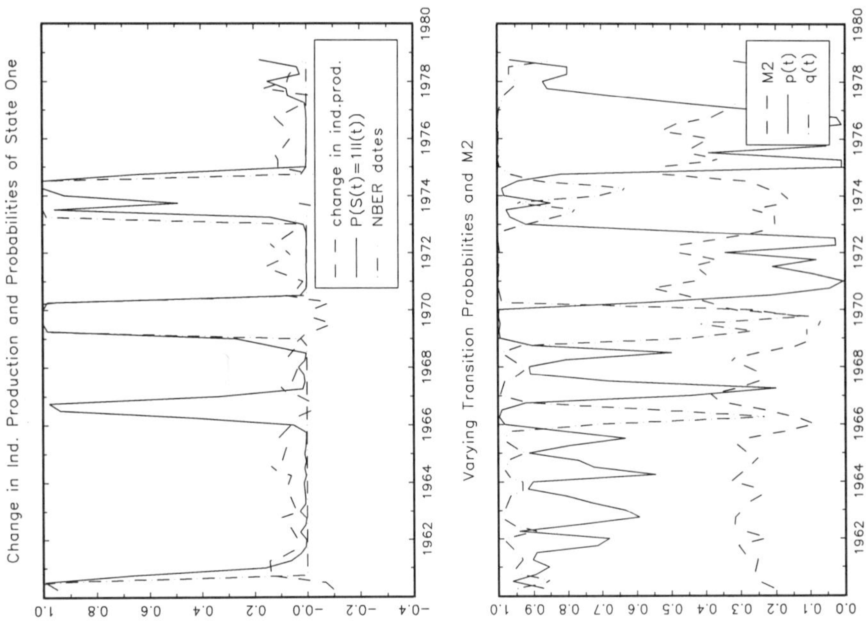

Figure 6b

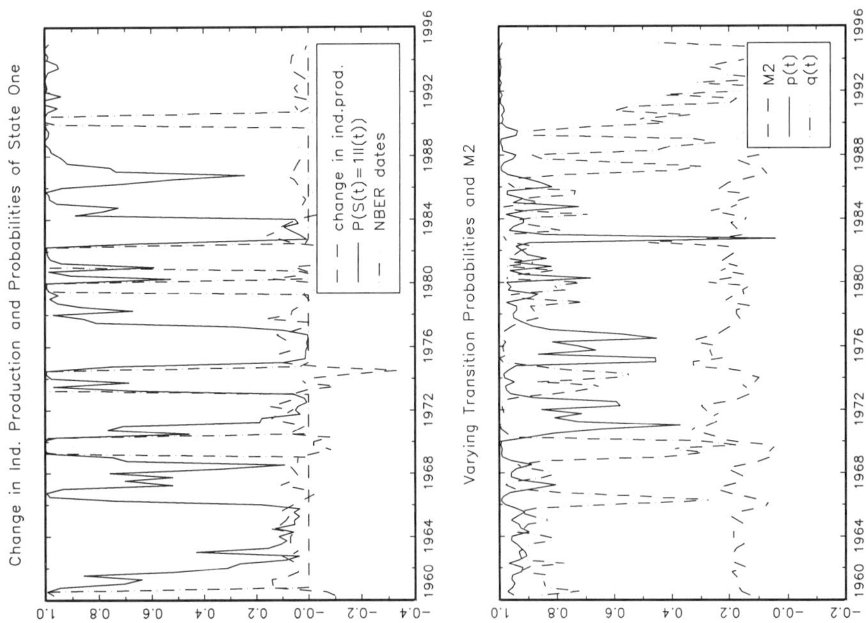

Figure 6a

increases.[6]

Finally, we report the results obtained using non-borrowed reserves as the variable affecting the transition probabilities. Column 6 in Table 3a shows that non-borrowed reserves have negligible effects on the probability of remaining both in a boom and in a recession ((β_0 and β_1 insignificantly different from zero). The bottom panel of Figure 7a shows that the major change in the transition probabilities takes place in 1984, and is clearly unrelated with either the booms or the recessions, and this result is therefore not surprising. This application gives us another example of the damaging effects that breaks in the stochastic process affecting the transition probabilities may have.

For the sub-sample we find that *NBR* significantly affects only the transition probabilities for the recession state (see column 7 in Table 3b). Figure 7b shows that the ends of the booms are associated with increases in *NBR*. A more detailed analysis of the graphs may help to understand this puzzling result. Consider for example the recession of the mid-1970s when changes in non-borrowed reserves experience the most dramatic variations of the sub-sample; *NBR* increases immediately before the recession, and therefore seems to predict it (the end of the boom). Once this recession starts, there is a big reduction in *NBR*, probably in an attempt to reduce inflation (see Figure 8b), which seems to have deepened the recession (the expected causality). Then, in 1974 there is a sharp increase in *NBR*, which was probably instrumental in ending the recession (once again, this is the expected causality), followed by a sharp decrease in *NBR*, probably directed to end the high inflation of the mid-seventies (see Figure 8b). Since the recession ended when the rate of expansion of *NBR* decreased (even though high), we get the counterintuitive result mentioned above.

In summary, the results for the monetary aggregates depend on the measure of the money stock and on the sample period. Clearly, the results seem far more intuitive when one restricts the sample period to the early part of the sample, indicating that the important changes in the conduct of monetary policy over the sample have probably modified the covariation between output and the monetary aggregates in the first part of the sample. An alternative interpretation is that the sensitivity to the sample period is linked to the problems associated with the use of standard monetary aggregates. Using Divisia indices, as advocated by Barnett, *et al.* (1995), would make an interesting extension of this analysis.

5.2 *Lagged Endogenous Variables*

We now turn to the results obtained when we use lagged endogenous variables affecting the transition probabilities. Column 1 in Table 4 presents the results when inflation is used as the variable affecting the transition probabilities. This variable is of special interest since periods of high interest rates seem to have triggered policy responses and which in turn have driven the economy into recession many times. Not surprisingly, we find that inflation is a useful nonlinear predictor for the change in industrial production (but recall also that one cannot reject causality from output to inflation).

Interestingly, inflation affects the transition probabilities in a fashion similar to the federal funds rate. We find that an increase in inflation significantly decreases the probability of staying in a boom (β_0<0). Figure 8a shows that the recessions of the

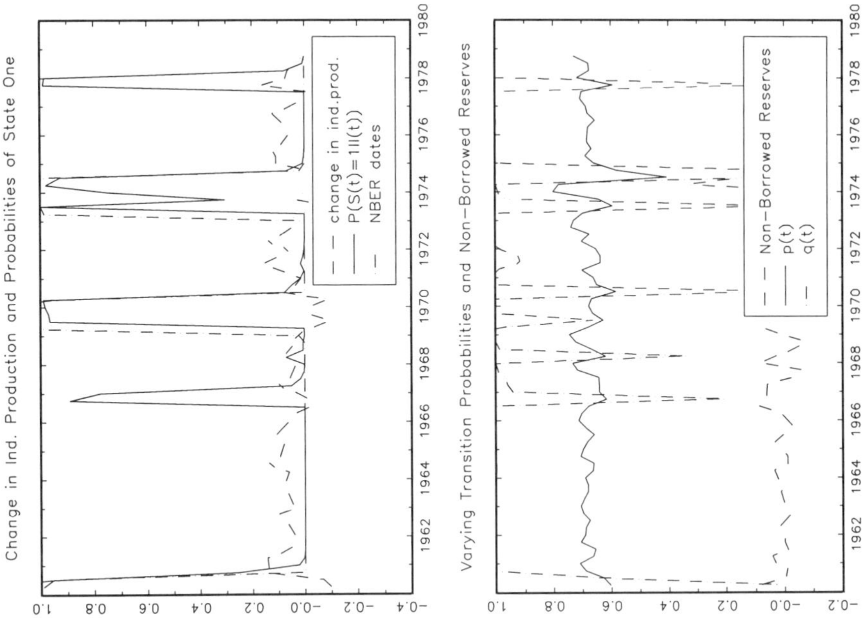

Figure 7b

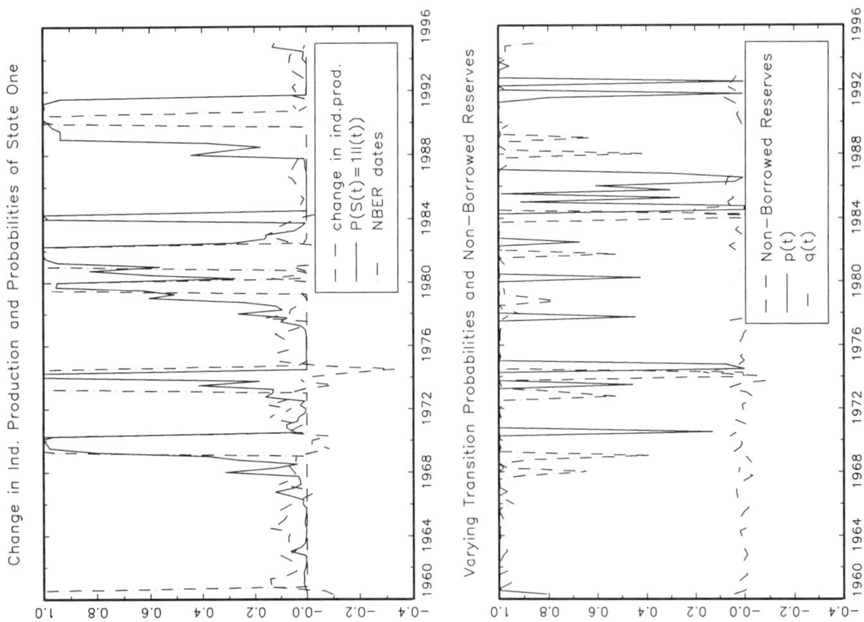

Figure 7a

Table 4. Maximum Likelihood Estimation Results With Lagged Endogenous Variables (Time-Varying Transition Probabilities).

	Lagged Endogenous Variables					
	INFL	$INFL_{ss}$	*RFFR*	$RFFR_{ss}$	$\Delta(M1/P)$	$\Delta(M1/P)_{ss}$
α_0	1.550 (0.304)	1.577 (0.191)	1.541 (0.137)	1.712 (0.123)	1.316 (0.129)	1.692 (0.118)
α_1	-1.463 (0.237)	-3.269 (0.846)	-1.602 (0.274)	-2.841 (0.715)	-2.154 (0.631)	-2.564 (0.553)
σ_0	1.181 (0.853)	1.063 (0.253)	1.041 (0.393)	1.048 0.332)	0.998 (0.153)	1.166 (0.487)
c_1	0.652 (1.415)	0.073 (1.054)	-1.701 (1.877)	-2.895 (1.524)	1.505 (0.930)	0.753 (0.888)
β_1	0.986 (1.097)	0.245 (1.185)	2.725 (1.403)	1.529 (0.805)	-2.499 (1.650)	-2.174 (1.691)
c_0	3.813 (1.628)	5.355 (2.530)	4.366 (1.780)	5.362 (1.974)	3.025 (1.303)	2.775 (1.448)
β_0	-1.799 (0.871)	-2.050 (1.009)	-2.262 (1.225)	-1.298 (0.622)	2.855 (1.475)	3.1061 (2.394)
φ_1	0.357 (0.131)	0.370 (0.103)	0.331 (0.061)	0.276 (0.115)	0.413 (0.092)	0.193 (0.112)
φ_2	-0.144 (0.097)	-0.161 (0.161)	-0.160 (0.072)	-0.194 (0.119)	-0.074 (0.098)	-0.139 (0.079)
φ_3	0.007 (0.042)	-0.057 (0.110)	0.010 (0.061)	-0.054 (0.086)	0.008 (0.099)	-0.067 (0.071)
φ_4	-0.142 (0.035)	-0.140 (0.081)	-0.158 (0.049)	-0.166 (0.074)	-0.133 (0.069)	-0.166 (0.075)
h	2.430 (1.160)	4.042 (2.011)	2.715 (1.013)	3.262 (1.723)	3.905 (1.583)	2.931 (1.543)
$\log \mathcal{L}$	-227.596	-126.229	-224.167	-122.711	-221.536	-122.692
$Q(1)$	2.552	0.158	0.297	0.392	0.047	0.072
$Q(5)$	7.423	2.890	6.721	4.582	7.432	4.196
$Q_2(1)$	1.483	1.687	2.648	2.104	3.419	1.746
$Q_2(5)$	3.261	3.738	4.410	3.086	7.429	3.040

Notes. Numbers in parentheses are asymptotic standard errors and $_{ss}$ denotes sub-sample.

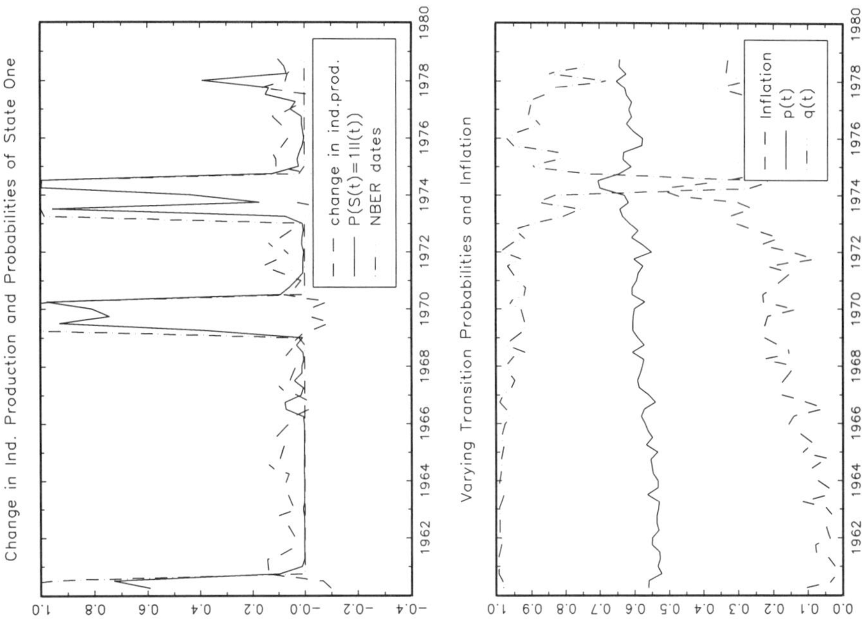

Figure 8b

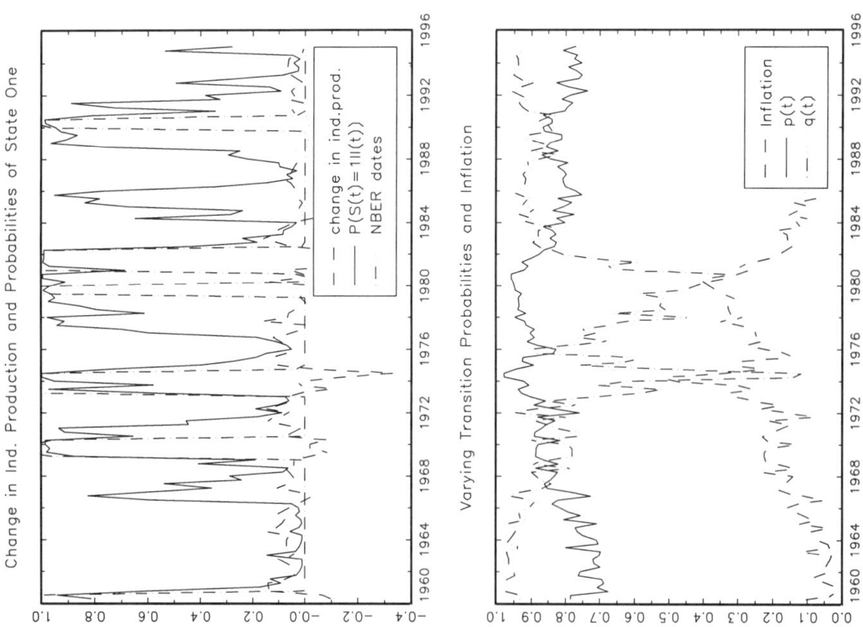

Figure 8a

mid-seventies and early eighties are preceded by two periods of very high inflation rates. For these periods the probability q_t of remaining in a boom decreases and the probability p_t of remaining in a recession increases (see Ravn and Sola (1995) for further evidence). A possible interpretation of this result is that it reflects policy responses to periods of high inflation, in which case these effects are fully revealed by the federal funds rates. These results are consistent with both the causality tests reported in Table 1b and the results and discussion of the effects of *M0*.

Next we investigate the effects of the *real* federal funds rate, the variable which potentially has the most important effects on the real economy and therefore should be a useful leading indicator for the switches between booms and recessions, as most theoretical models predict. Thus, we looked at the ex-post real interest rate constructed as the federal funds rate minus the inflation rate. The results presented in Table 4 differ from those for the nominal funds rate in that the real funds rate significantly affects both the persistence of the boom state and the persistence of the recession state. We find that β_0 and that β_1 is positive, and both are significantly different from zero.

The results imply that an increase in the real federal funds rate helps to predict the end of a boom state and to predict a higher probability of remaining in a recession state. Thus, a lower real federal funds rate is associated with a higher probability of switching from a recession to a boom and with a higher probability of staying in a boom as one would expect.

The estimated probabilities of the two states and the estimated transition probabilities are shown in Figure 9a. The identification of the two states is very similar to that for the nominal funds rate. The plots of the transition probabilities illustrate that the periods with significant movements in the real federal funds rate are associated with major movements in the estimated transition probabilities. The recession of the early seventies was preceded by a higher rel federal funds rate and was therefore associated with a significant increase in p_t. Furthermore, OPEC I was associated with both an increase in the probability of remaining in the recession state and a decrease in the probability of remaining in a boom state. We also find that the extraordinary high real interest rates of the early 1980s led to very high values of the transition probability associated with the recession state and extremely low values for the transition probability associated with the boom state. Finally, it is worth noting that the low real federal funds rate of the early 1990s only affected the time-varying probabilities associated with the recession state.

The results of the sub-sample seem to strengthen the conclusions for the full sample. Figure 9b clearly shows clearly that increases in the real interest rate preceded the recessions of the early and middle 1970s. It also shows that decreases in the real interest rate preceded the end of both recessions. The time-varying transition probabilities successfully capture this regularity, and are therefore successful predictors of the beginning and ending of booms and recessions. Thus, while the results for the nominal funds rate appear to be structurally unstable, this is not the case for the real funds rate, a results that may indicate that the real transmission mechanism of the economy remained robust to changes in the Fed's operating procedures.

The results presented in Table 4 using *real M1* show that an increase in the real amount of money help to predict the end of a recession ($\beta_1<0$ and significant at the 90% confidence level) while is also increases the probability of staying in a boom

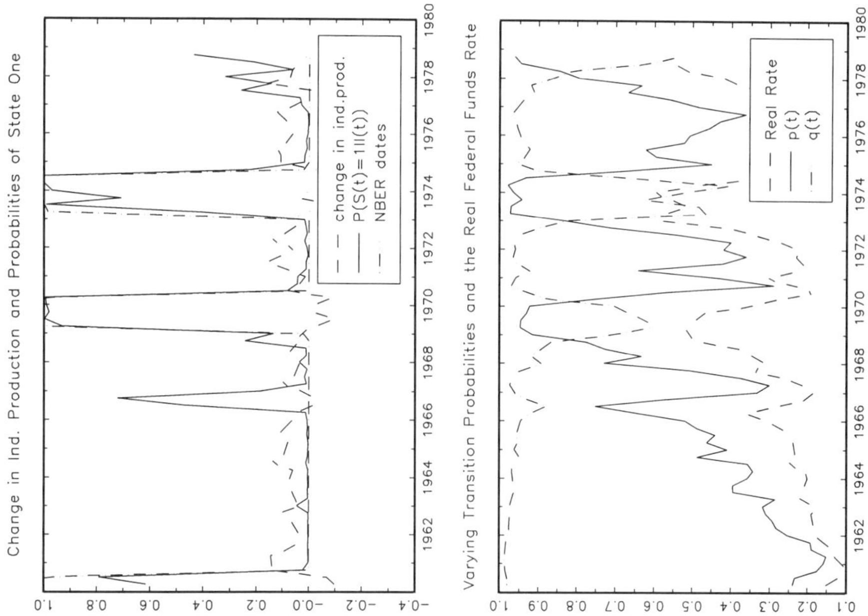

Figure 9b

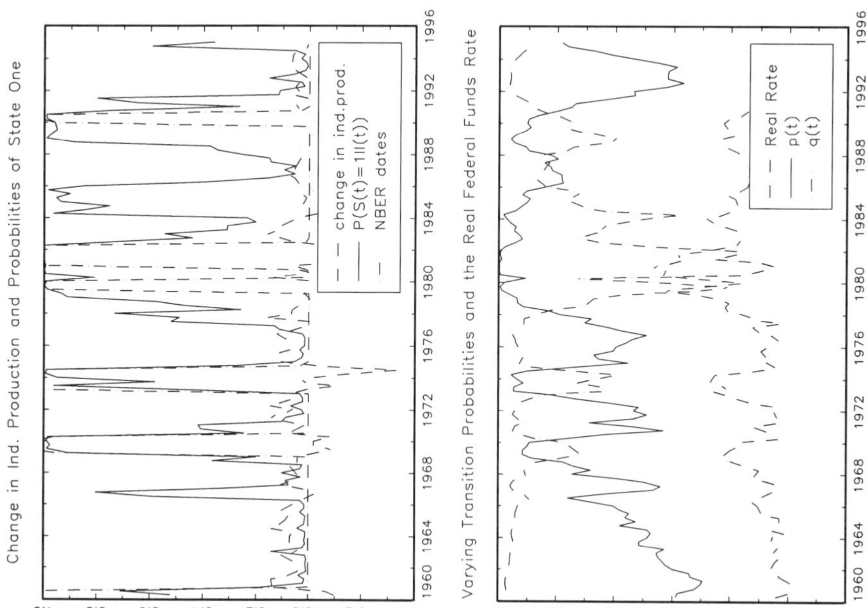

Figure 9a

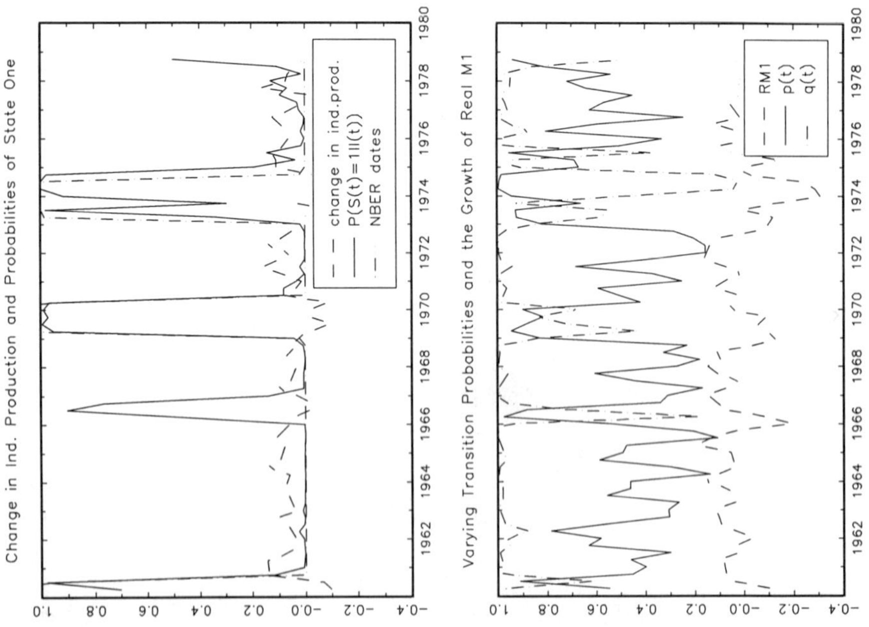

Figure 10b

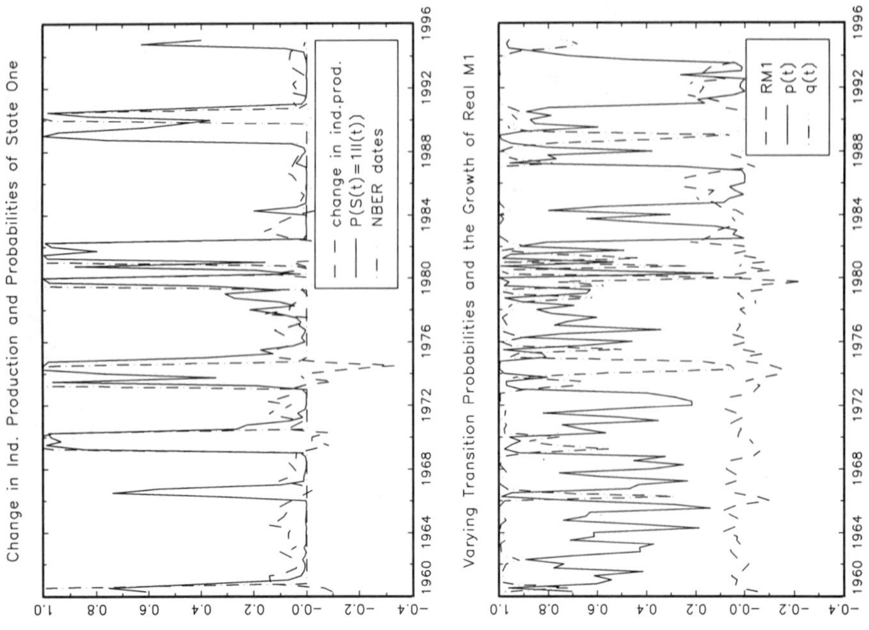

Figure 10a

(β_0>0 and significant). The top panel of Figure 10a shows a separation of the states similar to the *NBER* dates, while the bottom panel shows that even though the time-varying probabilities of the states are very (too) volatile, the model correctly predicts the beginning and ending of booms and recession. For example, there is a decrease in the real amount of money preceding each recession, while the real amount of money seems to increase before the end of each recession.

Thus, lagged endogenous variables are helpful predictors of the transition dynamics of output but one cannot, of course, make any causal interpretation of these results. Instead the results can be used to provide stylized facts for the business cycle literature.

6. Concluding Remarks

We have studied how macrovariables help to predict the transitional dynamics of output. We have looked at a discrete two-state model of the change in industrial production and examined whether macrovariables affect the transition probabilities of the model. This framework allows one to study whether there are asymmetric relationships between the variables and whether a given macrovariable helps to predict the beginning and ends of the phases of the business cycle. In this respect, we found that the time-varying transition probability model provides a very useful framework to conduct this analysis. We also showed that this model in some cases leads to an identification of the states of the economy which is practically identical to the business cycle chronology of the *NBER*.

We argued that in some cases the results should be interpreted as providing leading indicators of business cycle dynamics, while in other cases they have a much clearer causal interpretation. We have argued that it is important to look at the exogeneity of the regressors before one can decide on the proper interpretation of the results: finding that an endogenous variable helps predict transitions from one state of the cycle to another should not be given any causal interpretation. We also illustrated how time-varying transition probability models are affected by breaks in the relationship between the variables that enter in the transition probabilities and the change in industrial production. We showed that in those cases it may greatly affect inference about the states, and therefore may make any possible policy evaluation invalid.

We hope that the results of the analysis help to provide stylized facts that one might wish to address in future work. For example, the finding that ends of booms (recessions) were associated with higher (lower) levels of the federal funds rate for the early part of the sample might give further impetus to the literature on liquidity effects. Also, the result that the various monetary aggregates seem to be quite differently related to the output dynamics would be interesting to examine in further detail.

Acknowledgements

The authors wish to thank Fabio Canova, Haris Psaradakis, Howard Wall, and Ron Smith for comments. The usual disclaimer applies.

Notes

1. As an alternative to our approach of endogenizing the transition variables, one could instead look at models where the switching is dictated directly by other (observable) variables. This is the approach taken by researchers who have looked at transition regression models where the state variable is assumed to be observable. See Granger and Teräsvirta (1993) for a survey.
2. This squares well with many economic theories of the relationship between economic policy and the business cycle. One example is the line of research that has looked into financial market imperfections, e.g., Bernanke and Gertler (1992) or Greenwald and Stiglitz (1993). According to these types of models, monetary policy affects aggregate activity stronger in recessions than in booms because households and firms are more likely to be credit-constrained in recessions. Thus, the prediction is that monetary policy might be used to shorten recessions (thus negatively affecting p_t if state 1 is the low-mean state) while being close to neutral in booms.
3. In our discussion we use different expressions such as Granger causality, exogenous policy variables, response policy variables, and lagged endogenous variables. In order to analyze the effects of the business cycle it is not crucial to inquire whether a policy variable is Granger caused or not by another variable. For example, the government may increase interest rates in order to reduce inflation (inflation Granger causes interest rates) and by doing this, it drives the economy into a recession. The fact that interest rates are Granger caused by inflation (what we call a response variable) should not prevent us from interpreting the results as if the increase in interest rates caused the recession. In this example we may also say that an increase in inflation helps to predict the beginning of a recession. On the other hand, variables such as *M1* or *M2* may not be Granger caused by any other variables but may simply reflect the equilibrium between the demand for and supply of money. The best way to think about the distinction between exogenous and endogenous variables for interpretation of the results is in terms of how they enter in a macroeconomic model. In the remainder of the paper we stipulate whenever this interpretation changes.
4. Maximization of the likelihood function was carried out using a quasi-Newton algorithm that approximates the Hessian according to the Broyden-Fletcher-Goldfarb-Shanno update computed from numerical derivatives. Asymptotic standard errors were calculated from numerical second-order derivatives evaluated at the optimum.
5. This relative success of the univariate filter in classifying the states is partially due to the fact that we have allowed for fatter tails in the distribution of the errors, ε_t. This allows for outliers which would be classified as separate regimes when using a normal distribution.
6. Notice that the results for *M2* are inconsistent with the "credit view" interpretation, which implies that we should be cautious when interpreting the results for *M1*. It is not clear from this theory why such results should appear for *M1* but not for other monetary policy indicators.

References

Baba, Y., D. F. Hendry, and R. M. Starr (1992). "The Demand for *M1* in the US, 1960-88," *Review of Economic Studies*, 59, 25-61.

Barnett, W. A. (1980). "Economic Monetary Aggregates: An Application of Index Number and Aggregation Theory," *Journal of Econometrics*, 14, 11-48.

Barnett, W. A., M. Kirova, and M. Pasupathy (1995). "Estimating Policy-Invariant Deep Parameters in the Financial Sector When Risk and Growth Matter," *Journal of Money, Credit, and Banking*, 27, 1402-1429.

Bernanke, B. and M. Gertler (1989). "Agency Costs, Net Worth, and Business Fluctuations," *American Economic Review*, 79, 14-31.

Chari, V. V., L. J. Christiano, and M. Eichenbaum (1995). "Inside Money, Outside Money, and Short-Term Interest Rates," *Journal of Money, Credit, and Banking*, 27 1354-1386.

Christiano, L. J., M. Eichenbaum, and C. Evans (1996). "The Effects of Monetary Policy Shocks: Evidence From the Flow of Funds," *Review of Economics and Statistics*, 16-34.

Christiano, L. J., M. Eichenbaum, and C. Evans (1997). "Modeling Money," working paper, Department of Economics, Northwestern University.

Coleman, W. J., C. Gilles, and P. A. Labadie (1996). "A Model of the Federal Funds Market," *Economic Theory*, 7, 337-357.

Diebold, F. X., J. H. Lee, and G. C. Weinbach (1994), "Regime Switching with Time-Varying Transition Probabilities," Chp. 10 in Hargreaves, C. (ed.) *Nonstationary Time Series Analysis and Cointegration*. Oxford: Oxford University Press.

Diebold, F. X. and G. D. Rudebusch (1996), "Measuring Business Cycles: A Modern Perspective," *Review of Economics and Statistics*, 78, 67-77.

Filardo, A. J. (1994). "Business Cycle Phases and Their Transitional Dynamics," *Journal of Business and Economic Statistics*, 12, 299-308.

Ghysels, E. (1994). "On the Periodic Structure of the Business Cycle," *Journal of Business and Economic Statistics*, 12, 289-298.

Granger, C. W. J. and T. Teräsvirta (1993). *Modeling Non-Linear Economic Relationships*. Oxford: Oxford University Press.

Greenwald, B. C. and J. E. Stiglitz (1993), "Financial Market Imperfections and Business Cycles," *Quarterly Journal of Economics*, 108, 77-114.

Hamilton, J. D. (1989). "A New Approach to the Economic Analysis of Nonstationary Time Series and the Business Cycle," *Econometrics*, 57, 357-384.

Hamilton, J. D. (1990). "Analysis of Time Series Subject to Changes in Regime," *Journal of Econometrics*, 45, 39-70.

Hamilton, J. D. (1994). *Time Series Analysis*. Princeton: Princeton University Press.

Ljung, G. M. and G. E. P. Box (1978). "On a Measure of Lack of Fit in Time Series Models," *Biometrika*, 65, 297-303.

McCallum, B. T. (1983). "A Reconsideration of Sims's Evidence Regarding Monetarism," *Economics Letters*, 13, 161-171.

McLeod, A. I. and W. K. Li (1983). "Diagnostic Checking ARMA Time Series Models Using Squared-Residual Autocorrelations," *Journal of Time Series Analysis*, 4, 269-273.

Ravn, M. O. and M. Sola (1995). "Stylized Facts and Regime Changes: Are Prices Procyclical?" *Journal of Monetary Economics*, 36, 497-526.

Sims, C. A. (1986). "Are Forecasting Models Usable for Policy Analysis?" *Quarterly Review* (Winter), Federal Reserve Bank of Minneapolis.

Sims, C. A. and T. Zha (1994). "Does Monetary Policy Generate Recessions?" working paper, Department of Economics, Yale University.

Zarnowitz, V. (1992), *Business Cycles: Theory, History, Indicators, and Forecasting*, *NBER Studies in Business Cycles*, 27. Chicago: University of Chicago Press.

13 SEARCHING FOR THE SOURCES OF ARCH BEHAVIOR: TESTING THE MIXTURE OF DISTRIBUTIONS MODEL

Patrick de Fontnouvelle
General Electric

1. Introduction

Over the past decade, a large body of literature has evolved documenting changing volatility (ARCH) in financial data. For a review, see Bollerslev, Chou, Kroner (1992) and Bollerslev, Engle, Nelson (1994). This literature has generated numerous statistical methods for volatility forecasting, which are widely used in both the financial and the policy making sectors.

Despite the ubiquity of their application, current methods for forecasting volatility suffer from two serious drawbacks. The first is their poor performance: they seem to be largely incapable of forecasting large volatility shifts such as that which followed the crash of 1987. In addition, West and Cho (1995) have shown that a variety of popular forecast methods cannot consistently outperform (out of sample) a simple constant volatility model.

The second drawback is that current methods do not result from a deep understanding of the structural forces that generate ARCH behavior, but rather are purely statistical models that seem to fit financial data. Nelson (1992) shows that misspecified ARCH models provide good short term forecasts for a wide variety of true data generating processes. For accurate longer term forecasts, however, one must choose a more correctly specified model of volatility. Structural models provide a natural base for making his choice.

This paper investigates the structural foundations of ARCH behavior in the context of a well-known financial model: the Mixture of Distributions Model (MODM). We choose to analyze the MODM because it is the simplest model capable of explaining the volatility clustering effects seen in asset market data. In the MODM, these effects are generated by serial correlation in an exogenous information arrival process: markets are volatile when a lot of news arrives at the market and stable when little news arrives.

Various formulations of the MODM have been proposed by Andersen (1996), Clark (1973), Epps and Epps (1976), Harris (1986, 1987), Tauchen and Pitts (1983), among others. These authors have also provided evidence that many of the model's empirical implications are consistent with securities data. More recently, however, Hiemstra and Jones (1994), Lamoureux and Lastrapes (1994), and Richardson and Smith (1994) have uncovered empirical evidence against the MODM.

The issue of whether or not the MODM is consistent with the data is clouded by the fact that the existing empirical analysis relies on auxiliary assumptions that are not fundamental to the model. Although the MODM in its most general form places no restrictions on the joint process governing information arrival, volume, and volatility, many authors do specify this process in order to make the model empirically useful Because Hiemstra and Jones (1994), Lamoureux and Lastrapes (1994), and Richardson and Smith (1994) all rely on such auxiliary assumptions, they leave open the possibility that the MODM holds for alternate untested parameterizations.

We avoid such problems by observing that the MODM's main empirical prediction is that all intertemporal dependence between volatility and trading volume takes place through the information arrival process. In contrast to the empirical predictions tested by previous work, this prediction is invariant to any parametric assumptions one might make about the joint process for information arrival, volatility, and trading volume.

The initial motivation for this work was generated by the graphical results of Gallant, Rossi, Tauchen (GRT) (1993) and Tauchen, Zhang, Liu (TZL) (1996). Intuitively, one would expect that if all intertemporal dependence between volatility and trading volume takes place through the information arrival process (as the MODM requires), then these two time series should display a certain intertemporal symmetry. In particular, the effect of present volatility on future volume should be identical to the effect of present volume on future volatility. Even a cursory look at the results of GRT and TZL suggests that this is not so.

This paper follows up on the initial motivation by providing formal analysis and a testing procedure for the above intuition. With respect to the MODM, our results and those of GRT and TZL can then be viewed as complementary. We show that the departure seen in the data from the MODM's predictions is statistically significant, while the graphical results of GRT and TZL show that it is economically significant.

On a more technical level, our test of the MODM relies upon the concept of time reversibility. Heuristically speaking, a process is reversible if it "looks the same" going forwards and backwards in time. We show that if the information arrival process is reversible, then the MODM requires that time series for volatility and trading volume display a type of intertemporal symmetry which is not seen in the data. If, on the other hand, the information arrival process is irreversible, then the time series for returns and volume must also be irreversible. We find no evidence of such irreversibility in the data. Because any information arrival process must be either reversible or irreversibile, we interpret our results as statistical evidence against the MODM itself.

The key feature of our test is that it does not rely upon a specific parameterization of the joint stochastic process for price, volume, and information arrival, and therefore applies to all previously proposed versions of MODM. The evidence we find against the MODM cannot thus be attributed to auxiliary distributional assumptions, but rather suggests that structural factors outside the context of the MODM are at least partly responsible for determining financial market volatility.

Section 2 reviews the Mixture of Distributions Model, as well as previous empirical tests of the MODM. Section 3 presents a version of the MODM which does not rely on auxiliary assumptions about the relation between information arrival, volatility , and trading volume. Section 4 uses the results of Gallant, Rossi, Tauchen (1993) to provide motivation for our analysis. Section 5 describes the data. Section 6 develops theoretical restrictions on the data. Section 7 develops empirical tests based on these restrictions, and Section 8 extends the model to allow trading volume to display persistence that is unrelated to the rate of information arrival. Section 9 concludes and suggests directions for further research.

2. The Mixture of Distributions Model

Mixture of Distributions Models explain volatility shifts in financial markets by attributing them to changes in the rate of information arrival. When many pieces of information arrive at the market on a given day, the market moves through many price equilibria, and there is thus a much greater chance of a large price movement than on a day when few pieces of information arrive at the market. If one assumes that the expected trading volume generated at each price equilibrium is a constant, then total daily trading volume will also be positively related to the rate of information arrival. The MODM thus implies that volatility and trading volume are contemporaneously correlated. If the stochastic process representing the rate of information arrival is serially correlated, then the time series for volatility and trading volume will also be serially correlated.

Andersen (1996) develops an explicit structural model in which informational asymmetries and agents' liquidity needs generate trade in response to the arrival of new information. The rate of information arrival varies from day to day, thus generating variation in both price volatility and trading volume. Denote time t returns by r_t, time t volume by V_t, and the time t rate of information arrival by k_t. Andersen estimates the following empirical version of his MODM:

$$r_t = \bar{r} + Z_t\sqrt{k_t} \tag{1}$$

$$V_t \sim Po(m_0 + m_1 k_t), \qquad m_0 > 0 \tag{2}$$

where $\bar{r}$ denotes mean returns, "Po" denotes the Poisson distribution, and Z_t is i.i.d. $N(0,1)$ and independent of all other variables. Note that under this specification, the variance of returns and expected trading volume are both increasing functions of k_t, the rate of information arrival. Andersen uses GMM to test the joint unconditional moments for returns and volume implied by (1-2), and finds that the x^2 statistic for overidentifying restrictions does not indicate any misspecification.

Harris (1986) and Harris (1987) assume that on each day, k_t pieces of information arrive at the market. Price change and trading volume are assumed to be independent and identically distributed across information arrivals. If the number of information arrivals on each day is large, then the Central Limit Theorem implies the following version of the MODM:

$$r_t \sim N(ak_t, bk_t) \tag{3}$$

$$V_t \sim N(ck_t, dk_t) \tag{4}$$

$$Cov(r_t, V_t | k_t) = 0 \tag{5}$$

where *a, b, c, d* are constants. Harris (1986) finds empirical support for the cross-security implications of MODM (3-5), and Harris (1987) finds support for the transactions data implications of MODM (3-5).

Tauchen and Pitts (1983) use a similar motivation to derive the following MODM:

$$r_t = \sigma_1 z_{1t}\sqrt{k_t} \tag{6}$$

$$V_t = \mu_2 k_t + \sigma_2 z_{2t}\sqrt{k_t} \tag{7}$$

where $\{z_{1t}\}$ and $\{z_{2t}\}$ are i.i.d. $N(0,1)$ random variables, and z_{1t}, z_{2t}, and k_t are mutually independent. Here also, $Var(r_t|k_t) = \sigma_1^2 k_t$ and $E[V_t|k_t] = \mu_2 k_t$ are increasing functions of k_t. Tauchen and Pitts (1983) estimate (6-7) using maximum likelihood, and find that the parameter estimates have the correct signs and are of reasonable magnitude. Other papers investigating the MODM include Epps and Epps (1976), Gallant, Hsieh, and Tauchen (1991), Jain and Joh (1986), Lamoureux and Lastrapes (1990), Locke and Sayers (1992), Morgan (1976), and Westerfield (1977).

Despite the MODM's long history, only recently have authors begun to uncover statistical evidence against the model. After using an EGARCH[1] model to control for volatility persistence in returns, Hiemstra and Jones (1994) find that there is still significant causality from trading volume to returns. Because their version of the MODM is that returns and trading volume are related only through volatility effects, they view their results as evidence against MODM. We argue that their conclusions lack robustness in two ways. First, the general version of the MODM *does* require that returns and trading volume be related only through the rate of information arrival. This, however, does not rule out a relationship between trading volume and moments of returns other than the second. Second, the MODM imposes no restrictions on the parametric form of volatility persistence; using EGARCH to remove this persistence thus amounts to a nuisance parametric assumption. In other words, it is possible that they are rejecting the EGARCH specification of volatility rather than the MODM itself.

Richardson and Smith (1994) use Generalized Method of Moments (GMM) to test the joint moment restrictions on mean, variance, skewness, and kurtosis implied by the MODM (3-5). Richardson and Smith find that "at both the daily and transactions data level, the overidentifying restrictions test provides strong evidence against MODM."

Andersen (1996), however, suggests that Richardson and Smith (1994) are merely rejecting the auxiliary assumption that volume is distributed normally conditional on k_t. In order to investigate this possibility, he first re-estimates their model, and rejects it. Andersen then finds that when volume is assumed to be distributed Poisson as in (1-2), the MODM is no longer rejected.

Lamoureux and Lastrapes (1994) estimate a version of (6-7) in which $\{z_{2t}\}$ can display serial dependence, although it must still be independent of $\{z_{1t}\}$. They then substitute the estimated $\{k_t\}$ into (6-7) to calculate the estimated residuals $\{\hat{z}_{1t}\}$. Since (6-7) requires the true residuals to be i.i.d., and the estimated $\{\hat{z}_{1t}\}$ are heteroskedastic, Lamoureux and Lastrapes reject the model.

Andersen suggests that Lamoureux and Lastrapes (1994) also might be rejecting a nuisance assumption rather than the MODM itself. He notes that they implicitly restrict volume to be zero when information flow is zero. In the context of (1-2), this is equivalent to the restriction that $m_0 = 0$. Andersen finds that the null hypothesis $m_0 = 0$ is strongly rejected in a standard likelihood ratio test. Having thus dismissed these empirical rejections of the MODM, Andersen concludes that "the empirical evidence is strongly in favor of our 'modified' version of the MODM."

We conclude that although the MODM *per se* imposes no specific parameterization upon the joint process for information arrival, returns, and trading volume, previous empirical investigations have imposed restrictions on this process in order to test the MODM. It should thus not be too surprising that the results of these investigations depend strongly upon what parametric restrictions were chosen. We avoid this difficulty by testing a general MODM that makes minimal assumptions.

3. A General Mixture of Distributions Model

This section introduces a set of assumptions we use to define what constitutes an MODM. These assumptions are quite general; they simply state that all dependence in the returns and volume series must result from dependence in the rate of information arrival. It is a simple exercise to verify that all of the models mentioned in Section 2 satisfy these assumptions. The sole exception is the model of Lamoureux and Lastrapes (1994), who allow trading volume to display serial dependence that is unrelated to the rate of information arrival. Section 8 will extend the model presented here to allow for such dependence.

Information flow into the market is represented by a stationary random scalar $\{k_t\}$ with continuous forward transition density $f_k(k_t|k_{t-1})$ and backward transition density $f_k'(k_{t-1}|k_t)$. Returns are determined by the continuous conditional density function $i(r_t|k_t)$. Trading volume is determined by the continuous conditional density function $h(V_t|k_t)$. We define our version of the MODM by the following set of Assumptions:

A1: $\mathrm{E}[abs(r_t)|k_t] \equiv \int abs(r)\, i(r|k_t)\, dr$ exists and is monotonically increasing in k_t.

A2: $\mathrm{E}[V_t|k_t] \equiv \int V h(V|k_t)\, dV$ exists and is monotonically increasing in k_t.

A3: Define

$$I_\infty \equiv \{\ldots, k_{t-1}, r_{t-1}, V_{t-1}, k_t, r_t, V_t, k_{t+1}, r_{t+1}, V_{t+1}, \ldots\}.$$

Then $\Pr(r_t \in B|I_\infty / r_t) = \Pr(r_t \in B|k_t)$ and $\Pr(V_t \in B|I_\infty / V_t) = \Pr(V_t \in B|k_t)$ for any Borel set B in $\Re$.

Assumptions A1 and A2 say that expected volume and volatility both increase as the rate of news arrival increases. Assumption A3 says that all knowledge concerning r_t and V_t (short of their actual values) is contained in k_t. Although further technical assumptions will be necessary to test the MODM, these assumptions carry little economic significance. They are also extremely weak in that they are satisfied by a wide range of models, including the ones reviewed in Section 2.

4. Motivation for Our Tests

In this section, we provide a simple graphical motivation for our test of the Mixture of Distributions Model. This motivation comes from Gallant, Rossi, and Tauchen [GRT] (1993), who use semi-parametric methods to investigate the intertemporal dynamics of volatility and trading volume. Using daily S&P index data from 1928 to 1987, GRT use nonparametric techniques to calculate the nonlinear impulse response functions found in Figures 1 and 2.

Figure 1 presents the impulse response function of trading volume to volatility shocks. Figure 2 presents the impulse response function of volatility to trading volume shocks. In both figures, the dashed line corresponds to a positive shock, and the solid line to a negative shock. The heavy solid line corresponds to the baseline case, in

Figure 1: Impulse Response of Trading Volume to Price Shock
From Gallant, Rossi, Tauchen (1993), p. 895

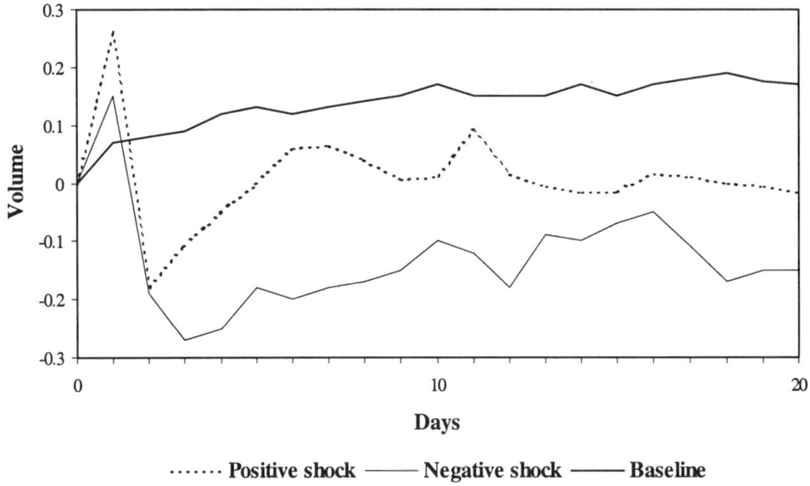

Figure 2: Impulse Response of Volatility to Volume Shock
From Gallant, Rossi, Tauchen (1993), p. 896

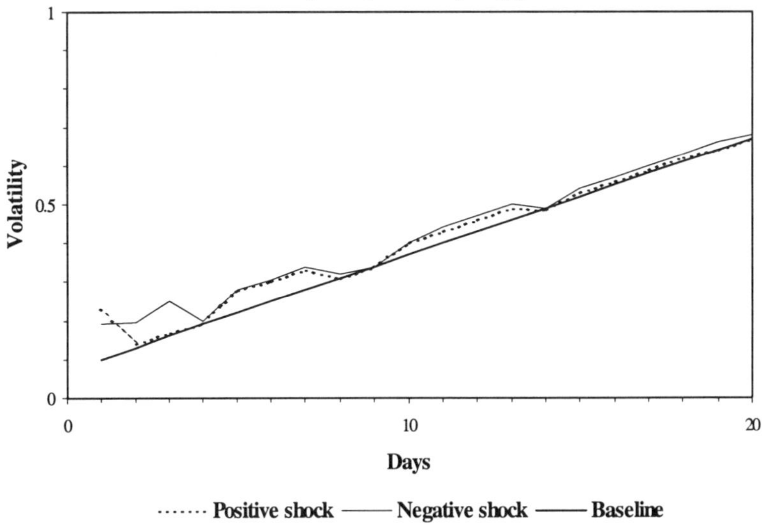

which there is no shock.

It is clear from these figures that the effect of present volume on future volatility is different from the effect of present volatility on future volume. In particular, shocks to present volatility have a strong effect on future volume, while shocks to volume have little effect on future volatility.

I argue that these graphs already provide intuitive evidence against the Mixture of Distributions Model. To see why, recall that MODM requires all intertemporal dependence between volume and volatility to occur through a common factor, the rate of information arrival. A strong relation between present volatility and future volume can occur only if there is strong serial correlation in the rate of information arrival. Serial correlation in the rate of information will invariably lead to a strong relation between present volume and future volatility. So under the MODM, shocks to present volatility can have a strong effect on future volume if and only if shocks to present volume also have a strong effect on future volatility. The MODM thus requires the time series for volatility and trading volume to satisfy a strong intertemporal symmetry restriction. The results of GRT show that this symmetry restriction does not hold empirically.

Tauchen, Zhang, and Liu (1996) have expanded GRT's analysis to the stocks of four individual companies on the NYSE. These authors find an intertemporal relationship between Volume and Volatility similar to that displayed in Figures 1 and 2. Examining hourly NYSE returns and trading volume, Jain and Joh (1986) also find that "although there is some evidence of (causality from volume to returns), it is quite weak compared to that from returns to volume and is essentially insignificant." The results of Tauchen, Zhang, and Liu (1996) and Jain and Joh (1986) lend credence to the claim that GRT's findings (with respect to the intertemporal relation between volatility and volume) are not unique to their data set, but instead constitute a general feature of financial time series. The analysis in this paper, which is carried out for 30 individual companies, provides additional empirical support for this claim.

Besides serving a didactic purpose, this graphical motivation for our analysis serves a substantive purpose as well: GRT's graphical results provide a useful gauge for the economic significance of our results. Because any model is by definition merely an approximation of reality, one would like to know if a statistical rejection indicates a substantive failure of the model or represents "approximation error." Figures 1 and 2 suggest that our results indicate not an "approximation error," but rather an economically significant rejection of the MODM's central empirical prediction.

5. Data

The data consist of daily returns and trading volume for the Dow Jones 30 firms from 1982 to 1993. Appendix A contains a list of these firms and their stock market symbols. The sample is split into two five year subsamples, 1982-1986 and 1989-1993, in order to avoid the crash of 1987. The data are adjusted for splits and dividends. The trading volume series used in the empirical analysis consists of the raw trading volume divided by the number of shares outstanding on each day.

6. Developing Restrictions on the Data

A stochastic process is time reversible if it has the same stochastic properties going forwards and backward in time. More formally, Liggett (1985) defines a process $\{x_t\}$ to be reversible if the following equivalent conditions hold:

R1: $E[f(x_t)g(x_{t+j})] = E[f(x_{t+j})g(x_t)]$ for all t, j, and continuous functions f and g.

R2: $\{x_t, -\infty<t<\infty\}$ and $\{x_{-t}, -\infty<t<\infty\}$ have the same joint distribution.

Rothman (1992) has proposed reversibility as a criterion for testing for linearity. In this paper, we use reversibility merely as a stepping stone for testing the MODM, and are interested in neither reversibility or linearity *per se*. Readers further interested in reversibility should refer to Kelly (1979).

Our approach will be to develop one set of restrictions on the data based on the premise that the information arrival process $\{k_t\}$ is reversible, and another set of restrictions based on the premise that $\{k_t\}$ is irreversible. Because the union of the set of reversible processes and the set of irreversible processes is the set of all stationary processes, we can assess the performance of the MODM for any stationary information arrival process $\{k_t\}$.

Proposition 1: *If $\{k_t\}$ is reversible, then*

$$E[V_t \text{abs}(r_{t+1})] = E[\text{abs}(r_t)V_{t+1}]. \tag{8}$$

Proof: *See Appendix.*

Proposition 2: *Suppose that $\{r_t, V_t, k_t\}$ satisfy the auxiliary assumptions A5 and A6. Then if $\{k_t\}$ is not reversible, $\{\text{abs}(r_t)\}$ and $\{V_t\}$ are not reversible.*

Proof: *See Appendix.*

Assumptions A5 and A6 are relegated to the appendix because they are violated only by processes for $\{r_t, V_t, k_t\}$ which are mathematical curiosities. To make empirical use of Proposition 2, we need the following:

Proposition 3: *If the probability density function $g(x_t, x_{t+1})$ is not symmetric on a set B such that $Pr(B)>0$, then there exist natural numbers i, j, N such that $M>N$ implies*

$$\mathrm{E}[\langle x_t,M\rangle^i\langle x_{t+1},M\rangle^j - \langle x_t,M\rangle^j\langle x_{t+1},M\rangle^i] \neq 0 \tag{9}$$

where $\langle x,y\rangle$ *denotes the minimum of x and y.*

Proof: *See Appendix.*

Proposition 3 lists a countable set of restrictions on the observable processes $\{abs(r_t)\}$ and $\{V_t\}$, one of which must be violated if $\{k_t\}$ is irreversible. The variable M is necessary for the proof of Proposition 3 because the Weierstrass approximation theorem requires the function being approximated to have bounded domain. This bound M will also serve to ensure that $\langle x_t,M\rangle$ has finite moments of all orders.

7. Testing Restrictions on the Data

First assume $\{k_t\}$ is reversible, and invoke Proposition 1. We wish to test the null hypothesis

$$H_0 : \mathrm{E}[x_t] = 0 \tag{10}$$

where

$$x_t \equiv abs(r_t)\,V_{t-1} - abs(r_{t-1})\,V_t. \tag{11}$$

In order to use a central limit theorem, we make the usual assumptions concerning mixing and moment conditions:

A4: $\{x_t\}$ is α-mixing. Furthermore, $\exists\,\rho$ such that $\alpha(m)$ is of size $\rho/(\rho-1)$ and $E\,|x_t|^{2\rho}<C<\infty$.

This mixing condition places restrictions on the amount of dependence in the $\{x_t\}$ process. For details, refer to White (1984). Under Assumption A4, we can use the White-Domowitz (1984) Central Limit Theorem to calculate the following statistic:

$$TS_1 \equiv \frac{1}{\sqrt{n}\sigma_n}\sum_{t=1}^{n} x_t \tag{12}$$

$$\sigma_n \equiv Var\left[n^{-1/2}\sum_{t=1}^{n} x_t\right] \tag{13}$$

Table 1. Tests of Hypothesis (10)

Co.	82-86	89-93	Co.	82-86	89-93
AA	* 0.146	* 0.070	JPM	* 0.280	* 0.100
ALD	0.027	0.036	KO	2.0E-4	0.001
AXP	8.4E-5	0.029	MCD	6.0E-4	0.010
BA	0.001	5.7E-5	MMM	4.6E-5	6.0E-4
BS	0.024	* 0.060	MO	0.007	0.001
CAT	0.007	0.025	MRK	9.0E-4	5.0E-4
CHV	5.9E-5	* 0.085	PG	0.006	0.005
DD	7.2E-6	0.003	S	6.0E-4	0.004
DIS	0.005	1.4E-5	T	0.002	0.014
EK	1.5E-6	8.0E-4	TX	* 0.054	* 0.110
GE	0.002	0.003	UK	* 0.260	0.008
GM	3.9E-6	2.4E-5	UTX	0.006	0.004
GT	0.028	2.0E-4	WX	0.008	* 0.520
IBM	8.8E-9	0.001	XON	4.0E-4	0.001
IP	3.0E-4	0.010	Z	0.010	* 0.050

where n denotes the sample size. The sample variance σ_n is calculated using a prewhitened Parzen kernel estimator with automatic bandwidth selection. See Andrews (1991) and Andrews and Monahan (1992) for details.

Table 1 presents the p-values of TS_1 for the Dow 30 companies over the two five year periods 1982-1986 and 1989-1993. An asterisk indicates that the null could not be rejected at the 5% level. The null is broadly and strongly rejected for almost all of the companies in both samples.

Recall that (10) holds only if the information arrival process $\{k_t\}$ is reversible. This rejection thus leads us to conclude that if our version of the MODM is valid, then $\{k_t\}$ must be irreversible. If $\{k_t\}$ is irreversible, then Proposition 2 implies that $\{|r_t|\}$ and $\{V_t\}$ must both be irreversible. In particular, there must be some i, j, k, l, M_r, M_v such that

$$\mathrm{E}[\langle |r_t|, M_r\rangle^i \langle |r_{t+1}|, M_r\rangle^j - \langle |r_t|, M_r\rangle^j \langle |r_{t+1}|, M_r\rangle^i] \neq 0 \tag{14}$$

$$\mathrm{E}[\langle V_t, M_v\rangle^k \langle V_{t+1}, M_v\rangle^l - \langle V_t, M_v\rangle^l \langle V_{t+1}, M_v\rangle^k] \neq 0. \tag{15}$$

We choose M_r so that

$$n^{-1} \sum_{t=1}^{n} I(abs(r_t) < M_r) = 0.99 \tag{16}$$

where I denotes the indicator function. We thus lose power against irreversibility that occurs only in the extreme tails of the distribution of $abs(r_t)$, but believe in any case that such an alternative would be highly improbable. The bound M_v is chosen in a similar manner.

If one is to use the irreversibility of $\{k_t\}$ as the basis for supporting the MODM despite its violation of the symmetry restriction (10), one would like to find some evidence of this irreversibility in either $\{abs(r_t)\}$ or $\{V_t\}$. Our procedure is to test the null hypotheses

$$\mathrm{E}[\langle |r_t|, M_r\rangle^i \langle |r_{t+1}|, M_r\rangle^j - \langle |r_t|, M_r\rangle^j \langle |r_{t+1}|, M_r\rangle^i] = 0 \tag{17}$$

$$\mathrm{E}[\langle V_t, M_v\rangle^k \langle V_{t+1}, M_v\rangle^l - \langle V_t, M_v\rangle^l \langle V_{t+1}, M_v\rangle^k = 0 \tag{18}$$

for all $i, j, k, l \leq 10$. Because $\langle |r_t|, M_r\rangle$ and $\langle V_t, M_u\rangle$ have finite moments of all orders, we can construct as before a test statistic using the White-Domowitz (1984) Central Limit Theorem.

Table 2 presents the results of testing the null hypotheses (17-18), which must be violated for some (i, j) if $\{k_t\}$ is irreversible. The figures presented are the minima of the p-values taken over all $i, j \leq 10$. For all but two of the companies (Phillip Morris and AT&T), we cannot reject either (17) or (18) at the 5% level for either of the five year periods.

Note that because this procedure takes the minimum across many nulls, the reported p-value is biased downwards. Note also that our test is a generalization of that proposed by Ramsey and Rothman (1996) and Rothman (1990, 1991, 1992), who proposed testing

$$E[x_t^2 x_{t-1}] - E[x_t x_{t-1}^2] = 0 \tag{19}$$

Table 2. Tests for Reversibility of $\{\langle |r_t|, M_r \rangle\}$ and $\{\langle V_t, M_v \rangle\}$

Co.	82-86	89-93	Co.	82-86	89-93
AA	0.11, 0.42	0.13, 0.32	JPM	0.16, 0.32	0.32, 0.32
ALD	0.26, 0.32	0.27, 0.31	KO	0.10, 0.08	0.34, 0.50
AXP	0.31, 0.09	0.09, 0.18	MCD	0.09, 0.50	0.32, 0.10
BA	0.27, 0.32	0.23, 0.34	MMM	0.15, 0.42	0.32, 0.22
BS	0.27, 0.32	0.21, 0.29	MO	0.04, 0.36	0.16, 0.32
CAT	0.69, 0.32	0.32, 0.34	MRK	0.22, 0.32	0.37, 0.21
CHV	0.17, 0.28	0.32, 0.32	PG	0.16, 0.35	0.21, 0.67
DD	0.07, 0.32	0.54, 0.32	S	0.05, 0.24	0.28, 0.30
DIS	0.16, 0.32	0.32, 0.24	T	0.04, 0.22	0.31, 0.32
EK	0.13, 0.37	0.10, 0.16	TX	0.32, 0.32	0.16, 0.32
GE	0.09, 0.75	0.15, 0.22	UK	0.32, 0.16	0.17, 0.32
GM	0.29, 0.18	0.10, 0.26	UTX	0.28, 0.32	0.35, 0.18
GT	0.23, 0.54	0.31, 0.28	WX	0.32, 0.32	0.23, 0.24
IBM	0.11, 0.38	0.28, 0.32	XON	0.37, 0.44	0.05, 0.70
IP	0.14, 0.16	0.50, 0.31	Z	0.26, 0.68	0.33, 0.67

to test for reversibility of the time series $\{x_t\}$. Because our test procedure always includes the test of (19) for $i = 1$ and $j = 2$, our test will have at least as much power as the Ramsey-Rothman TR test against any time irreversible alternative.[2]

We have thus shown that if the information arrival process $\{k_t\}$ is reversible, then the MODM requires that time series for volatility and trading volume obey the intertemporal symmetry restriction (10-11). Table 1 shows that the observed time series for returns and trading volume strongly violate this restriction. We have also shown that if the information arrival process is irreversible, then the MODM requires that $\{\langle |r_t|, M_r \rangle\}$ and $\{\langle V_t, M_v \rangle\}$ must also be irreversible. Table 2 suggests no evidence of such irreversibility in the data. Because any information arrival process $\{k_t\}$ must be either reversible or irreversible, we interpret these statistical results as evidence against the MODM itself.

8. Extending the Model

This section extends the MODM presented in Section 3 to allow for the idiosyncratic component of volume to display serial dependence, as in Lamoureux and Lastrapes (1994). Let $\{l_t\}$ denote this idiosyncratic component. One might imagine that $\{l_t\}$ represents liquidity trading which is not motivated by information arrival and does not affect asset prices. We are thus allowing the amount of liquidity trading to display time dependence. We define the extended MODM by the following replacement assumptions:

A2': $E[V_t | k_t, l_t] \equiv \int Vh(V | k_t, l_t) dV$ exists and is monotonically increasing in k_t and l_t.

A3': Let

$$I_\infty \equiv \ldots, k_{t-1}, l_{t-1}, r_{t-1}, V_{t-1}, k_t, l_t, r_t, V_t, k_{t+1}, l_{t+1}, r_{t+1}, V_{t+1}, \ldots$$

Then $\Pr(r_t \varepsilon B | I_\infty / r_t) = \Pr(r_t \varepsilon B | k_t)$, $\Pr(V_t \varepsilon B | I_\infty / V_t) = \Pr(V_t \varepsilon B | k_t, l_t)$ for any Borel set B in $\Re$.

By following the logic of the proofs for Propositions 1 and 2, it is easy to show that if $\{k_t\}$ is reversible, then

$$E[V_t abs(r_{t+1})] = E[abs(r_t) V_{t+1}], \tag{20}$$

and if $\{k_t\}$ is irreversible, then $\{\langle |r_t|, M_r \rangle\}$ must also be irreversible.[3] Table 1, however, provides strong evidence that (20) does not hold for most companies in the Dow 30, and Table 2 presents little evidence against the reversibility of $\{\langle |r_t|, M_r \rangle\}$. We conclude that the extended MODM of Lamoureux and Lastrapes (1994) does not appear consistent with the data either.

9. Conclusions and Future Directions

This paper has proposed a general version of the Mixture of Distributions Model which encompasses all previously proposed MODM's. We have argued that a common requirement of these MODM's is that all intertemporal dependence between volatility and trading volume must occur through the information arrival process $\{k_t\}$.

If $\{k_t\}$ is reversible, then this common requirement implies that the time series for volatility and trading volume must obey a certain intertemporal symmetry restriction. In Section 7 we tested and rejected this restriction using daily data for 30 companies over two separate five year periods.

If $\{k_t\}$ is irreversible, we have shown that the time series for volatility and trading volume must also be irreversible. The data, however, provide no evidence against the reversibility of these two series. Because the union of the sets of reversible and irreversible processes is the set of all stationary processes, our results suggest that the MODM is inconsistent with the data for any information arrival process $\{k_t\}$.

We conclude that the intertemporal dependence between volatility and trading volume must occur through two or more latent common factors. One could thus imagine constructing an augmented MODM with two information arrival processes, each affecting volatility and trading volume in a different way. At the level of generality maintained in this paper, however, this augmented MODM would not constitute a scientifically satisfactory theory: because any empirically observed behavior for volatility and volume can *always* be constructed from two latent common factors, this two factor MODM could never be falsified.

In order to construct an empirically falsifiable theory for volatility and trading volume, it will be necessary to use structural models to generate further restrictions on the data. This is the approach taken by Brock and LeBaron (1996) and their references. We expect that in future multi-factor models of volatility and trading volume, the information arrival process will come to play much the same role that the Solow residual plays in growth models: variations in volatility and volume that cannot be accounted for by structural shifts in the market will be attributed to variation in the rate of information arrival. Construction and testing of these multi-factor models remains an important area for future research.

Appendix

A. Definition of Company Symbols

Dow Jones Firms		Dow Jones Firms	
AA	Alum. Co. of America	JPM	J.P. Morgan
ALD	Allied Signal	KO	Coca Cola
AXP	American Express	MCD	McDonalds
BA	Boeing Company	MMM	Minn. Mining & Mfg.
BS	Bethlehem Steel	MO	Phillips Morris
CAT	Caterpillar	MRK	Merck
CHV	Chevron	PG	Proctor & Gamble
DD	Dupont	S	Sears
DIS	Walt Disney	T	AT&T
EK	Eastman Kodak	TX	Texaco

Continuation of Definition of Company Symbols

Dow Jones Firms		Dow Jones Firms	
GE	General Electric	UK	Union Carbide
GM	General Motors	UTX	United Technologies
GT	Goodyear	WX	Westinghouse
IBM	Intl. Bus. Machines	XON	Exxon
IP	International Paper	Z	Woolworth

B. Proof of Proposition 1

Adding and subtracting the same term to V_t and $\text{abs}(r_{t+1})$ yields

$$\begin{aligned} E[V_t abs(r_{t+1})] &= E[((V_t - E[V_t | k_t]) + E[V_t | k_t]) \\ &\quad \times ((abs(r_{t+1}) - E[abs(r_{t+1}) | k_{t+1}]) + E[abs(r_{t+1}) | k_{t+1}])] \\ &= E[(V_t - E[V_t | k_t])(abs(r_{t+1}) - E[abs(r_{t+1}) | k_{t+1}])] \\ &\quad + E[(V_t - E[V_t | k_t])E[abs(r_{t+1}) | k_{t+1}])] \\ &\quad + E[E[V_t | k_t](abs(r_{t+1}) - E[abs(r_{t+1}) | k_{t+1}])] \\ &\quad + E[E[V_t | k_t]\, E[abs(r_{t+1}) | k_{t+1}]] \end{aligned} \tag{21}$$

The first three terms are zero by assumptions A1-A3. This yields

$$E[V_t abs(r_{t+1})] = E[E[V_t | k_t]\, E[abs(r_{t+1}) | k_{t+1}]] \tag{22}$$

By similar argument,

$$E[abs(r_t) V_{t+1}] = E[E[abs(r_t) | k_t]\, E[V_{t+1} | k_{t+1}]] \tag{23}$$

Because $\mathrm{E}[V_t abs(r_t) \mid k_t]$ and $\mathrm{E}[V_t \mid k_t]$ are continuous functions of k_t, the definition of reversibility implies that the right hand sides of the two previous expressions are equal, and thus

$$\mathrm{E}[V_t abs(r_{t+1})] = \mathrm{E}[abs(r_t) V_{t+1}]. \tag{24}$$

C: Proof of Proposition 2

We will say that a function *crosses* the X axis at a point x_c if $\exists\, \delta>0$ such that $0<\varepsilon<\delta \Rightarrow (x_c - \varepsilon)(x_c + \varepsilon)<0$. The following are the two auxiliary assumptions for Proposition 2:

> **A5:** Define $F(y,x) \equiv f(y \mid x) - f'(y \mid x)$. For each y, $F(y,x)$ crosses the X axis at only a finite number of points x. Call the greatest of these points $x^*(y)$. For each x, $F(y,x)$ crosses the Y axis at only a finite number of points.
>
> **A6:** For all x, there exists V^* such that $\frac{d}{dV}\Pr(k>x \mid V^*)>0$.

Recall that f is the forward transition density for $\{k_t\}$ and f' is the backward transition density for $\{k_t\}$. We now prove Proposition 2 for $\{V_t\}$; the proof for $\{\,|r_t|\,\}$ is similar. Let g denote the transition density function for $\{V_t\}$, then by the law of iterated expectations,

$$\begin{aligned} g(V_{t+1} \mid V_t) &= \int h(V_{t+1} \mid k_{t+1}) \left[\int f(k_{t+1} \mid k_t) h'(k_t \mid V_t)\, dk_t \right] dk_{t+1} \\ &= \int h(V_{t+1} \mid y) \left[\int f(y \mid x) h'(x \mid V_t)\, dx \right] dy \end{aligned} \tag{25}$$

By similar argument,

$$g'(V_t \mid V_{t+1}) = \int h(V_t \mid y) \left[\int f'(y \mid x) h'(x \mid V_{t+1})\, dx \right] dy \tag{26}$$

Define

$$G(a,b) \equiv g(a \mid V_t = b) - g'(a \mid V_{t+1} = b) = \int h(a \mid y) [F(y,x) h'(x \mid b)\, dx]\, dy \tag{27}$$

Since the bivariate density f is continuous, there exists a set A of positive measure such that $F(y,x) \neq 0$ for all $(y,x) \in A$. (This follows from the definition of reversibility.) Fix y and consider $F(y, x)$ as a univariate function of x. Then by Proposition 4, there exists a set B such that $\Pr(B)>0$ and $b \in B$ implies

$$\Gamma(y,b) \equiv \int F(y,x) h'(x \mid b) dx \neq 0. \tag{28}$$

Now consider

$$\Gamma'(y,b) \equiv \Gamma(y,b)/h(y). \tag{29}$$

From assumption A5, it follows that $\Gamma'(y,b)$ crosses the Y axis at a finite number of points, and thus by Proposition 4, there exists a set A such that $\Pr(A)>0$ and $\forall a \in A$,

$$\int \Gamma'(y,b) h'(y \mid a) dy \neq 0 \Rightarrow \tag{30}$$

$$\int \Gamma(y,b) \frac{H(y,a)}{h(y)h(a)} dy \neq 0 \Rightarrow \tag{31}$$

$$\int \Gamma(y,b) h(a \mid y) dy \neq 0, \tag{32}$$

which implies

$$G(a,b) \neq 0 \tag{33}$$

where H is the bivariate density function for $\{V_t, k_t\}$. In summary, we have shown that $\exists B \in \Re^2$ such that $\Pr(B)>0$ and $G(a,b) \neq 0 \; \forall (a,b)$ in B. Thus $g(V_{t+1} \mid V_t)$ is not symmetric.

Proposition 4: *Suppose that the random variables x and b are such that the conditional density $h'(x|b)$ is continuous, and $\mathrm{E}[x|b]$ is monotonically increasing in b. Suppose $F: \Re \to \Re$ is continuous and is non-zero on a set A where $\Pr(x \in A)>0$. Suppose that $F(x)$ crosses the X axis at only a finite number of points, the greatest of which is x^+. Suppose also that there exists b^* such that $d/db \Pr(x>x^+ \mid b^*)>0$. Then there exists B such that $\Pr(b \in B)>0$ and $\mathrm{E}[F(x) \mid b] \neq 0$ whenever $b \in B$.*

Proof: If $\mathrm{E}[F(x) \mid b^*] \neq 0$, we can use continuity of F to immediately show the result. So suppose that $\mathrm{E}[F(x) \mid b^*] = 0$. Without loss of generality,[4] assume that $F(x) \geq 0$ for $x > x^+$. Then write

$$\mathrm{E}[F(x) \mid b^*] = \mathrm{E}[F(x) I(x<x^+) \mid b^*] + \mathrm{E}[F(x) I(x>x^+) \mid b^*]. \tag{34}$$

Because $F(x) \geq 0$ for all $x > x^+$ and $F(x) > 0$ for x in some neighborhood $(x^+, x^+ + \delta)$, it follows that

$$\mathrm{E}[F(x) I(x>x^+) \mid b^*] > 0 \text{ and } \mathrm{E}[F(x) I(x<x^+) \mid b^*] < 0. \tag{35}$$

Using $F(x^+) = 0$ and assumptions A1 and A2, we get

$$d/db\, \mathrm{E}[F(x) I(x>x^+) \mid b^*] > 0, \; d/db\, \mathrm{E}[F(x) I(x<x^+) \mid b^*] > 0 \Rightarrow \tag{36}$$

$$d/db\, \mathrm{E}[F(x) \mid b^*] > 0. \tag{37}$$

By continuity, we conclude the existence of B such that $b \in B$ implies $\mathrm{E}[F(x) \mid b] > 0$.

D: Proof of Proposition 3

By continuity of g, we can choose bounded sets B_1 and B_2 so that

$$a \in B_1 \text{ and } b \in B_2 \Rightarrow (a,b) \in B \tag{38}$$

$$\Pr(x_t \in B_1 \text{ and } x_{t+1} \in B_2) > 0 \tag{39}$$

$$\Pr(x_t \in B_1 \text{ and } x_{t+1} \in B_2) \neq \Pr(x_{t+1} \in B_1 \text{ and } x_t \in B_2). \tag{40}$$

This last expression can be rewritten

$$\mathrm{E}[I(x_t, B_1) I(x_{t+1}, B_2) - I(x_t, B_2) I(x_{t+1}, B_1)] \neq 0, \tag{41}$$

where I denotes the indicator function. Since these indicator functions can be approximated arbitrarily well by continuous functions, it follows that there exist bounded continuous functions r,s of bounded non-negative support such that

$$E(r(x_t)s(x_{t+1}) - s(x_t)r(x_{t+1})) \neq 0. \tag{42}$$

The proof is concluded by contradiction. Suppose that $\forall i,j$,

$$E[\langle x_t,M\rangle^i \langle x_{t+1},M\rangle^j - \langle x_t,M\rangle^j \langle x_{t+1},M\rangle^i] = 0. \tag{43}$$

Because we can choose M above the support of r and s, it follows

$$E(r(x_t)s(x_{t+1}) - s(x_t)r(x_{t+1})) = \tag{44}$$

$$E(r(\langle x_t,M\rangle)s(\langle x_{t+1},M\rangle) - s(\langle x_t,M\rangle)r(\langle x_{t+1},M\rangle)) \tag{45}$$

By the Weierstrass approximation theorem, there exist $\{a_i\}$ and $\{b_i\}$ such that

$$E(r(\langle x_t,M\rangle)s(\langle x_{t+1},M\rangle) - s(\langle x_t,M\rangle)r(\langle x_{t+1},M\rangle)) = \tag{46}$$

$$E[(\sum_{i=0}^{\infty} \langle x_t,M\rangle^i a_i)(\sum_{i=0}^{\infty} \langle x_{t+1},M\rangle^i b_i) - (\sum_{i=0}^{\infty} \langle x_t,M\rangle^i b_i)(\sum_{i=0}^{\infty} \langle x_{t+1}, M\rangle^i a_i)]. \tag{47}$$

Expanding the summation terms and invoking the bounded convergence theorem yields

$$E(r(x_t)s(x_{t+1}) - s(x_t)r(x_{t+1})) = \tag{48}$$

$$\lim_{N\to\infty} E[\sum_{i,j\le N} \langle x_t,M\rangle^i \langle x_{t+1},M\rangle^j a_i b_j - \sum_{i,j\le N} \langle x_t,M\rangle^i \langle x_{t+1},M\rangle^j a_j b_i] \tag{49}$$

But this last expression is zero by (43), contradicting (42), and thus completing the proof.

Acknowledgments

I would like to thank the Iowa State University Department of Economics for supporting this research while I was an Assistant Professor there. I also thank William Brock, Blake LeBaron, and Steven Durlauf for their many contributions.

Notes

1. See Nelson (1991).
2. This is true except, as mentioned before, in the unlikely case in which irreversibility occurs only in the extreme tails of the distribution of $|r_t|$ or V_t.
3. Note that in this extended version of MODM, $\{V_t\}$ can be reversible even if $\{k_t\}$ is not.
4. The proof proceeds in a similar manner under the assumption that $F(x) \leq 0$ for $x > x^+$.

References

Andersen, T. (1995). "Return Volatility and Trading Volume: An Information Flow Interpretation of Stochastic Volatility," *Journal of Finance*, 51, 169-204.

Andrews, D. and J. Monahan (1992). "An Improved Heteroskedasticity and Autocorrelation Consistent Covariance Matrix Estimator," *Econometrica*, 60, 953-966.

Andrews, D. (1991). "Heteroskedasticity and Autocorrelation Consistent Covariance Matrix Estimation," *Econometrica*, 59, 817-858.

Apostol, T. (1974). *Mathematical Analysis*. Menlo Park: Addison-Wesley.

Bollerslev, T., R. Chou, and K. Kroner (1992). "ARCH Modelling in Finance," *Journal of Econometrics*, 52, 5-59.

Bollerslev, T., R. Engle, and D. Nelson (1994). "A Dynamic Structural Model for Stock Return Volatility and Trading Volume," in Engle, R. and D. McFadden (eds.), *Handbook of Econometrics*, Volume 4. Amsterdam: Elsevier Science.

Brock, W. and B. Lebaron (1996). "A Dynamic Structural Model for Stock Return Volatility and Trading Volume," *Review of Economics and Statistics*, 78, 94-110.

Epps, T. and M. Epps (1976). "The Stochastic Dependence of Security Price Changes and Transactions Volumes: Implications for the Mixture of Distributions Hypothesis," *Econometrica*, 44, 305-321.

Gallant, A., D. Hsieh, and G. Tauchen (1991). "On Fitting a Recalcitrant Series: The Pound/Dollar Exchange Rate, 1974-83," in Barnett, W., J. Powell, and G. Tauchen (eds.) *Nonparametric and Semiparametric Methods in Econometrics and Statistics*. Cambridge: Cambridge University Press.

Gallant, A., P. Rossi, and G. Tauchen (1993). "Nonlinear Dynamic Structures," *Econometrica*, 61, 871-907.

Harris, L. (1986). "Cross Security Tests of the Mixture of Distributions Hypothesis," *Journal of Financial and Quantitative Analysis*, 21, 39-46.

Harris, L. (1987). "Transaction Data Tests of the Mixture of Distributions Hypothesis," *Journal of Financial and Quantitative Analysis*, 22, 127-141.

Hiemstra, C. and J. Jones (1994). "Testing for Linear and Nonlinear Granger Causality in the Stock Price-Volume Relation," *Journal of Finance* 49, 1639-1664.

Jain, P. and G. Joh (1986). "The Dependence Between Hourly Price Changes and Trading Volume." *Journal of Financial and Quantitative Analysis,* 21, 39-46.

Kelly, F. (1979). *Reversibility and Stochastic Networks*. New York: Wiley.

Lamoureux, C. and W. Lastrapes (1990). "Heteroskedasticity in Stock Return Data: Volume versus GARCH Effects," *Journal of Finance*, 45, 221-229.

Lamoureux, C. and W. Lastrapes (1994). "Endogenous Trading Volume and Momentum in Stock Return Volatility," *Journal of Business and Economic Statistics*, 12, 253-260.

Liggett, T. (1985). *Interacting Particle Systems*. New York: Springer-Verlag.

Morgan, I. (1976). "Stock Prices and Heteroskedasticity," *Journal of Business*, 49, 496-508.

Nelson, D. (1991). "Conditional Heteroskedasticity in Asset Returns: a New Approach," *Econometrica*, 59, 347-370.

Nelson, D. (1992). "Filtering and Forecasting with Misspecified ARCH Models I: Getting the

Right Variance with the Wrong Model," *Journal of Econometrics*, 52, 61-90.

Locke, P. and C. Sayers (1992). "Intra-day Futures Price Volatility: Information Effects and Variance Persistence," *Journal of Applied Econometrics*, 8, 15-30.

Morgan, I. (1976). "Stock Prices and Heteroskedasticity," *Journal of Business*, 49, 496-508.

Ramsey, J. and P. Rothman (1996). "Time Irreversibility and Business Cycle Asymmetry," *Journal of Money, Credit, and Banking*, 28, 1-21.

Richardson, M. and T. Smith (1994). "A Direct Test of the Mixture of Distributions Hypothesis: Measuring the Information Flow Throughout the Day," *Journal of Financial and Quantitative Analysis*, 29, 101-116.

Rothman, P. (1990). "Characterization of the Time Irreversibility of Economic Time Series," Ph.D. dissertation, Department of Economics, New York University.

Rothman, P. (1991). "Higher Order Residual Analysis for Simple Bilinear and Threshold Autoregressive Models with the TR Test," Department of Economics, University of Delaware.

Rothman, P. (1992). "The Comparative Power of the TR Test against Simple Threshold Models," *Journal of Applied Econometrica*, 7, S187-S195.

Tauchen, G. and M. Pitts (1983). "The Price Variability-Volume Relationship on Speculative Markets," *Econometrica*, 51, 485-505.

Tauchen, G., H. Zhang, and M. Liu (1996). "Volume, Volatility, and Leverage: A Dynamic Analysis," *Journal of Econometrics*, 74, 177-208.

West, K. and D. Cho (1995). "The Predictive Ability of Several Models of Exchange Rate Volatility," *Journal of Econometrics*, 69, 367-392.

Westerfield, R. (1977). "The Distribution of Common Stock Price Changes: An Application of Transactions Time and Subordinated Stochastic Models," *Journal of Financial and Quantitative Analysis*, 12, 743-765.

White, H. (1984). *Asymptotic Theory for Econometricians*. New York: Academic Press.

White, H. and I. Domowitz (1984). "Nonlinear Regression with Dependent Observations," *Econometrica*, 52, 143-161.

14 IMPROVED TESTING AND SPECIFICATION OF SMOOTH TRANSITION REGRESSION MODELS

Alvaro Escribano
Universidad Carlos III de Madrid
Oscar Jordá
University of California, Davis

1. Introduction

This chapter provides tools for the empirical practitioner interested in modelling Smooth Transition Regression (STR) models. Nonlinear time series models are being used more frequently in empirical applications, leaving the researcher with a virtual infinity of models and specification from which to choose. However, STR models are a general class of state-dependent, reduced form, nonlinear time series models in which the transition between states is generally endogenously generated. They encompass as particular cases the Smooth Transition Autoregressive (STAR), the Exponential Autoregressive (EAR), the Threshold Autoregressive (TAR), and the SETAR models.[1] Together with Hamilton's Regime Switching model[2] (where the transition between regimes is exogenously generated by a Markov chain), state-dependent models have proven useful in modelling the asymmetric behavior of economic fluctuations.[3]

Even after restricting attention to a certain class, the rich parameterization and flexibility of these models makes the task of specifying the models complicated. Model building usually starts by performing a nonlinearity test. If there is not enough evidence of nonlinearity, there is no reason to pursue a model that is much more difficult to specify, estimate, and evaluate than a linear model is. Chan and Tong

(1986) discuss the possibility of using likelihood ratio test statistics for testing linearity against SETAR models. The drawback of this approach is that the distribution of the statistic has to be determined by simulation for each application. Based on work by Tsay (1986), Luukkonen *et al.* (1988a) introduce a set of Lagrange Multiplier (*LM*) type tests that have asymptotic χ^2 distributions. Saikkonen and Luukkonen (1988) considered *LM* tests against bilinear and EAR alternatives. Teräsvirta (1994) uses these procedures in several stages of the specification of STAR models.

This chapter introduces two new main results, which rely on Taylor series approximations to the transition function (between states) around the scale parameter.[4] The results, derived in the general context of STR models (which encompass STAR-based tests as a particular case), are an extension of previous work in Escribano and Jordá (1997). First, we introduce a new specification strategy extended to include the choice between logistic and exponential STR models. Our selection procedure has higher correct selection frequencies of the right model, avoids the pitfalls of the rules proposed in Teräsvirta (1994), and is much simpler to apply. Second, we suggest that in some cases, tests of the null hypothesis of linearity against STR-type nonlinearity[5] should be augmented to include up to fourth-order terms. Including these terms is necessary to gain power against alternatives where an exponential STR might be involved. We complement this analysis with an exhaustive study of the dynamics of STR models and the shapes of transition functions in practice through simulations. The chapter also provides ample Monte Carlo evidence in support of these claims (providing several additional cases to those available in Escribano and Jordá (1997)) and shows how these procedures work in practice with a different empirical example, a brief study of the dynamics of the U.S. unemployment rate.

This chapter is organized as follows: Section 2 briefly defines and introduces the basic properties of STR models; Section 3 presents the tests of nonlinearity and their properties; Section 4 discusses the selection procedure proposed by Teräsvirta (1994) and then introduces the new alternative, Section 5 presents Monte Carlo simulations of the new tests; Section 6 applies the new techniques to the U.S. unemployment rate; and Section 7 concludes.

2. Smooth Transition Regression (STR) Models

Consider the following STR model:

$$y_t = \pi' \underline{x}_t + \theta' \underline{x}_t F(z_{t-d}, \gamma, c) + u_t \tag{1}$$

where y_t is a scalar,

$$\underline{x}_t = (1, y_{t-1}, \ldots, y_{t-r}; \underline{w}_{t-1}, \ldots, \underline{w}_{t-q})' = (1, \tilde{\underline{x}}_t)',$$

with $\underline{w}_t$ a vector of exogenous variables, and z_{t-d} is a scalar, although in general it could be a be a vector. Usually $z_{t-d} = y_{t-d}$ where d is the delay parameter which satisfies $1 \leq d \leq p$ for $p = \max(r,q)$ and is assumed known.[6] Note that when $z_{t-d} = y_{t-d}$ and $\underline{w}_t = \underline{0}$, we have a conventional STAR model. Further,

$$\pi' = (\pi_0, \pi_1, ..., \pi_p) = (\pi_0, \tilde{\pi}'); \quad \text{and}$$

$$\theta' = (\theta_0, \theta_1, ..., \theta_p) = (\theta_0, \tilde{\theta}').$$

u_t is a martingale difference sequence with constant variance[7] and y_t is assumed to be stationary and ergodic. The function $F(z_{t-d}, \gamma, c)$ is at least fourth-order, continuously differentiable with respect to the scale parameter, γ.

The transition function, $F(\cdot)$, is traditionally chosen to be either a logistic or an exponential distribution function.[8] The exponential function is:

$$F(z_{t-d}, \gamma, c) = \left[1 - \exp\left\{-\gamma(z_{t-d} - c)^2\right\}\right]. \tag{2}$$

When $\theta_0 = c$, $\underline{w}_t = \underline{0}\ \forall t$, and $z_{t-d} = y_{t-d}$, the corresponding ESTR model is reduced to the exponential autoregressive model (EAR). The logistic transition function[9] is:

$$F(z_{t-d}, \gamma, c) = \left[\left\{1 + \exp(-\gamma(z_{t-d} - c)\right\}^{-1} - 1/2\right]. \tag{3}$$

In order to illustrate the properties and shapes of the logistic and exponential transition functions, we simulate two specifications from Teräsvirta (1994). This exercise will provide a better understanding of the details involved in the specifications and nonlinearity testing of STR models. Consider the following simulation:

$$y_t = 1.8 y_{t-1} - 1.06 y_{t-2} + (\theta_0 - 0.9 y_{t-1} + 0.795 y_{t-2}) F(y_{t-1}, \gamma, c) + u_t \tag{4}$$

where:

$$F(z_{t-d}, \gamma, c) = \left[1 + \exp\left(-1000 (y_{t-1} - c)^2\right)\right] \qquad \text{(ESTAR)}$$

and

$$F(z_{t-d}, \gamma, c) = \left[1 + \exp(-100 (y_{t-1} - c))\right]^{-1} \qquad \text{(LSTAR)}$$

Figure 1. STR Simulated Models

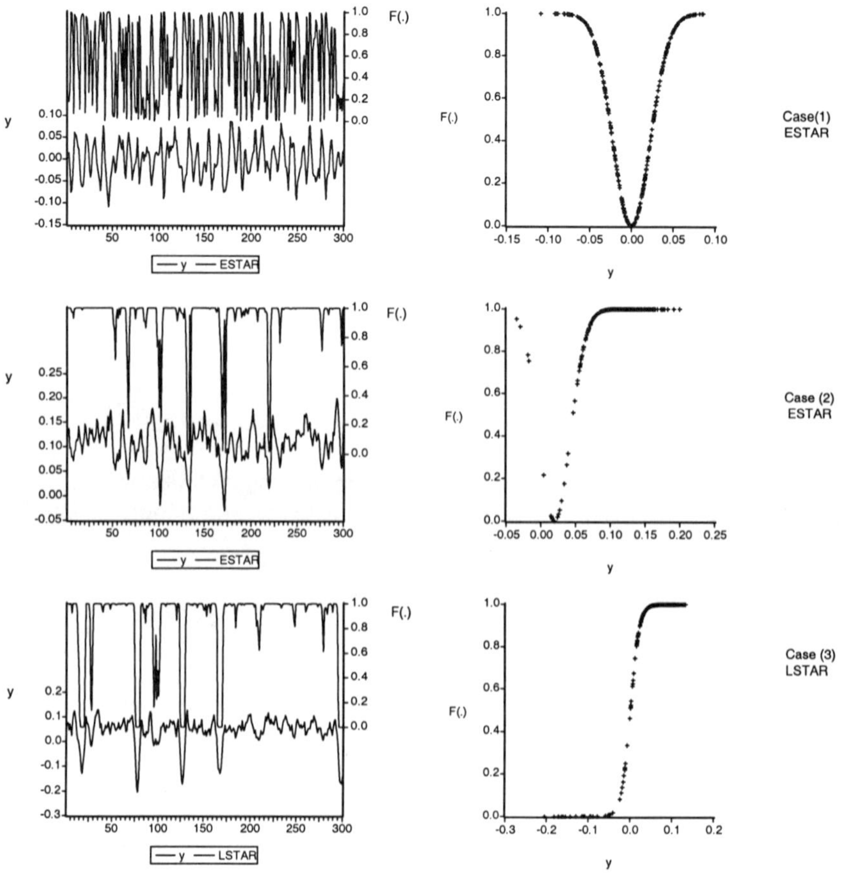

with u_t i.i.d. $N(0,0.02^2)$ and $T=300$. Figure 1 depicts three graphs of the plots of the data, y_{t-1}, and $F(y_{t-1},\gamma,c)$ for the following cases: (1) $\theta_0=c=0$ and the ESTAR specification; (2) $\theta_0=0.04$, $c=0.02$, and the ESTAR specification; and (3) $\theta_0=0.02$, $c=0$, and the LSTAR specification. The plots in the first column depict the time series of the data and the values of the transition function. The plots in the second column depict the scatter plots of the data versus the corresponding values of the transition function, thus illustrating its shape.

Consider case (1). The data are symmetrically distributed at each side of the threshold c. When $|y_{t-1}|\gg c$, then $F(y_{t-1},\gamma,c)\simeq 1$, and the dynamic behavior of the data is characterized by the "upper" linear regime. When $|y_{t-1}|$ is "close" to c, then $F(y_{t-1},\gamma,c)\simeq 0$, and the dynamics of the "lower" regime dominate. The speed with which the transition from one regime to the other takes place is set by the value of the scale parameter, γ. The shape of the plot of the values of the transition function and the data that we obtain in case (1) therefore represent those of the usual ESTAR model.

Now consider case (3), the graph of the typical transition function for an LSTAR model. When $y_{t-1} \ll c$, then $F(y_{t-1},\gamma,c) \simeq 0$ (what we designated as the "lower" regime). Conversely, when $y_{t-1} \gg c$, then $F(y_{t-1},\gamma,c) \simeq 1$ (the "upper" regime). γ again determines how smooth the transition between regimes is. In the extreme case in which $\gamma \to \infty$, we get a particular case of LSTAR, the Threshold Autoregressive (TAR) model.

Finally, the most interesting scenario is that of case (2). The model is of the ESTAR class but notice that the data y_{t-1} tend to cluster around the "upper" regime, to the right of the threshold c. In fact, the graphs for cases (2) and (3) are rather similar although the dynamic behavior of each model is quite different. This asymmetric behavior in the cases when $\theta_0 \neq 0$ and/or $c \neq 0$ will influence both the ability to specify and test for a STAR models as we shall see in the next sections.

3. Testing Linearity Against STR Models

Testing linearity against STR-type nonlinearity implies testing the null hypothesis $H_0{:}\,\theta' = \underline{0}$ in equation (1). However, under the null hypothesis, the parameters γ and c are not identified, that is, they can take on any value. Alternatively, we could choose as our null hypothesis $H_0{:}\,\gamma = 0$, in which case neither c nor θ' would be identified. Davies (1977) showed that conventional maximum likelihood theory is not directly applicable to this problem. A solution proposed in Luukkonen *et al.* (1988a) and adopted in Teräsvirta (1994) is to replace $F(z_{t-d},\gamma,c)$ in equation (1) with a suitable Taylor series approximation. Under the null hypothesis of linearity, the *LM* test is shown to possess asymptotically the usual χ^2 distribution.[10] In practice, the test is performed on the following auxiliary regression:

$$y_t = \pi'\underline{x}_t + \left[\theta'\underline{x}_t\{\gamma F_\gamma(z_{t-d},\gamma=0,c)\}\right] + \upsilon_{1t}, \tag{5}$$

where $F_\gamma(\cdot)$ indicates the first derivative of F with respect to γ; $\gamma F_\gamma(\cdot)$ is obviously the first term of the Taylor approximation of F around $\gamma = 0$. In particular, $F_\gamma(\cdot)$ for the logistic transition function is:

$$F_\gamma(z_{t-d},\gamma=0,c) = \frac{1}{4}(z_{t-d}-c), \tag{6}$$

which substituted into (5) yields:

$$y_t = \pi'\underline{x}_t + \frac{1}{4}\gamma\theta'\underline{x}_t(z_{t-d}-c) + \upsilon_{1t}. \tag{7}$$

The parameters under this specification cannot be identified. After combining terms, we get the final version of the auxiliary regression:

$$y_t = \delta_0 + \tilde{\delta}'\tilde{\underline{x}}_t + \beta_1'\tilde{\underline{x}}_t z_{t-d} + \upsilon_{1t}, \text{ where} \tag{8}$$

$$\delta_0 = \left(\pi_0 - \frac{1}{4}\gamma c \theta_0 \right),$$

$$\tilde{\delta}' = \left(\tilde{\pi}' - \frac{1}{4}\gamma c \tilde{\theta}' \right), \text{ where the } d^{th} \text{ term is } \delta_d = \left(\pi_d + \frac{1}{4}\gamma c (\theta_0 - \theta_d) \right),$$

$$\text{and } \beta_1{}' = \frac{1}{4}\gamma\tilde{\theta}'.$$

The null hypothesis of linearity therefore becomes $H_0'\!:\beta_1' = \underline{0}$. Note that equation (8) is explosive and generally not a meaningful time series model (see Granger and Andersen (1978)[11]). We will refer to this test as "NL1." Luukkonen *et al.* (1988a) realized that this test would have low power against alternatives for which $\tilde{\theta}'$ is "small" and θ_0 is "large" in absolute value since β_1' does not include the θ_0 coefficient, the "detecting a shift in the mean" problem.

To overcome this difficulty, they proposed to approximate the transition function with a third order Taylor series expansion, that is:

$$y_t = \pi'\underline{x}_t + \frac{1}{4}\gamma\theta'\underline{x}_t(z_{t-d} - c) + \frac{1}{48}\gamma^3\theta'\underline{x}_t(z_{t-d} - c)^3 + \upsilon_{3t}, \tag{9}$$

where it is important to note that the square powers of the approximation are identically zero. Recombining in terms of identified parameters, we obtain:

$$y_t = \delta_0 + \tilde{\delta}'\underline{x}_t + \beta_1'\tilde{\underline{x}}_t z_{t-d} + \beta_2'\tilde{\underline{x}}_t z_{t-d}^2 + \beta_3'\tilde{\underline{x}}_t z_{t-d}^3 + \upsilon_{3t}, \tag{10}$$

where now $\beta_{2d} = \frac{1}{48}\gamma^3\theta_0$ and thus we avoid the "detecting a shift in the mean" problem.[12]

The null hypothesis now becomes $H_0'''\!:\beta_1' = \beta_2' = \beta_3' = \underline{0}$. Following Saikkonen and Luukkonen (1988), Teräsvirta *et al.* (1994), and Teräsvirta (1994), a convenient procedure for computing the *LM* statistic by OLS consists of estimating the auxiliary regression (10) under the null hypothesis and computing the sum of square residuals, SSR_0. Next, (10) is estimated under the alternative hypothesis and the corresponding sum of squared residuals is computed, SSR_1. Finally, the statistic $T(SSR_0 - SSR_1)/SSR_0$ has asymptotically a χ^2 distribution under the null. It is usually recommended to use the approximation given by the F distribution because of the good size and power properties of the test in small samples. We will call the nonlinearity test based on equation (10) "NL3."

Note that when the model is an ESTR, we have that $F_{\gamma}(z_{t-d}, \gamma = 0, c) = (z_{t-d} - c)^2$, which, after substituting into (5) becomes:

$$y_t = \gamma c^2 \theta_0 + \pi' \underline{x}_t - 2\gamma c \tilde{\theta}' \underline{\tilde{x}}_t z_{t-d} + \gamma \tilde{\theta}' \underline{\tilde{x}}_t z_{t-d}^2 + \upsilon_{2t}, \tag{11}$$

which in terms of identified parameters becomes:

$$y_t = \delta_0 + \tilde{\delta}' \underline{\tilde{x}}_t + \beta_1' \underline{\tilde{x}}_t z_{t-d} + \beta_2' \underline{\tilde{x}}_t z_{t-d}^2 + \upsilon_{2t}, \tag{12}$$

where in particular $\beta_{1d} = \gamma \theta_0$ and therefore we do not need to pursue further terms of the expansion since we are able to identify shifts in the mean via the term β_{1d}. As a result, equation (10) is a valid nonlinearity test for either LSTR or ESTR alternatives.

3.1 Properties of the Taylor Approximation

The transition function of a STR model exhibits two important features. First, the logistic function (see equation (3), Figure 1 case (3)) has a single inflection point while the exponential function (see equation (2), Figure 1, cases (1) and (2)) has two inflection points. Second, the even powers of the Taylor expansion of a logistic function are all zero. Meanwhile, all the odd powers of the Taylor expansion of an exponential function are zero. The first feature suggests ways to improve the nonlinearity test NL3 of the previous section. In Section 4 we use the second feature to introduce a new specification procedure to choose between LSTR and ESTR models.

The immediate consequence resulting from the difference in shape between the logistic and the exponential function is that we need a second-order Taylor series expansion to be able to capture its two inflection points. This is particularly the case when γ is "small" in absolute value (smooth transition) and/or the variance of the residuals, u_t, is "large" such that a reasonable number of observations are driven to the "upper" regime. A solution to this problem is to expand the auxiliary regression (10) to include the terms resulting from the second-order expansion of the exponential. In particular:

$$y_t = \pi' \underline{x}_t + \theta' \underline{x}_t \left[(z_{t-d} - c)^2 (\gamma + \gamma^2) - \frac{1}{2} (z_{t-d} - c)^4 \gamma^2 \right] + \upsilon_{4t}, \tag{13}$$

which rewritten in terms of identified parameters[13] simply means augmenting the auxiliary regression in (10) as follows:

$$y_t = \delta_0 + \tilde{\delta}'\tilde{\underline{x}}_t + \beta_1'\tilde{\underline{x}}z_{t-d} + \beta_2'\tilde{\underline{x}}_t z_{t-d}^2 + \beta_3'\tilde{\underline{x}}_t z_{t-d}^3 + \beta_4'\tilde{\underline{x}}_t z_{t-d}^4 + \upsilon_{4t}. \quad (14)$$

Now the null hypothesis becomes $H_0^{IV}: \beta_1' = \beta_2' = \beta_3' = \beta_4' = \underline{0}$. We call this test "NL4."[14] The computation of the test is parallel to that of NL3, where either the χ^2 or the F version of the test can be used.

An alternative to the above *LM* tests (namely NL3 and NL4) is to use the Wald test of Hansen (1996). This procedure approximates unknown limiting distributions by generating *p*-values based on simulation methods. Pesaran and Potter (1997) adapt this strategy in an interesting application to their floor and ceiling model. However, this alternative involves additional elements of complexity which are easily avoided with the type of *LM* testing discussed here.

Testing in practice involves several important steps such as choice of the lag length of the linear ARX model,[15] choice of the delay parameter d,[16] and others which are all well documented in Teräsvirta (1994). However, an important empirical question is to realize that NL4 requires p extra regressors with respect to NL3, and that NL3 requires $2p$ extra regressors with respect to NL1. Lack of parsimony is particularly important with small sample sizes and/or when the order of the ARX polynomial, p, is high, since it reduces the power of the test. Luukkonen *et al.* (1988a) recognized this lack of parsimony and suggested an augmented first-order procedure based on equation (8). This consists of using the following auxiliary regression instead of (10):

$$y_t = \delta_0 + \tilde{\delta}'\tilde{\underline{x}}_t + \beta_1'\tilde{\underline{x}}_t z_{t-d} + \beta_{2d} z_{t-d}^3 + \beta_{3d} z_{t-d}^4 + \upsilon_{at}, \quad (15)$$

which we will call "NL3A." In the same spirit, we can augment the first-order procedure further to take into account the results that led us to equation (14). This means including the term $\beta_{4d} z_{t-d}^5$ in the auxiliary regression (15) to obtain the equivalent augmented first-order version of the test which we will call "NL4A." The null hypothesis simply becomes $H_O^{IVA}: \beta_1' = \beta_{2d} = \beta_{3d} = \beta_{4d} = 0$, which only requires $p+3$ terms. The intuition behind this test is that the parameters β_{jd} for $j = 2,3,4$ collect the effects of a shift in the mean in the nonlinear regime.

4. Choosing Between LSTR or ESTR

4.1 Teräsvirta's (1994) Selection Procedure

Upon rejecting the null hypothesis of linearity (with any of the tests suggested in the previous section), one might consider using a STR model as a useful nonlinear alternative; recall that the tests for nonlinearity have power against forms of nonlinearity other than the STR-type. Teräsvirta (1994) introduced the following model selection procedure (which we will call "TP"), based on equation (10), and reproduced here for convenience:

$$y_t = \delta_0 + \tilde{\delta}'\tilde{\underline{x}}_t + \beta_1'\tilde{\underline{x}}_t z_{t-d} + \beta_2'\tilde{\underline{x}}_t z_{t-d}^2 + \beta_3'\tilde{\underline{x}}_t z_{t-d}^3 + \upsilon_{3t}.$$

1. Test the null hypothesis $H_{03}: \beta_3' = \underline{0}$ versus the alternative $H_{13}: \beta_3' \neq \underline{0}$ with an F-test (F_3). According to Teräsvirta, rejection of this null would imply rejection of the ESTR specification since cubic powers of z_{t-d} in a first-order approximation of $F(z_{t-d}, \gamma, c)$ are 0.

2. Test the null hypothesis $H_{02}: \beta_2' = \underline{0} | \beta_3' = \underline{0}$ with an F-test (F_2). Teräsvirta's reasoning is that the terms z_{t-d}^2 of a first-order approximation to a logistic function are zero when $c = \theta_0 = 0$ (see equation (6)). However, these terms will be non-zero in the ESTR case (except in the unlikely case that $\tilde{\theta}' = \underline{0}$). Failure to reject this null is taken as evidence in favor of an LSTR model. Nevertheless, rejection of H_{02} is not very informative one way or the other.

3. Test the null hypothesis $H_{01}: \beta_1' = \underline{0} | \beta_2' = \beta_3' = \underline{0}$ with an F-test (F_1). Following Teräsvirta, failing to reject H_{01} after rejecting H_{02} points to an ESTR model. On the other hand, rejecting H_{01} after failing to reject H_{02} supports the choice of an LSTR model.

4. Note which hypotheses are rejected and compare the relative strengths of the rejections. If the model is LSTR, typically H_{01} and H_{03} are rejected more strongly than H_{02}. Therefore, if the p-value of F_2 is the smallest of F_1, F_2, F_3, then select the ESTR specification, otherwise select LSTR.

To discuss the problems and pitfalls with the TP we reproduce equation (14) for an LSTR and ESTR model in terms of the Taylor expansion to the transition function, namely:

$$y_t = \pi_0 + \tilde{\pi}'\tilde{x}_t + \psi_1' x_t (z_{t-d} - c) + \psi_3' x_t (z_{t-d} - c)^3 + \upsilon_{3t}, \tag{16}$$

for the logistic STR third-order expansion, and:

$$y_t = \pi_0 + \tilde{\pi}'\tilde{x}_t + \psi_1' x_t (z_{t-d} - c)^2 + \psi_4' x_t (z_{t-d} - c)^4 + \upsilon_{4t}, \tag{17}$$

for the second-order expansion of the exponential STR. The parameters ψ_j' are not directly identifiable but make the discussion easier to follow.

Consider the following problems with the TP. First, whenever $c \neq 0$, expansion of the term $\psi_4' x_t (z_{t-d} - c)^4$ will yield non-zero $\tilde{x}_t z_{t-d}^3$ terms in the auxiliary regression (10) when the model is ESTR. This is particularly problematic if π_0 and θ_0 are also non-zero. Additionally, when the variance of the error term is "large" (causing the data to be distributed asymmetrically around c as we showed in Figure 1, case (2)), false detection of an LSTR model will become more frequent.

The second source of problems is intrinsic to the design of the rule: the three F-tests suggested in the TP are nested. This feature becomes troublesome again when $c \neq 0$. For example, if the true model is LSTR, expansion of the term $\psi_3' x_t (z_{t-d} - c)^3$ yields non-zero $\tilde{x}_t z_{t-d}^2$ terms in the auxiliary regression (10). In addition, by conditioning on $\beta_3' = \underline{0}$ in H_{02}, the terms $\tilde{x}_t z_{t-d}^2$, irrespective of whether $c \neq 0$, or not, are left to approximate the transition function; an approximation that the cubic terms presumably were successfully capturing. It is therefore unclear whether F_3 or F_2 will have the smallest p-value.

4.2 A New Selection Procedure

Consider the following example. Suppose that $c = 0$. It is clear that (based on equation (16)) if the model is LSTR, the terms $\tilde{x}_t z_{t-d}^j$ for $j = 2, 4, 6, \ldots$, are exactly zero (i.e., $\beta_2' = \beta_4' = \underline{0}$ in equation (14). Conversely, if the model is ESTR, based on equation (17) the terms $\tilde{x}_t z_{t-d}^j$ for $j = 1, 3, 5, \ldots$, are exactly zero (i.e., $\beta_1' = \beta_3' = \underline{0}$ in equation (14). This suggests the following selection procedure (which we will call "EJP") based on NL4, conditional on prior rejection of linearity:

1. Test the null $H_{0E}: \beta_2' = \beta_4' = \underline{0}$ with an F-test (F_E).

2. Test the null $H_{0L}: \beta_1' = \beta_3' = \underline{0}$ with an F-test (F_L).

3. Compare the relative strength of the rejection of each hypothesis. If the minimum p-value corresponds to F_L, select LSTR, otherwise, if it corresponds to F_E, select ESTR.

Note that when $c \neq 0$, the test is still effective since we rely on testing the joint significance of linear and cubic terms relative to the joint significance of quadratic and fourth order terms, without conditioning. In addition, the EJP provides information regarding non-zero thresholds, c. Linear and cubic terms are exactly zero when $c = 0$ and the model is ESTR. Quadratic and fourth-order terms are exactly zero when $c = 0$ and the model is LSTR. Therefore, rejecting H_{0L} and failing to reject H_{OE} suggests an LSTR model with $c = 0$. Conversely, rejecting H_{0E} and failing to reject H_{0L} suggests an ESTR model with $c = 0$. This feature provides useful starting values in the estimation stage of the model.

Table 1. Relative Correct Selection Frequencies for LSTAR Models in Luukkonen *et al.* (1988a) Figure 2, p. 496, Cases (a) and (b).

		Selection Procedure				
Sample Size	θ_1	TP	GTP	EJP	C = 0	Power NL4
50	-0.4	0.500	0.406	0.594	0.868	0.064
	0.0	-	-	-	-	0.039
	0.5	0.736	0.597	0.736	0.868	0.072
	1.0	0.853	0.776	0.871	0.838	0.170
	1.5	0.904	0.889	0.936	0.787	0.467
100	-0.4	0.459	0.377	0.811	0.798	0.122
	0.0	-	-	-	-	0.043
	0.5	0.811	0.724	0.724	0.793	0.127
	1.0	0.952	0.916	0.936	0.867	0.498
	1.5	0.963	0.960	0.978	0.917	0.883

Notes: The DGP is

$$y_t - 0.5 y_{y-1} + (\theta_1 y_{t-1}) F(z_t) = \varepsilon_t$$

with $\varepsilon_t \sim N(0, 25.0)$, $c = 0$, $d = 1$, $\gamma = 0.5$.

5. Monte Carlo Evidence

This section examines how accurate the EJP is with respect to the TP (and a generalization of the TP for completeness) in choosing the correct STR specification (logistic or exponential). In addition, we analyze the power properties of NL2 versus NL4. The models that we simulated here are not proposed by us but rather are based on previous experiments in the literature, in particular Luukkonen *et al.* (1988a, b), Saikkonen and Luukkonen (1988), and Teräsvirta (1994). The specifics and technical details of the experiments are provided in the Appendix. Here we give a general overview and concentrate on the results.

The first 100 observations of each simulated series were dropped to avoid initialization problems. Each experiment was replicated 1,000 times. Some specifications allowed for values of the variance of the residuals that complement those originally proposed in the literature. Whenever the specification proposed in the literature was restricted to one type of model (an ESTR model, for example), we

Table 2. Relative Correct Selection Frequencies for LSTAR Models in Luukkonen *et al.* (1988a) Figure 2, p. 496, Cases (c) and (d).

		Selection Procedure				
Sample Size	θ_1	TP	GTP	EJP	C = 0	Power NL4
50	-1.4	0.568	0.537	0.947	0.787	0.322
	-1.0	0.872	0.788	0.872	0.785	0.203
	-0.5	0.802	0.630	0.630	0.667	0.081
	0.0	-	-	-	-	0.047
	0.5	0.841	0.735	0.690	0.782	0.113
100	-1.4	0.594	0.585	0.978	0.823	0.744
	-1.0	0.941	0.916	0.935	0.858	0.491
	-0.5	0.898	0.814	0.814	0.854	0.118
	0.0	-	-	-	-	0.037
	0.5	0.919	0.871	0.860	0.791	0.272

Notes: The DGP is

$$y_t + 0.5y_{t-1} + (\theta_1 y_{t-1})F(z_t) = \varepsilon_t,$$

with $\varepsilon_t \sim N(0, 25.0)$, $c = 0$, $d = 1$, $\gamma = 0.5$.

constructed its counterpart (an LSTR model, continuing with the example) with the same choice of parameters.[17] The next section examines the accuracy of the selection procedures while Section 5.2 examines the power properties of the nonlinearity tests.

5.1 Accuracy of the EJ Selection Procedure

This section will compare the performance of the EJP versus the TP. In addition, and for the sake of completeness, we generalize the TP to account for the fact that the EJP is based on equation (14) and therefore uses information not available to the TP. The generalized version of the TP will be called the "GTP" and is described as follows. Conditional on rejecting linearity, use the following sequence of nested tests:

1. Test the null hypothesis $H_{04}^G\colon \beta_4' = \underline{0}$ with an F-test (F_4^G).

2. Test the null hypothesis $H_{03}^G\colon \beta_3' = \underline{0} \mid \beta_4' = \underline{0}$ with an F-test (F_3^G).

Table 3. Relative Correct Selection Frequencies for Models 4.1 and 4.6 in Teräsvirta (1994).

LSTAR Model							
			Selection Procedure				
Sample Size	c	π_{20}	TP	GTP	EJP	C = 0	Power NL4
100	0.0	0.0	0.975	0.965	0.997	0.984	0.950
	0.0	0.02	0.932	0.909	0.907	0.913	0.636
	0.02	0.04	0.859	0.827	0.577	0.611	0.156
300	0.0	0.0	1.000	1.000	1.000	1.000	1.000
	0.0	0.02	0.998	0.996	0.985	0.979	0.993
	0.02	0.04	0.959	0.925	0.671	0.691	0.386
ESTAR Model							
			Selection Procedure				
Sample Size	c	π_{20}	TP	GTP	EJP	C = 0	Power NL4
100	0.0	0.0	0.885	0.922	0.971	0.963	0.729
	0.0	0.02	0.673	0.698	0.670	0.863	0.838
	0.02	0.04	0.488	0.569	0.350	0.816	0.606
300	0.0	0.0	0.983	0.989	1.000	1.000	0.999
	0.0	0.02	0.876	0.883	0.850	0.995	1.000
	0.02	0.04	0.332	0.378	0.569	0.961	0.982

Notes: The DGP is

$$y_t = 1.8 y_{t-1} - 1.06 y_{t-2} + \left(\pi_{20} - 0.9 y_{t-1} + 0.795 y_{t-2}\right) F(z_t) + u_t$$

with $d = 1$, $u_t \sim N(0, 0.0004)$, $\gamma = 100$ (LSTAR), 1000 (ESTAR).

Table 4. Relative Correct Selection Frequencies for ESTAR Models in Luukkonen *et al.* (1988b) Table 4, p. 172.

		Selection Procedure				
Sample Size	μ	TP	GTP	EJP	C = 0	Power NL4
50	0.0	0.632	0.698	0.792	0.833	0.106
	0.3	0.472	0.552	0.736	0.739	0.125
	1.0	0.129	0.252	0.624	0.466	0.210
100	0.0	0.805	0.836	0.898	0.904	0.256
	0.3	0.522	0.609	0.830	0.768	0.317
	1.0	0.095	0.142	0.704	0.403	0.493
200	0.0	0.899	0.914	0.963	0.936	0.616
	0.3	0.659	0.686	0.923	0.753	0.692
	1.0	0.039	0.053	0.765	0.510	0.881

Notes: The DGP is

$$y_t - 0.3y_{t-1} - \left\{\exp(-y_{t-1}^2) - 1\right\}0.9y_{t-1} = \mu + \varepsilon_t,$$

with $\varepsilon_t \sim N(0, 0.36)$.

3. Test the null hypothesis $H_{02}^G: \beta_2' = \underline{0} \mid \beta_3' = \beta_4' = \underline{0}$ with an F-test (F_2^G).

4. Test the null hypothesis $H_{01}^G: = \beta_1' = \underline{0} \mid \beta_2' = \beta_3' = \beta_4' = \underline{0}$ with an F-test (F_1^G).

5. If the minimum p-value from these sequence of tests corresponded to either F_4^G or F_2^G, select ESTR. Otherwise, if it corresponds to either F_3^G or F_1^G, select LSTR.

The simulations were done using the general guidelines introduced in the previous section and available in the Appendix. We used NL4 as our test of nonlinearity (we also used NL3 but obtained similar results). Conditional on rejecting linearity, we then applied the EJP, TP, and GTP, each. Furthermore, conditional on having found the correct model, we then looked at the accuracy of being able to determine whether the threshold is zero. The results of these exercises are reported in Tables 1 - 5. In all, we tested three LSTR models and three ESTR models for different specifications.

The results of the experiments indicate that the EJP is much more accurate than

Table 5. Relative Correct Selection Frequencies for ESTAR Models in Saikkonen and Luukkonen (1988), Table 4.3, p. 64.

			Selection Procedure				
Sample Size	a	θ	TP	GTP	EJP	C = 0	Power NL4
50	0.9	0.3	0.667	0.746	0.702	0.925	0.114
	0.9	0.6	0.826	0.852	0.852	0.836	0.264
	-0.3	-0.9	0.747	0.783	0.816	0.768	0.217
	-0.6	-0.9	0.720	0.752	0.858	0.700	0.218
	-0.6	0.9	0.379	0.485	0.591	0.667	0.066
100	0.9	0.3	0.820	0.860	0.847	0.894	0.222
	0.9	0.6	0.958	0.967	0.973	0.969	0.669
	-0.3	-0.9	0.854	0.858	0.913	0.820	0.528
	-0.6	-0.9	0.751	0.757	0.941	0.829	0.666
	-0.6	0.9	0.600	0.650	0.750	0.475	0.080
200	0.9	0.3	0.924	0.940	0.946	0.954	0.485
	0.9	0.6	0.993	0.994	0.998	1.000	0.974
	-0.3	-0.9	0.959	0.959	0.982	0.961	0.903
	-0.6	-0.9	0.805	0.806	0.980	0.931	0.966
	-0.6	0.9	0.805	0.842	0.858	0.840	0.190

Notes: The DGP is

$$y_t + \{a + \theta\exp(-y_{t-1}^2)\}y_{t-1} = u_t,$$

with $u_t \sim N(0, 0.09)$.

either the TP or the GTP. The EJP's success rate always increases as the sample size increases (a highly desirable "consistency" feature- the result of the particular design of the procedure). In contrast, both the TP and the GTP lack this feature in some cases. For example, consider the $\mu = 1$ case in Table 4. The TP's correct selection frequency drops form 12.9% to 9.5% and to 3.9% as the sample size *increases* from 50 to 100 and to 200 observations, respectively. Compare these numbers to the EJP's selection frequency for the same case: 62.4%, 70.4%, and 78.5%.

Table 6. Power Simulations for ESTAR and LSTAR Models in Luukkonen *et al.* (1988b), Table 4, p. 172.

		$\sigma=0.3$		$\sigma=0.6$		$\sigma=0.9$	
Sample Size	Model	NL3 Power	NL4 Power	NL3 Power	NL4 Power	NL3 Power	NL4 Power
	$\mu=0.0$						
50	ESTAR	0.183	0.168	0.106	0.138	0.054	0.071
	LSTAR	0.125	0.103	0.203	0.173	0.242	0.202
100	ESTAR	0.405	0.411	0.225	0.292	0.107	0.152
	LSTAR	0.242	0.225	0.374	0.347	0.458	0.436
200	ESTAR	0.725	0.746	0.459	0.623	0.157	0.263
	LSTAR	0.486	0.425	0.683	0.636	0.828	0.792
	$\mu=0.3$						
50	ESTAR	0.232	0.201	0.130	0.139	0.077	0.090
	LSTAR	0.127	0.107	0.200	0.173	0.264	0.233
100	ESTAR	0.439	0.433	0.291	0.339	0.102	0.142
	LSTAR	0.266	0.238	0.393	0.361	0.521	0.461
200	ESTAR	0.823	0.828	0.502	0.621	0.178	0.303
	LSTAR	0.487	0.423	0.707	0.653	0.822	0.793
	$\mu=1.0$						
50	ESTAR	0.384	0.339	0.181	0.172	0.087	0.078
	LSTAR	0.114	0.102	0.222	0.203	0.281	0.248
100	ESTAR	0.702	0.661	0.321	0.319	0.155	0.169
	LSTAR	0.240	0.206	0.419	0.385	0.529	0.442
200	ESTAR	0.963	0.946	0.629	0.603	0.246	0.246
	LSTAR	0.490	0.442	0.714	0.754	0.850	0.802

Notes: LSTAR models constructed for the same values of γ and c as in ESTAR models.

Table 7. Power Simulations for ESTAR and LSTAR Models in Saikkonen and Luukkonen (1988), Table 4.3, p. 64.

		$\sigma = 0.3$		$\sigma = 0.6$		$\sigma = 0.9$	
		NL3 Power	NL4 Power	NL3 Power	NL4 Power	NL3 Power	NL4 Power
Sample Size = 50							
$A = 0.9$	ESTAR	0.220	0.201	0.085	0.095	0.073	0.072
$\theta = 0.3$	LSTAR	0.994	0.994	0.997	0.997	0.996	0.996
$A = 0.9$	ESTAR	0.462	0.478	0.144	0.158	0.074	0.077
$\theta = 0.6$	LSTAR*	0.045	0.035	0.045	0.047	0.048	0.059
$A = -0.3$	ESTAR	0.387	0.352	0.308	0.306	0.173	0.167
$\theta = -0.9$	LSTAR	0.043	0.042	0.038	0.038	0.053	0.053
$A = -0.6$	ESTAR	0.363	0.348	0.338	0.321	0.154	0.167
$\theta = -0.9$	LSTAR†	0.774	0.769	0.829	0.811	0.830	0.814
$A = -0.6$	ESTAR	0.081	0.078	0.160	0.162	0.100	0.122
$\theta = 0.9$	LSTAR	0.069	0.057	0.108	0.104	0.169	0.143
Sample Size = 100							
$A = 0.9$	ESTAR	0.354	0.368	0.117	0.115	0.070	0.067
$\theta = 0.3$	LSTAR	1.000	1.000	1.000	1.000	1.000	1.000
$A = 0.9$	ESTAR	0.787	0.856	0.218	0.265	0.084	0.079
$\theta = 0.6$	LSTAR*	0.681	0.933	0.679	0.929	0.701	0.930
$A = -0.3$	ESTAR	0.742	0.703	0.566	0.578	0.274	0.331
$\theta = -0.9$	LSTAR	0.053	0.053	0.065	0.065	0.078	0.078
$A = -0.6$	ESTAR	0.790	0.761	0.677	0.685	0.270	0.316
$\theta = -0.9$	LSTAR†	0.567	0.553	0.761	0.741	0.778	0.771
$A = -0.6$	ESTAR	0.153	0.141	0.359	0.355	0.209	0.273
$\theta = 0.9$	LSTAR	0.084	0.081	0.205	0.174	0.400	0.351

Table 7, Continued

		$\sigma = 0.3$		$\sigma = 0.6$		$\sigma = 0.9$	
		NL3 Power	NL4 Power	NL3 Power	NL4 Power	NL3 Power	NL4 Power
Sample Size = 200							
$A = 0.9$	ESTAR	0.610	0.643	0.148	0.162	0.072	0.071
$\theta = 0.3$	LSTAR	1.000	1.000	1.000	1.000	1.000	1.000
$A = 0.9$	ESTAR	0.969	0.969	0.340	0.470	0.094	0.108
$\theta = 0.6$	LSTAR*	0.981	0.997	0.975	0.987	0.978	0.991
$A = -0.3$	ESTAR	0.970	0.967	0.869	0.895	0.531	0.658
$\theta = -0.9$	LSTAR	0.085	0.085	0.073	0.071	0.091	0.090
$A = -0.6$	ESTAR	0.980	0.978	0.917	0.957	0.445	0.577
$\theta = -0.9$	LSTAR†	0.576	0.557	0.724	0.703	0.786	0.774
$A = -0.6$	ESTAR	0.312	0.278	0.699	0.735	0.448	0.599
$\theta = 0.9$	LSTAR	0.170	0.152	0.466	0.410	0.730	0.685

Notes: LSTAR models constructed for the same values of γ and c as original ESTAR models, with the following exceptions. '*' indicates that $\gamma/10$ was used and '†' indicates that $\gamma/3000$ was used.

Teräsvirta (1994) recognized that the TP works well when the LSTR and ESTR specifications are not close substitutes. However, the TP is less effective when the two models are close substitutes and the model is ESTR. It is remarkable that the most impressive gains of using the alternative the EJP occur precisely in this situation. The results are fairly conclusive: the EJP outperforms the TP (and the GTP); it is simpler to construct (it requires two simple F-tests and a straightforward choice); and it is "consistent" in the sense mentioned above.

5.2 Power Properties of the NL4 Test

The key question we examine here is whether the gains in power from adding the terms $\tilde{x}_t z_{t-d}^4$ in NL4 outweigh the losses from including p extra regressors in the auxiliary regression (14). It is clear that if the DGP is an LSTR model, we will lose power because we are including redundant regressors. If the model is ESTR we should expect to perform well whenever $c = 0$ and the data are symmetrically distributed

Table 8. Power Simulations for ESTAR and LSTAR Models (4.1) and (4.6) in Teräsvirta (1994).

	$\sigma = 0.01$		$\sigma = 0.02$		$\sigma = 0.04$	
	NL3 Power	NL4 Power	NL3 Power	NL4 Power	NL3 Power	NL4 Power
Sample Size = 100						
ESTAR	0.817	0.824	0.612	0.722	0.245	0.392
LSTAR	0.875	0.828	0.962	0.951	0.981	0.975
ESTAR	1.000	1.000	0.983	0.997	0.724	0.924
LSTAR	0.061	0.058	0.691	0.656	0.946	0.931
ESTAR	0.038	0.041	0.611	0.623	0.424	0.466
LSTAR	0.028	0.035	0.157	0.139	0.883	0.870
Sample Size = 200						
ESTAR	0.927	0.911	0.825	0.835	0.358	0.414
LSTAR	1.000	1.000	1.000	1.000	1.000	1.000
ESTAR	1.000	1.000	1.000	1.000	0.876	0.937
LSTAR	0.148	0.148	0.993	0.993	1.000	1.000
ESTAR	0.113	0.114	0.984	0.993	0.903	0.947
LSTAR	0.030	0.043	0.378	0.373	1.000	1.000

between the upper and lower regimes (recall Figure 1, case (1)). If $c = 0$, the benefits of including extra regressors will depend on each particular case. Tables 6 - 8 report this Monte Carlo exercise.

The simulations indicate that with large sample sizes ($T \geq 300$) there is little to no loss or gain in using NL4 rather than NL3. Both tests detect nonlinearity appropriately with the power approaching 1 in most cases. However, for small sample sizes, NL4 performs better, particularly when the variance is "large" and/or c and θ_0 are non-zero (asymmetric cases mentioned in Section 2, Figure 1). On the other hand, there are no significant losses of power when the true model is LSTR. One should view these results with caution. While NL4 is able to capture some nonlinearities that are hard to detect with NL3, it involves p additional regressors compared to NL3. When the lag length of the model, p, is long and the sample size is small, parsimony becomes an issue regarding the power of these tests. A parsimonious alternative in these cases is the test NL4A that we introduced in Section 3.1.

6. Unemployment Dynamics in the U.S.

This section will apply the techniques developed above in an empirical study of the dynamics of the U.S. unemployment rate; the aim is to illustrate how the previous tests work in practice rather than produce a detailed analysis of the U.S. unemployment rate. The existence of asymmetries in this unemployment rate series has been a debated topic in recent years. Neftçi (1984) used the theory of Markov chains and applied it to the series of the signs of the first differences of the unemployment rates. Symmetry was thus defined as a situation in which next period's probability of moving from a positive to a negative sign is equal to the probability of moving from a negative sign to a positive sign. Sichel (1989), after correcting some errors in Neftçi (1984), found no evidence of asymmetry using a second-order Markov chain, while Rothman (1991), using a first-order Markov chain concluded that the unemployment rate series was indeed asymmetric. DeLong and Summers (1986) defined asymmetry as non-zero skewness in a detrended series and found the U.S. unemployment rate to be asymmetric.

The issue of asymmetry is crucial in both theoretical and empirical studies of unemployment behavior. Theoretical explanations for why unemployment might be asymmetric vary widely. Non-competitive theories can generate asymmetries as a result of nominal and real rigidities. In Layard, *et al.* (1991), asymmetries are generated by insider-outsider mechanisms in wage setting. Labor turnover costs (a difference between hiring and firing costs) may also induce asymmetry as in Burgess (1988), Burgess and Dolado (1989), and Pfann and Palm (1993). According to these theories, adjustment costs depend on the size of unemployment, labor market legislation, and union power. Search-theoretic models can also predict asymmetries through a reduction of search effectiveness and loss of skills induced by long-term unemployment (see Pissarides (1992)).

Following the previous discussion, it is natural to test the unemployment rate series for nonlinearities. Before we proceed, it is useful to consider the raw data and its properties. Figure 2 depicts the U.S. monthly seasonally adjusted unemployment rate from January 1948 to July 1997. Perhaps the first concern that arises from observing the graph is whether the series is stationary (recall that the nonlinearity tests are based on the assumption of stationarity and ergodicity). Both the Dickey-Fuller test and the Phillips-Perron test clearly rejected the hypothesis of a unit root. This result coincides with previous studies in the literature. However, the majority of these studies incorporate a time trend, an approach that is hard to justify on theoretical grounds.

Regarding the asymmetric behavior of unemployment, it is perhaps best to begin with a simple inspection of the graph of the series and comment on its most notable characteristics. Two features are most striking. First, recessions are characterized by sharp increases in the unemployment rate within a few months. After reaching its peak the unemployment rate drops quickly at first, but its descent to the original level

Figure 2. U.S. Unemployment Rate, 1948:01 - 1997:07.

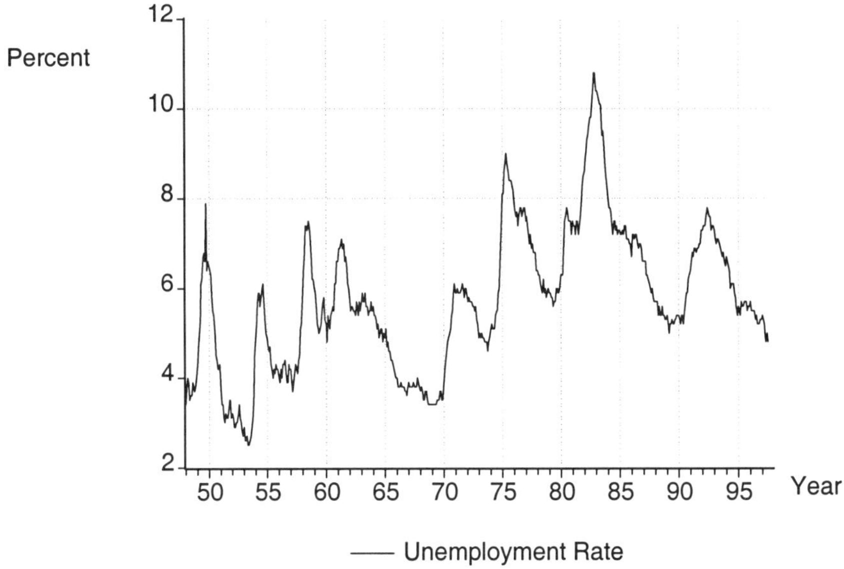

becomes protracted and slow. Second, related to the previous property, the duration between trough to peak is much shorter than the duration from peak to trough (note that we are referring to the unemployment rate, a counter-cyclical series, not the business cycle itself). If is therefore natural to consider a two-state model such as the STAR model to analyze this series.

6.1 Nonlinearity and Model Selection

The first step is to construct the linear model from which to build the nonlinearity tests and from which to compare the performance of the nonlinear alternatives. The usual information criteria (Akaike's and Schwarz's) select an AR(6). However, upon inspection of the autocorrelogram of the residuals and the Ljung-Box statistic, there is evidence of left-over autocorrelation corresponding to seasonal lags 12, 13, 24, and 25 (despite the fact that we used seasonally adjusted data). Recall that omitting terms in the specification of the linear model could lead to false rejections of linearity. Therefore, the linear model is expanded to include these lags as well.

Based on this model, we then perform tests for nonlinearity that have been presented in the previous sections, namely, NL3, NL4, and their augmented versions, NL3A and NL4A. Table 9 reports the results of these tests as well as the choice of delay parameter, d, and model selection with the EJP and TP. All four nonlinearity tests (NL3, NL4, NL3A, and NL4A) clearly detected nonlinearity for $d=6$. With

Table 9. Nonlinearity Tests, Choice of Delay Parameter, and Model Selection Tests for Seasonally Adjusted U.S. Unemployment Rate, 1948:01 - 1997:07.

Test	p-value	Delay Parameter
NL3A	0.02984	6
NL3	0.00012	6
NL4A	0.01926	6
NL4	0.00009	6

Selection Procedure	Model Selected
TP	LSTAR
EJP	ESTAR with $c \neq 0$

respect to model choice, the EJP clearly selected ESTAR with a non-zero threshold while the TP unequivocally selected the LSTAR specification.

6.2 Estimation and Evaluation

Following the results of the previous section, we proceeded to build and estimate ESTAR and LSTAR specifications for this time series to compare their performance. Following Teräsvirta (1994), estimation is performed by nonlinear least squares. Under certain conditions, including stationarity and ergodicity of the series, the estimators of the parameters are consistent and asymptotically normal.[18] The LSTAR alternative produced models that were either not superior to the linear alternative or did not converge. As a result we do not report these estimation results. The ESTAR alternative did produce models that had better fits than the linear model. The results of the preferred model are reported in Table 10.

An important result is worth noting. The ratio of the residual variance of the nonlinear model to that of the linear model was 0.97, clearly above the 0.90 level proposed by Granger and Andersen (1978) as a guideline to avoid spurious results. This indicates that while the ESTAR model has a better fit than its linear counterpart, the improvement is not very significant and therefore caution should be exercised in not putting too much weight on this specification. Given this caveat, there is perhaps still some interesting information that can be extracted from the model.

Figure 3 plots the transition function versus the data for this series, as we did in Section 2. This exponential function shows that the transition between regimes is very smooth; perhaps indicating why NL4 achieved the lowest p-value of all the

Table 10. ESTAR Estimates and Summary Statistics for Seasonally Adjusted U.S. Unemployment Rate, 1948:01 - 1997:07.

Coefficient	Estimate	Standard Error	*t*-Statistic
π_0	0.0320	0.1031	0.3101
π_1	0.9333	0.0789	11.823
π_2	0.2044	0.0929	2.2013
π_6	-0.1887	0.0404	4.6681
π_{12}	-0.1727	0.0429	4.0284
π_{13}	0.2286	0.0431	5.3080
π_{24}	-0.1380	0.0387	3.5697
π_{25}	0.1262	0.0367	3.4427
θ_0	0.0956	0.1262	0.7573
θ_1	0.2888	0.1316	2.1948
θ_2	-0.3490	0.1575	2.2156
θ_6	0.1521	0.0818	1.8599
θ_{13}	-0.1062	0.0494	2.1476
γ	1.1735	1.0422	1.1260
c	6.8691	0.2808	24.467
Summary Statistics			
R-Squared	0.986316	ARCH Test (*p*-val.)	0.012627
S.S.R.	19.64620	Jarque-Bera Res.	0.000000
Log-Likelihood	151.0145	Skewness Res.	0.361244
Durbin-Watson	2.002727	Kurtosis Res.	4.286392

Figure 3. Transition Function from ESTAR Model Estimated for U.S. Unemployment Rate.

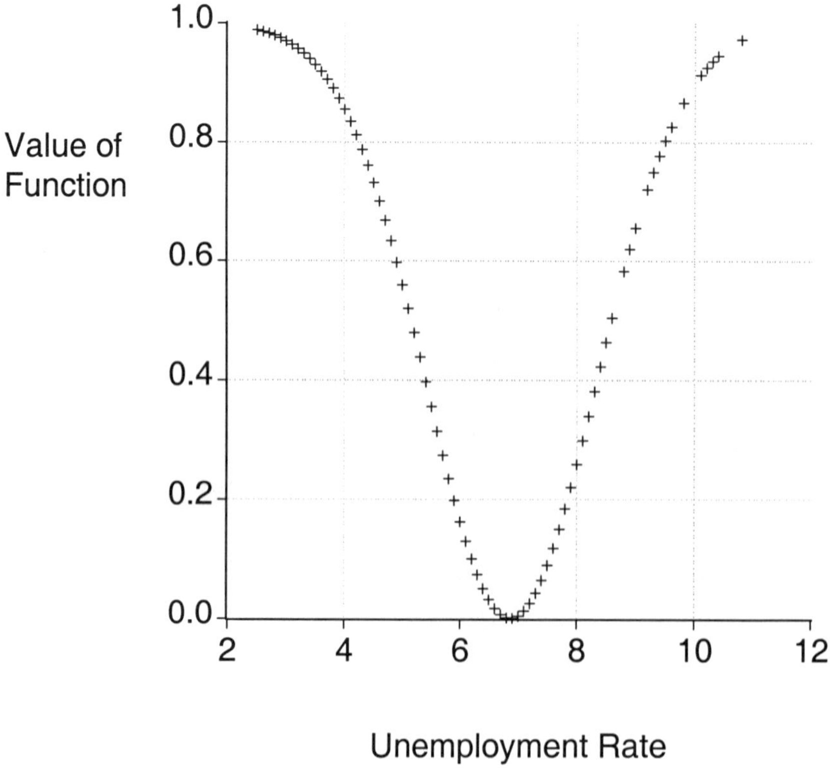

nonlinearity tests and why the TP might have selected the LSTAR class. Figure 4 plots the impulse response functions for each of the linear regimes in the estimated ESTAR model. For both regimes, the impulse response functions reveal very strong persistence. The upper regime exhibits a sinusoidal decay, characteristic of polynomials with complex roots, with a period of approximately 40 months. The lower regime is far more persistent (a 1% increase takes over 20 years to die out). Complementing these plots, Figure 5 shows a graph of the unemployment rate and the values of the transition function at each point. An important aspect of this graph is the following. Prior to approximately 1975, recessions (periods of high unemployment) are associated with periods in which the lower regime dominates while periods of low unemployment are associated with the upper regime. Starting around 1975 these dynamics reverse themselves. In fact, a Chow test for a break point in 1975 performed on the residuals of the ESTAR model rejects the null hypothesis of no-break.[19]

Figure 4. Impulse Response Functions of Linear Regimes from ESTAR Model Estimated for U.S. Unemployment Rate.

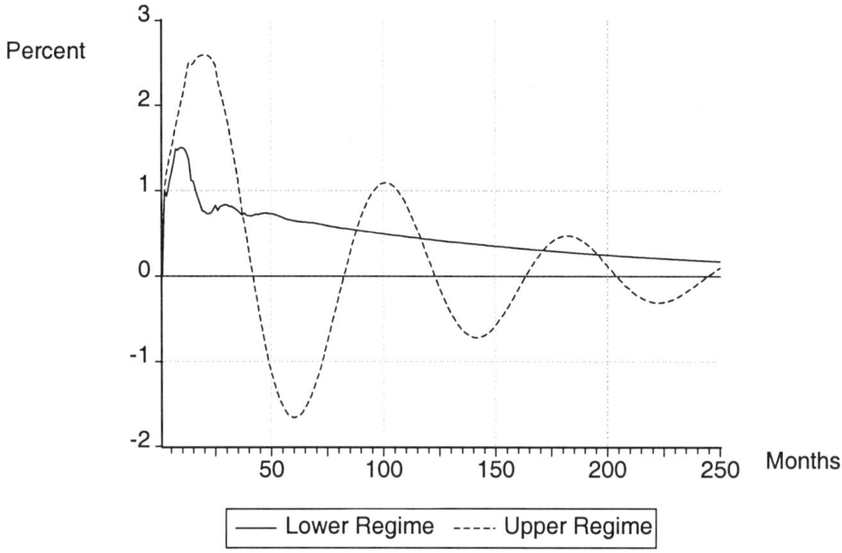

How successful is the ESTAR model in picking up asymmetries? Aside from the differences in dynamic behavior explained above, there seems to be ample scope to improve the modeling stage. The residuals exhibit ARCH, the Jarque-Bera test rejects the null of normality, and there is left-over skewness in the data (asymmetry defined in the sense of DeLong and Summers (1986)). This suggests that the ESTAR model, while it provided some useful insights, was not a good model in itself. For our purposes, however, we did learn a few things. The ESTAR model was clearly more helpful than its LSTAR alternative in describing some features of the data. The nonlinearity tests that we introduced had lower p-values than those available in the literature and our selection rule did suggest using what turned out to be the most useful specification within the STR class, the exponential STR, unlike the rule proposed by Teräsvirta (1994).

7. Conclusion

This chapter provides a variety of useful suggestions and testing strategies for empirical analysis with STR models. Our analysis is based on the properties of Taylor approximations to the transition function and the construction of different *LM* tests based on auxiliary regressions that use these approximations. The most significant contribution is the introduction of a new specification strategy to choose between

Figure 5. U.S. Unemployment Rate and Transition Function from ESTAR Model.

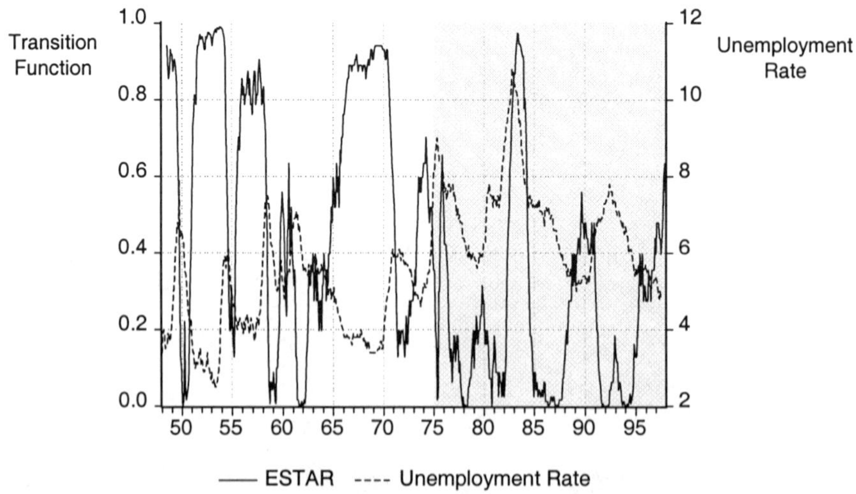

exponential and logistic STR models, the EJP. Monte Carlo evidence showed that this procedure generates much higher correct selection frequencies, it is consistent unlike its predecessor, the TP, and is straightforward to apply. Another important result was the demonstration that nonlinearity testing can be improved. By augmenting the existing tests with fourth-order terms, we are able to approximate salient features of the exponential function that increase the power of the test. While this result is not general, since there is not an equivalent gain when the alternative is an LSTR model, Monte Carlo evidence suggests that the gains in power are significant. Moreover, it constitutes the natural test in view of our selection procedure.

Understanding the nature of the Taylor approximations and the testing strategies is useful in adapting the testing procedures to empirical practice of each particular case. For example, we showed that in situations where parsimony is at stake, one can construct simplified versions of the test, namely, NL4A. Additionally, if one had transition functions other than the exponential and the logistic in mind, specific procedures can be readily constructed from the derivations in the chapter. Finally, the paper expands specification and testing for the STR family of models. Extensions to multivariate analysis are straightforward generalizations of the equations proposed except for a few obvious details.

Appendix

A1. Derivations of NL3 and NL4

Recall equation (9):

$$y_t = \pi'\underline{x}_t + \frac{1}{4}\gamma\theta'\underline{x}_t(z_{t-d} - c) + \frac{1}{48}\gamma^3\theta'\underline{x}_t(z_{t-d} - c)^3 + \upsilon_{3t}, \tag{A1}$$

from which we derived equation (10):

$$y_t = \delta_0 + \tilde{\delta}'\tilde{\underline{x}}_t + \beta_1'\tilde{\underline{x}}_t z_{t-d} + \beta_2'\tilde{\underline{x}}_t z_{t-d}^2 + \beta_3'\tilde{\underline{x}}_t z_{t-d}^3 + \upsilon_{3t}. \tag{A2}$$

The reparameterization from equation (9) to (10) is as follows:

$$\delta_0 = \pi_0 - \frac{1}{4}\gamma\theta_0 c\left[1 + \frac{1}{12}\gamma^2 c^2\right]$$

$$\tilde{\delta}' = \tilde{\pi}' - \frac{1}{4}\gamma c\left[1 + \frac{1}{12}\gamma^2 c^2\right]\tilde{\theta}'$$

$$\delta_d = \pi_d + \frac{1}{4}\gamma c[\theta_0 - \theta_d] + \frac{1}{48}\gamma^3 c^3[3\theta_0 - \theta_d]$$

$$\beta_1' = \frac{1}{4}\gamma\left[1 + \frac{1}{4}\gamma^2 c^2\right]\tilde{\theta}'$$

$$\beta_{1d} = \frac{1}{4}\gamma\theta_d + \frac{1}{16}\gamma^3 c(c\theta_d - \theta_0)$$

$$\beta_2' = -\frac{1}{16}\gamma^3 c\tilde{\theta}'$$

$$\beta_{2d} = \frac{1}{48}\gamma^3(\theta_0 - 3c\theta_d)$$

$$\beta_3' = \frac{1}{48}\gamma^3\theta_0$$

Recall equation (13):

$$y_t = \pi'\underline{x}_t + \theta'\underline{x}_t\left[(z_{t-d}-c)^2(\gamma+\gamma^2) - \frac{1}{2}(z_{t-d}-c)^4\gamma^2\right] + \upsilon_{4t}, \tag{A3}$$

from which we derived equation (14):

$$y_t = \delta_0 + \tilde{\delta}'\underline{\tilde{x}}_t + \beta_1'\underline{\tilde{x}}_t z_{t-d} + \beta_2'\underline{\tilde{x}}_t z_{t-d}^2 + \beta_3'\underline{\tilde{x}}_t z_{t-d}^3 + \beta_4'\underline{x}_t z_{t-d}^4 + \upsilon_{4t}.$$

This transformation is done as follows:

$$\delta_0 = \pi_0 + \theta_0(\gamma+\gamma^2)c^2 - \frac{1}{2}\theta_0\gamma c^4$$

$$\tilde{\delta}' = \tilde{\pi}' + c^2(\gamma+\gamma^2)\tilde{\theta}' + \frac{1}{2}\gamma c^4$$

$$\delta_d = \pi_d + c(\gamma+\gamma^2)(c\theta_d - 2) + \frac{1}{2}\gamma c^3(c\theta_d - 4\theta_0)$$

$$\beta_1' = 2c\gamma(c^2 - (1+\gamma))\tilde{\theta}'$$

$$\beta_{1d} = (\gamma+\gamma^2)(\theta_0 - 2c\theta_d) + \gamma c^2(2c\theta_d - 3\theta_0)$$

$$\beta_2' = \left[(\gamma+\gamma^2) - 3\gamma c^2\right]\tilde{\theta}'$$

$$\beta_{2d} = (\gamma+\gamma^2)\theta_d + \gamma c(2\theta_0 - 3c\theta_d)$$

$$\beta_3' = 2\gamma c\tilde{\theta}'$$

$$\beta_{3d} = \gamma\left(2cd - \frac{1}{2}\theta_0\right)$$

$$\beta_4' = \frac{1}{2}\gamma\tilde{\theta}'$$

A.2 Description of the Monte Carlo Experiments

The series in this study were generated according to each of the models. The first 100 observations from each of the series were dropped. Each experiment was replicated 1,000 times. The χ^2 version of the test was used for the simulations to determine the empirical power for the NL3 and the NL4 tests. The experiments were not size-corrected. Size-correction showed that the empirical values were close to the asymptotic values. The simulations of the decision rules were conditioned on prior rejection of linearity with NL4 in its *F*-version. Similar results were obtained by conditioning on NL3. The models used and the details of each simulation are described in Tables 1 - 5. Tables 6 - 8 are based on the same models and their specifics are therefore not repeated.

Acknowledgements

The first author acknowledges support from Spanish DGCIYT, PB95-0298. The second author acknowledges support from "La Caixa" and the Bank of Spain. We wish to thank R. F. Engle, J. Gonzalo, C. W. J. Granger, J. D. Hamilton, M. H. Pesaran, and T. Teräsvirta for useful comments and suggestions.

Notes

1. See Granger and Teräsvirta (1993), Teräsvirta (1994), Haggan and Ozaki (1981), Tsay (1989), and Chan and Tong (1986).
2. See Hamilton (1989).
3. A variety of recent studies justify this assertion. See Neftçi (1984), Sichel (1989), and Rothman (1991), for example.
4. Based on Davies (1977), Luukkonen *et al.* (1988a) and Teräsvirta (1994) introduced this solution for Smooth Transition Autoregressive models.
5. Note this alternative is an assumption imposed by the analysis. The test has power against alternatives other than STR.
6. See Teräsvirta (1994) when *d* is unknown.
7. This assumption is useful to derive the *LM* tests below. See White (1984).
8. In general, it could be any continuously differentiable function such that $0 \leq F(z_{t-d}, \gamma, c) \leq 1$ for all z_{t-d}, $1 \leq d \leq p$, $\gamma \neq 0$, and c. However, in practice the logistic and the exponential permit the necessary degree of generality.
9. The term 1/2 is added here for convenience but does not affect the results.
10. The delay parameter, *d*, is usually unknown. Based on Tsay (1989), Teräsvirta (1994) proposes choosing *d* so as to minimize the *p*-value of the nonlinearity test.
11. Also note that the alternative now includes models other than STR.
12. The details of the correspondence between the parameters in equations (9) and (10) are derived in the Appendix.
13. The details of how the parameters in (13) are transformed into (14) are available in the Appendix.
14. When $z_{t-d} = y_{t-d}$, this test is similar to a higher-order RESET test.
15. Too few terms can cause false rejections of linearity, excessive terms can undermine the power of the nonlinearity test.
16. We mentioned this in the previous section.
17. However, some of the values of γ had to be rescaled to obtain sensible models.
18. See Tong (1990) Chapter 5.

19. This might constitute another reason why the LSTAR model was a poor alternative to the ESTAR model.

References

Burgess, S. M. (1988). "Employment Adjustment in U.K. Manufacturing," *Economic Journal*, 98, 81-103.

Burgess, S. M. and J. J. Dolado (1989). "Intertemporal Rules with Variable Speed of Adjustment: An Application to U.K. Manufacturing Employment," *Economic Journal*, 99, 347-365.

Chan, K. S. and H. Tong (1986). "On Estimating Thresholds in Autoregressive Models," *Journal of Time Series Analysis*, 7, 179-190.

Davies, R. B. (1977). "Hypothesis Testing When a Nuisance Parameter is Present Only Under the Alternative," *Biometrika*, 64, 247-254.

DeLong, J. B. and L. H. Summers (1986). "Are Business Cycles Asymmetrical?" in Gordon, R.G. (ed.) *The American Business Cycle: Continuity and Change. NBER Studies in Business Cycles*, 25. Chicago: University of Chicago Press.

Escribano, A. and O. Jordá (1997). "Testing Nonlinearity: Decision Rules for Selecting Between Logistic and Exponential STAR Models," working paper, Department of Econometrics and Statistics, Universidad Carlos III de Madrid.

Granger, C. W. J. and A. P. Andersen (1978). *An Introduction to Bilinear Time Series Models*. Göttingen: Vandenhoeck and Ruprecht.

Granger, C. W. J. and T. Teräsvirta (1993). *Modelling Nonlinear Economic Relationships*. New York: Oxford University Press.

Haggan, V. and T. Ozaki (1981). "Modelling Nonlinear Random Vibrations Using an Amplitude-Dependent Autoregressive Time Series Model," *Biometrika*, 68, 189-196.

Hamilton, J. D. (1989). "A New Approach to the Economic Analysis of Non-Stationary Time Series and the Business Cycle," *Econometrica*, 57, 357-384.

Hansen, B. E. (1996). "Inference When a Nuisance Parameter is not Identified Under the Null Hypothesis," *Econometrica*, 64, 413-430.

Layard, R., S. Nickell, and R. Jackman (1991). *Unemployment: Macroeconomic Performance and the Labour Market*. Oxford: Oxford University Press.

Luukkonen, R., P. Saikkonen, and T. Teräsvirta (1988a). "Testing Linearity Against Smooth Transition Autoregressive Models," *Biometrika*, 75, 491-499.

Luukkonen, R., P. Saikkonen, and T. Teräsvirta (1988b). "Testing Linearity in Univariate Time Series Models," *Scandinavian Journal of Statistics*, 15, 161-175.

Neftçi, S. N. (1984). "Are Economic Time Series Asymmetric Over the Business Cycle?" *Journal of Political Economy*, 92, 307-328.

Pesaran, M. H. and S. M. Potter (1997). "A Floor and Ceiling Model of U.S. Output," *Journal of Economic Dynamics and Control*, 21, 661-696.

Pfann, G. A. and F. C. Palm (1993). "Asymmetric Adjustment Costs in Non-Linear Labour Demand Models for the Netherlands and U.K. Manufacturing Sectors," *Review of Economic Studies*, 60, 397-412.

Pissarides, C. (1992). "Loss of Skills During Unemployment and the Persistence of Employment Shocks," *Quarterly Journal of Economics*, 107, 1371-1392.

Rothman, P. (1991). "Further Evidence on the Asymmetric Behavior of Unemployment Rates Over the Business Cycle," *Journal of Macroeconomics*, 13, 291-298.

Saikkonen, P. and R. Luukkonen (1988). "Lagrange Multiplier Tests for Testing Nonlinearities in Time Series Models," *Scandinavian Journal of Statistics*, 15, 55-68.

Sichel, D. (1989). "Are Business Cycles Symmetric? A Correction," *Journal of Political Economy*, 97, 1055-1060.

Sichel, D. (1993). "Business Cycle Asymmetry: A Deeper Look," *Economic Inquiry*, 31, 224-236.

Teräsvirta, T. and H. M Anderson (1992). "Characterizing Nonlinearities in Business Cycles Using Smooth Transition Autoregressive Models," *Journal of Applied Econometrics*, 7, S119-S136.

Teräsvirta, T. (1994). "Specification, Estimation, and Evaluation of Smooth Transition Autoregressive Models," *Journal of the American Statistical Association*, 89, 208-218.

Teräsvirta, T., D. Tjøstheim, and C. W. J. Granger (1994). "Aspects of Modelling Nonlinear Time Series," in Engle, R.F. and D. McFadden (eds.) *Handbook of Econometrics, 4*. Amsterdam: Elsevier.

Tong, H. (1990). *Non-Linear Time Series: A Dynamical System Approach*. Oxford: Clarendon Press.

Tsay, R. (1986). "Non-Linearity Tests for Time Series," *Biometrika*, 73, 461-466.

Tsay, R. (1989). "Testing and Modelling Threshold Autoregressive Processes," *Journal of the American Statistical Association*, 84, 231-240.

White, H. (1984). *Asymptotic Theory for Econometricians*. London and San Diego: Academic Press.

15 SPECULATIVE BEHAVIOR, REGIME-SWITCHING, AND STOCK MARKET CRASHES

Simon van Norden
École des Hautes Études Commercial
Huntley Schaller
Carleton University

1. Introduction

Stock market crashes have presented a perennial challenge to our understanding of financial markets. The fact that there are sometimes abrupt changes in asset prices with little "news" about economic fundamentals is difficult to reconcile with simple models of asset pricing. Attempts to deal with this fact have tended to follow one of two approaches. The first allows asset prices to be determined by factors that do not affect their fundamental values. The second approach maintains that the market is "efficient," but suggests that the relationship between economic fundamentals and asset prices is highly nonlinear, so that relatively minor pieces of news sometimes have exceptionally large effects.

Stock market crashes have long fascinated observers of financial markets, but they are hard to analyze using conventional econometric techniques. One key problem is that crashes may involve sharp breaks in the relationships among variables. This suggests using techniques that explicitly recognize such breaks. This paper demonstrates that models of stock market crashes have empirical implications which can be tested using switching-regression econometrics. We show that switching regressions reveal new patterns in the data which go beyond existing stylized facts.

The first model that we consider is designed to capture several common features of historical accounts of "manias and panics."[1] We refer to this as a model of speculative behavior. The stylized pattern is an accelerating upswing in asset prices, followed by a precipitous reversal. Historical accounts suggest that an asset price crash becomes more likely as the relationship between asset prices and their fundamental value grows more extreme. In Section 2, we show how the model of speculative behavior has testable implications for the presence of regime switching in stock market returns. In particular, we show that the model predicts a correlation between stock returns and the deviation from fundamental price, but that this correlation should vary across regimes. It also suggests that the size of the deviation from fundamental price should help predict the probability of a crash.

Regime switching has been linked to speculative behavior in a number of previous studies. van Norden and Schaller (1993), and van Norden (1996) use similar methods to look for evidence consistent with such behavior in the Toronto Stock market and in the foreign exchange market. Hall and Sola (1993) and Funke, Hall, and Sola (1994) use a univariate regime-switching framework to look for such evidence in other markets.[2] Several other studies (such as Turner, Startz, and Nelson (1989) or Hamilton and Susmel (1994)) have applied regime-switching models to U.S. stock market data, but these typically feature a univariate framework and do not try to link regime-switching to speculative behavior.

As pointed out by Flood and Hodrick (1990), any empirical evidence consistent with the presence of speculative behavior can be re-interpreted as evidence of misspecified fundamentals. We examine this possibility in detail. In the case of switching regressions, evidence of the kind implied by the model of speculative behavior could also be consistent with a model where agents rationally anticipate changes in fundamentals. Barsky and DeLong (1993) and Donaldson and Kamstra (1996) argue that investors try to estimate a dividend growth rate that may vary over time and that therefore even small changes in the dividend growth rate can lead to substantial price changes. If new information causes investors to lower their expectations of future dividend growth rates, the stock price might suddenly plunge. Cecchetti, Lam, and Mark (1990, 1993) formalize a model of occasional shifts in the dividend growth rate between a high and a low growth regime. They have shown that such a model is capable of generating some of the previously documented features of U.S. stock market returns. To the best of our knowledge, we are the first to explore formally the possibility that stock market crashes are linked to regime changes in economic fundamentals.

In Section 2 of the paper, we develop the models of speculative behavior and switching fundamentals. Section 3 shows the relationship between speculative behavior and a switching-regression specification and tests the null hypothesis that stock market returns are unrelated to the deviation from fundamental price using parametric restrictions on the switching-regression specification. Section 4 presents parameter estimates of the model of switching fundamentals and discusses to what extent shifts in fundamentals explain the switching-regression results. Section 5 analyzes how well the probabilities of collapse generated by each model accord with actual stock market crashes. Section 6 offers conclusions and poses questions for future research.

2. Parametric Models of Stock Market Crashes

In this section, we develop the models of speculative behavior and switching fundamentals, each of which has the potential to account for extraordinary movements in stock market prices. Both the model of speculative behavior and the model of switching fundamentals are departures from the basic Lucas (1978) asset pricing model, which is discussed in Section 2.1. The distinctive feature of the model of speculative behavior, which is described in Section 2.2, is that it allows actual prices to diverge from the present value of future dividends. The distinctive feature of the model of switching fundamentals (outlined in Section 2.3) is that there are regime shifts in the endowment process, which might correspond to booms or recessions in the aggregate economy.

2.1 Equilibrium Asset Prices in the Lucas Model

This section briefly reviews key features of the Lucas (1978) exchange-economy asset pricing model; a more detailed derivation is left for the Appendix (which is available in van Norden and Schaller (1996)). The model is deliberately simple; in particular, the process we assume for dividends implies that in this model the observed price-dividend ratio should be constant. The relationship between ex post returns and variations in the price-dividend ratio then becomes a key feature to be explained by either a model of switching due to speculative behavior, or a model of switching in the growth of dividends. In the empirical work which follows, we show how the results are affected by relaxing the constant price-dividend ratio assumption in various ways. We find that, for the alternatives we examine, the conclusions are fairly robust. Therefore, we maintain this assumption here as a way of simplifying the exposition.

In the Lucas model, there are a large number of identical, infinitely-lived agents and a fixed number of assets that produce units of the non-storable consumption good. Since agents are identical, per capita consumption is equal to per capita dividends (D). This and the assumption of Constant Relative Risk Aversion (CRRA) utility gives the following stochastic difference equation for equilibrium prices

$$P_t \cdot D_t^{\gamma} = \beta \cdot \mathrm{E}_t D_{t+1}^{\gamma} \cdot (P_{t+1} + D_{t+1}) \tag{1}$$

which in turn yields the following equation for fundamental price

$$P_t^* = D_t^{-\gamma} \cdot \sum_{k=1}^{\infty} \beta^k \cdot \mathrm{E}_t D_{t+k}^{1+\gamma} \tag{2}$$

where P is the value-weighted stock market index adjusted for population size, β is the subjective discount factor, $0 < \beta < 1$, E_t is the mathematical expectation conditioned on information available at time t, and γ is the coefficient of relative risk aversion.

The expression for fundamental price depends on future dividends, which are not observable. One way to express the fundamental price in terms of observables is to make an assumption about the stochastic process for dividends. A common assumption (with some empirical justification) is that log dividends are a random walk with constant drift. This leads to a simple solution in which the fundamental price is a multiple of current dividends. Formally, dividends are

$$d_t = \alpha_0 + d_{t+1} + \varepsilon_t \tag{3}$$

where d_t is the logarithm of dividends, α_0 is the drift parameter, and ε_t is a sequence of independent, identically distributed normal random variables with mean zero and variance σ^2. This gives

$$\rho = \frac{\beta \cdot e^{\alpha_0(1+\gamma) + (1+\gamma)^2\sigma^2/2}}{1 - \beta \cdot e^{\alpha_0(1+\gamma) + (1+\gamma)^2\sigma^2/2}} \tag{4}$$

as the fundamental price-dividend ratio. The equilibrium gross return is:

$$\begin{aligned} R_t &\equiv \frac{P_t + D_t}{P_{t-1}} \\ &= \left(\frac{1+\rho}{\rho}\right) \cdot e^{\alpha_0 + \varepsilon_t} \end{aligned} \tag{5}$$

Note that the effect of an increase in the expected rate of dividend growth depends on whether γ is greater or less than -1. When $\gamma > -1$, increases in the dividend growth rate raise the price-dividend ratio; when $\gamma < -1$, the reverse is true.

2.2 A Model of Speculative Behavior

The model presented in this subsection introduces speculative behavior into the model of the previous subsection. Our goal is to develop a model which links switching-regression econometrics to some of the characteristics of speculative behavior which have been noted in historical accounts, experimental economics, and surveys of market participants[3]. The main result of this section is a switching-regression econometric specification (derived from the model of speculative behavior) which provides a framework for interpreting stock market crashes.

We define the size of the deviation from fundamental price as $B_t = P_t - P^*$. If the deviation from fundamental price is rational, then:

$$B_t \cdot D_t^{\gamma} = \beta \cdot \mathrm{E}_t[D_{t+1}^{\gamma}] \cdot \mathrm{E}_t[B_{t+1}] \tag{6}$$

where we use the fact that D_t will be independent of B_t (i.e., B_t is an extrinsic bubble). We can then use the assumed process for dividends given in (3) to show that

$$\begin{aligned} \frac{\mathrm{E}_t[B_{t+1}]}{B_t} &= \frac{D_t^{\gamma}}{\beta \cdot \mathrm{E}_t[D_{t+1}^{\gamma}]} \\ &= \beta^{-1} \cdot e^{-\gamma(\alpha_0 + \gamma\sigma^2/2)} \\ &\equiv M \end{aligned} \tag{7}$$

so the expected growth rate of the speculative component corresponds to the rate of return on the fundamental component. Blanchard (1979) and Blanchard and Watson (1982) propose a specific solution to (7) with two states of nature [4]. In one state (state C), the speculative component completely collapses, so $\mathrm{E}_t[B_{t+1}] = 0$. In the other state (state S), the speculative component survives. The probability of being in state S next period is assumed to be some constant, q. In this case, (7) implies

$$B_t \cdot M = \mathrm{E}_t[B_{t+1}] = q \cdot \mathrm{E}_t[B_{t+1} | S] + (1-q) \cdot 0 \tag{8}$$

$$\therefore \mathrm{E}_t[B_{t+1} | S] = \frac{M \cdot B_t}{q} \tag{9}$$

As noted by Kindleberger (1989) and others, historical accounts of "manias and panics" suggest that these episodes share many common features. One such feature is that a crash becomes more likely as the speculative component becomes a larger proportion of the price. To reflect the possibility that crashes are more likely to occur when the speculative component accounts for a large proportion of the price, we allow q, the probability of survival, to depend on the relative size of the speculative component

$$q \equiv q(b_t) \tag{10}$$

where $b_t = B_t / P_t$ and[5]

$$\frac{dq(b_t)}{d|b_t|} < 0 \tag{11}$$

Neither historical experience nor experimental evidence suggest that all deviations from fundamental price collapse in a single period. To take a prominent recent example, many observers interpreted the fall in the Tokyo stock exchange in the months following January 1990 as the gradual unwinding of a speculative episode. We therefore allow the expected value of b_{t+1} conditioned on collapse to be non-zero[6]. If we want to interpret state C as the collapse of a speculative component, then the expected value of b_{t+1} in state C must increase less than proportionately with b_{t+1}. To incorporate this, we define the expected value of b_{t+1} in state C as $u_t \cdot P_t$ and make u_t a function of b_t:

$$\mathrm{E}_t[B_{t+1}|C] = u(b_t) \cdot P_t \tag{12}$$

where $u(\cdot)$ is a continuous and everywhere differentiable function, $u(0) = 0$, and $1 \geq u' \geq 0$[7]. Imposing (7) then gives

$$\mathrm{E}_t[B_{t+1}|S] = \frac{M}{q(b_t)} \cdot B_t - \frac{1 - q(b_t)}{q(b_t)} \cdot u(b_t) \cdot P_t \tag{13}$$

The assumptions on $q(b_t)$ and $u(b_t)$ are intended to capture the spirit of speculative behavior as an explanation of crashes in a way that can empirically be translated into a switching-regression econometric specification. These restrictions are not imposed on the data. Rather, they help to give the model of speculative behavior empirically testable content.

As shown in the appendix, we can use the above equations to solve for the expected returns in each regime as a function of b_t. To guarantee that the estimates of q will be bounded between 0 and 1, we adopt the same approach used in Probit models by imposing the functional form $q = \Phi(\beta_{q0} + \beta_{qb} \cdot |b_t|)$ where Φ is the standard normal cumulative density function (CDF). After linearizing equations (12) and (13), we obtain the following three-equation model which we estimate below.

$$\begin{aligned} E(R_{t+1}|S) &= \beta_{S0} + \beta_{Sb} b_t \\ E(R_{t+1}|C) &= \beta_{C0} + \beta_{Cb} b_t \\ q &= \Phi(\beta_{q0} + \beta_{qb} \cdot |b_t|) \end{aligned} \tag{14}$$

As shown in the appendix, the model of speculative behavior has a number of testable implications. First, equation (11) implies that $\beta_{qb}<0$, so that as the deviation from fundamental price grows, so does the probability of a collapse. Second, the model implies that β_{Cb} should be negative. Intuitively, this is because a larger (positive) speculative component implies a larger capital loss if it collapses. Third, the model implies that $\beta_{Sb}>\beta_{Cb}$. Intuitively, a larger speculative component means a larger difference between returns in the surviving and collapsing states.

2.3 *Switching in Dividends*

The preceding section has shown how speculative behavior could generate regime switching behavior in stock returns. However, as we note in the introduction, regime switching in fundamentals might also lead to switching in stock market returns. In the remainder of this section, we consider a variation of the Lucas asset pricing model which explicitly allows for switching in the dividend process. A dividend process that meets this requirement is:

$$d_t = d_{t-1} + \alpha_0 + \alpha_1 S_{t-1} + \nu_t \tag{15}$$

where d_t is the logarithm of dividends, ν_t is a sequence of independent, identically normally distributed random variables with mean zero and variance σ^2 and S_t is the sequence of Markov random variables with state space $\{0,1\}$ and transition matrix:

$$\boldsymbol{P} \equiv \begin{bmatrix} P_{00} & P_{01} \\ P_{10} & P_{11} \end{bmatrix} \tag{16}$$

This is a Markov switching model. As noted in Cecchetti, Lam, and Mark (1990) (CLM), it is able to capture features of stock market returns not captured by other models, including ARIMA, ARCH, and GARCH models. In particular, CLM show that a Markov switching process in dividends can account for the evidence of mean reversion found by Fama and French (1988a) and Poterba and Summers (1988). Cecchetti, Lam, and Mark (1993) show that switching in the endowment process can also account for the first moment of the equity premium and the risk-free rate[8].

CLM show that the above equations lead to a relation between fundamental prices and dividends of the form:

$$P_t = \rho(S_t)D_t \tag{17}$$

In other words, the price-dividend ratio will take on one of two values, depending only on whether the economy is in the high or the low dividend growth state. Stock market crashes could then be caused by a transition from the high ρ to the low ρ state, implying large movements in stock prices[9]. The equilibrium gross return can be obtained from the relationship between current prices and dividends and the dividend process, so

$$R_t = \left(\frac{1 + \rho(S_t)}{\rho(S_{t-1})} \right) e^{(\alpha_0 + \alpha_1 S_{t-1} + v_t)} \tag{18}$$

3. Estimates of the Model of Speculative Behavior

This section is divided into four subsections. Part 3.1 explains the link between the model of speculative behavior presented in Section 2.2 and the switching-regression econometric specification, as well as how the model of speculative behavior can be tested. Part 3.2 reviews the data series that we use and introduces our empirical measure of the deviation from fundamental price. Part 3.3 presents, tests, and interprets parameter estimates for the model of speculative behavior. Part 3.4 explores the robustness of these results by re-estimating the model over various sub-periods and by using an alternative measure of the fundamental price which relaxes the assumption that the fundamental price-dividend ratio is constant.

3.1 Econometric Technique

To estimate the model of speculative behavior, we may rewrite (14) as:

$$R_{S,t+1} = \beta_{S0} + \beta_{Sb} b_t + \varepsilon_{S,t+1} \tag{19}$$

$$R_{C,t+1} = \beta_{C0} + \beta_{Cb} b_t + \varepsilon_{C,t+1} \tag{20}$$

$$q = \Phi(\beta_{q0} + \beta_{qb} \cdot |b_t|) \tag{21}$$

where $R_{S,t+1}$ and $R_{C,t+1}$ are the returns from period t to period $t+1$ conditional on survival and collapse respectively, and b_t is a measure of the deviation from fundamental price in period t. We assume $\varepsilon_{S,t+1}$ and $\varepsilon_{C,t+1}$ are mean zero i.i.d. normal random variables.

The three equations (19), (20), and (21) form a standard switching-regression model of the type described by Goldfeld and Quandt (1976) and Hartley (1978). Given normality of $(\varepsilon_{S,t+1}, \varepsilon_{C,t+1})$, estimates of the β's can be found by maximizing the likelihood function

$$\prod_{t=1}^{T}\left[\Phi(\beta_{q0}+\beta_{qb}|b_t|)\cdot\varphi\left(\frac{R_{t+1}-\beta_{S0}-\beta_{Sb}b_t}{\sigma_S}\right)\cdot\sigma_S^{-1} + \left\{1-\Phi(\beta_{q0}+\beta_{qb}|b_t|)\right\}\cdot\varphi\left(\frac{R_{t+1}-\beta_{C0}-\beta_{Cb}b_t}{\sigma_C}\right)\cdot\sigma_C^{-1}\right] \tag{22}$$

where φ is the standard normal probability density function (pdf) and σ_S, σ_C are the standard deviations of $\varepsilon_{S,t+1}, \varepsilon_{C,t+1}$. Note that this estimation technique not only allows us to recover consistent and efficient estimates of the parameters in both states but it does not require assumptions about which regime generated a given observation. Instead, it considers the probability that either regime may have generated a given observation and gives an optimal classification of observations into the underlying regimes. The probability of being in regime *i* at time *t+1* is defined as the probability conditioning on all relevant information available at the end of period *t*, namely b_t. This is determined solely by the classifying equation (21) and is given by the formula $\Phi(1(\mathrm{i})\cdot(\beta_{q0}+\beta_{qb}\cdot|b_t|)) \equiv P_i^A$, where 1(i) equals 1 in the surviving state and -1 in the collapsing state[10].

In Section 2.2, we note that the model of speculative behavior has three testable implications: $\beta_{Cb}<0$, $\beta_{Sb}>\beta_{Cb}$, and $\beta_{qb}<0$. Aside from the model of speculative behavior, however, the switching regression that we estimate also nests a variety of stylized facts about stock market behavior. To better understand the extent to which the data support or fail to support the model, it is useful to consider first whether the switching regression finds evidence of anything besides the existing stylized facts.

In particular, it would be interesting to know whether deviations from fundamental price have explanatory power for returns[11]. We therefore consider the extent to which our switching regression is able to reject three simpler specifications of return behavior, each of which captures a different stylized fact.

First, ARCH and related models have focused considerable attention on the changing volatility of stock market returns. Under the null hypothesis that b_t has no effect on returns, the switching regression can capture this fact (which we refer to as "volatility regimes") by imposing the restrictions $\beta_{S0}=\beta_{C0}=\beta_0$, $\beta_{Sb}=\beta_{Cb}=\beta_{qb}=0$ but allowing $\sigma_S \neq \sigma_C$, so

$$R_{t+1} = \beta_0 + \varepsilon_{t+1} \tag{23}$$

where

$$\begin{aligned} \varepsilon_{t+1} &\sim N(0,\sigma_S) \text{ with prob } q \\ \varepsilon_{t+1} &\sim N(0,\sigma_C) \text{ with prob } 1-q \end{aligned} \tag{24}$$

A similar specification has been estimated by Schwert (1989), who finds evidence of volatility regimes.

One fact that the volatility regime's null fails to capture is that periods of high volatility are more likely to occur during stock market declines while periods of low volatility tend to be associated with stock market increases. This is sometimes referred to as the "leverage effect." Therefore, we might wish to maintain the assumption that expected returns in each regime are constant but allow these constants to differ across regimes. This is the special case of the switching regression where $\beta_{Sb} = \beta_{Cb} = \beta_{qb} = 0$, so b_t has no effect. This implies that returns are well characterized by a mixture of normal distributions with different means and variances, which can be expressed as

$$\begin{aligned} R_{t+1} &\sim N(\beta_{SO},\sigma_S) \text{ with prob } q \\ R_{t+1} &\sim N(\beta_{CO},\sigma_C) \text{ with prob } 1-q \end{aligned} \tag{25}$$

for some constant q.

Another possibility we explore is that returns may be linearly predictable, but that mean returns do not differ across regimes. Testing this amounts to setting $\beta_{SO} = \beta_{CO} = \beta_0$, $\beta_{Sb} = \beta_{Cb} = \beta_1$, and $\beta_{qb} = 0$, giving

$$R_{t+1} = \beta_0 + \beta_1 b_t + \varepsilon_{t+1} \tag{26}$$

where[12]

$$\begin{aligned} \varepsilon_{t+1} &\sim N(0,\sigma_s) \text{ with prob } q \\ \varepsilon_{t+1} &\sim N(0,\sigma_c) \text{ with prob } 1-q \end{aligned} \tag{27}$$

We refer to this as the "mean-reversion" model, since it corresponds to the regression test for mean reversion in stock prices in Cutler, Poterba, and Summers (1991), except that we allow more flexibility for volatility by allowing the variances of returns to be drawn from high and low volatility distributions[13].

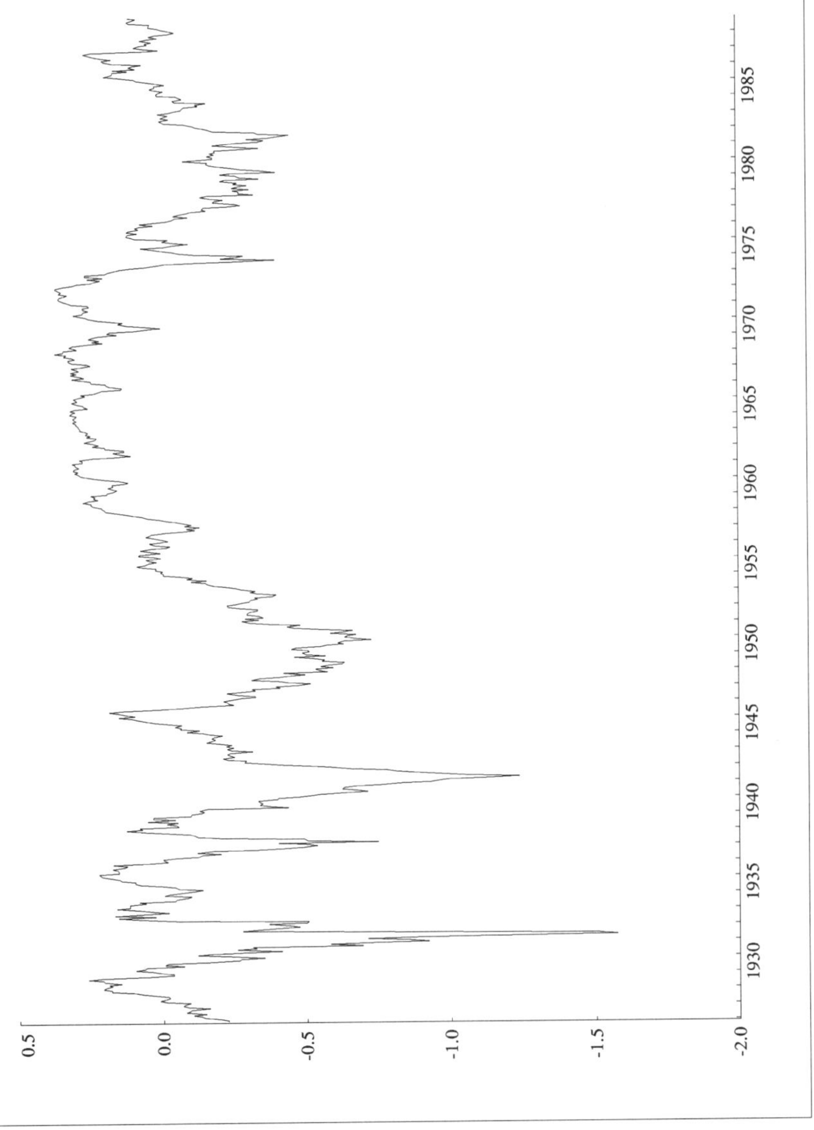

Figure 1 - Deviation of Actual Price from Fundamental Price

Table 1. The Model of Speculative Behavior

Parameters	Estimates
β_{SO}	1.007 (0.002)
β_{Sb}	-0.006 (0.006)
β_{CO}	0.976 (0.039)
β_{Cb}	-0.111 (0.071)
β_{q0}	2.098 (0.294)
β_{qb}	-1.560 (0.510)
σ_S	0.044 (0.002)
σ_C	0.170 (0.028)
Likelihood Ratio Tests	
Volatility Regimes	16.29 (0.003)
Mixture of Normals	16.21 (0.001)
Mean Reversion	14.79 (0.002)

Notes: The model of speculative behavior is equations (19) - (21) in the text. Figures in parentheses indicate standard errors for parameter estimates and p-values for likelihood ratio tests. The volatility regimes test imposes the restrictions $\beta_{SO} = \beta_{CO}$, $\beta_{Sb} = \beta_{Cb} = \beta_{qb} = 0$. The mixture of normals test imposes the restrictions $\beta_{Sb} = \beta_{Cb} = \beta_{qb} = 0$. The mean reversion test imposes the restrictions $\beta_{SO} = \beta_{CO}$, $\beta_{Sb} = \beta_{Cb}$, $\beta_{qb} = 0$.

3.2 Data and Measurement of Fundamental Price

The data we examine for evidence of speculative behavior are drawn from the Center for Research on Security Prices (CRSP) database. We use their monthly value-weighted price (P) and dividend (D) indices for all stocks from January 1926 to December 1989. We convert P and D to real per capita terms[14]. We use the All Items Consumer Price Index (CPI) to deflate nominal returns. Since dividends display strong seasonal fluctuations, we follow Fama and French (1988b) in using an average over the twelve-month period ending in the given month.

The first measure of deviations from fundamental price that we use (b_t^A) is tied to the model of equilibrium asset prices outlined in Section 2.1, where the fundamental price is $P_t^* = \rho D_t$. Under the hypothesis that actual prices may deviate from the fundamental price, we can measure the proportional deviation from fundamental price as:

$$b_t^A = \frac{B_t}{P_t} = \frac{P_t - P_t^*}{P_t} = 1 - \frac{\rho D_t}{P_t} \tag{28}$$

where we use the mean price-dividend ratio as ρ[15]. The solid line in Figure 1 shows how b_t moves over the sample period. Interestingly, two periods when b_t is rising and relatively large are 1929 and 1987. The value of b_t is unusually low in 1932, 1938, and 1942.

3.3 Estimation and Test Results

The estimated parameters for the model of speculative behavior (19)-(21) are presented in the top half of Table 1. The lower part of that table presents likelihood-ratio (LR) test statistics for the restrictions implied by the three "stylized fact" models of stock returns. We will begin by considering the latter.

As shown in the first column of the table, the LR statistic for the Volatility Regimes specification is 16.29 and has a p-value of 0.003. Our rejection of this null implies that the regimes must differ in more than their variances. This means that either the information contained in b_t helps to determine which regime prevails, or that the regimes have different expected conditional returns, or both.

In the second column, we see the LR statistic of 16.21 for the Normal Mixture specification has an even lower p-value (0.001). From this we conclude that there is more in the data than just the "leverage effect". Put another way, this tells us that deviations from the fundamental price have significant predictive power for the distribution of stock returns.

The LR statistic for the mean reversion model (14.79) allows a similar rejection (p-value = 0.002), implying that the previous rejections cannot be explained by simple linear predictability alone. It therefore seems that the relationship between b_t and R_t is highly significant, but nonlinear.

Turning now to the three coefficient restrictions implied by the model of speculative behavior, the first is $\beta_{Cb}<0$. As Table 1 illustrates, the point estimate of β_{Cb}

is negative; the standard error implies that we can reject the hypothesis that $\beta_{Cb}=0$ at just above the 0.05 level in a one-sided test. Second, β_{Sb} should be greater than β_{Cb}. The point estimates of the two parameters are consistent with this implication (*p*-value = 0.145). Third, β_{qb} should be negative. Again, this implication is consistent with the estimates reported in Table 1. Here, however, the statistical evidence is stronger: the *p*-value for the hypothesis that $\beta_{qb}=0$ is 0.001. Accordingly, we conclude that the estimated model is consistent with the restrictions implied by our model of speculative behavior.

To better understand the behavior that the estimated model is capturing, it is useful to consider the estimates in greater detail. Expected returns in the surviving regime for $b_t=0$ (the mean value of b_t) are given by β_{S0}. The point estimate of β_{S0} is 1.007, implying a monthly rate of return of 0.7% (8.7% on an annual basis). In contrast, the point estimate of β_{C0} is 0.976, implying a monthly rate of return of -2.4% (-25.3% on an annual basis) in the collapsing regime. The difference in returns is more dramatic in periods when b_t is relatively large. For example, in September 1929, b_t=.216; this implies an expected rate of return of 7% per year in the surviving regime and -44.6% in the collapsing regime[16].

The parameter β_{q0} can be used to calculate the probability that the collapsing regime will occur. When $b_t=0$, the probability of the collapsing regime is $1-\Phi(2.098)=0.018$. The point estimate of β_{qb} implies that in a period when $|b_t|$ is relatively large, the probability of the collapsing regime increases. For example, at the end of September 1929, the point estimates imply that the probability that the collapsing regime will occur in the following period is more than twice as large as when $b_t=0$.

3.4 Robustness

The results discussed in the previous section are generally supportive of the predictions of the model of speculative behavior. However, such econometric results are sometimes sensitive to the time period over which the estimation is done. It would also be useful to know how sensitive the results are to the assumption of a constant fundamental price-dividend ratio. We now examine both of these questions in turn.

We divide the sample into three subperiods: 1926-1954, 1954-74, 1974-89. The first sub-period includes the Great Depression, World War II and the Korean War. The division between the second and third sub-periods coincides with the first OPEC shock.

Results are shown in Table 2, and are generally similar across periods. As in the estimates for the full period, we always find $\beta_{S0}>1.0$ and $\beta_{C0}<1.0$. Some tests based on U.S. stock market data (such as mean reversion tests) show more evidence of a potential departure from efficient markets in time periods which include the 1929 crash and the Great Depression[17]. In contrast, the likelihood ratio tests presented in Table 2 actually show stronger rejection of the "stylized fact" null hypotheses in the periods 1954-1974 and 1974-1989 than in the period which includes the 1929 crash and the Great Depression. In all periods we reject the various null hypotheses at the 0.05 level[18].

The evidence on the testable implications of the model of speculative behavior also seems to be robust. In all sub-periods, the point estimate of β_{Cb} is negative; in

Table 2. The Model of Speculative Behavior (Sub-Periods)

Parameters Estimates	1926-1954	1954-1974	1974-1989
β_{S0}	1.009 (0.004)	1.035 (0.005)	1.006 (0.003)
β_{Sb}	0.007 (0.014)	-0.140 (0.017)	0.011 (0.025)
β_{C0}	0.996 (0.035)	0.997 (0.003)	0.956 (0.021)
β_{Cb}	-0.150 (0.119)	-0.036 (0.016)	-0.532 (0.071)
β_{q0}	1.794 (0.355)	-0.483 (0.299)	4.600 (1.080)
β_{qb}	-1.673 (0.768)	-3.726 (1.689)	-13.289 (4.034)
σ_S	0.050 (0.003)	0.012 (0.003)	0.044 (0.003)
σ_C	0.187 (0.035)	0.037 (0.002)	0.041 (0.015)
Likelihood Ratio Tests			
Volatility Regimes	9.40 (0.052)	19.02 (0.001)	14.98 (0.005)
Mixture of Normals	9.23 (0.026)	10.53 (0.015)	14.91 (0.002)
Mean Reversion	9.35 (0.025)	15.30 (0.002)	14.03 (0.003)

Notes: See Table 1 for notes.

most cases, the hypothesis that $\beta_{Cb}=0$ can be rejected at conventional levels. Except for the 1954-74 sub-period, $\beta_{Sb}>\beta_{Cb}$; the p-values for the first and third sub-periods are 0.156 and 0.000, respectively. In all sub-periods β_{qb} is negative; the hypothesis that $\beta_{qb}=0$ is strongly rejected in all cases.

Under the assumption that the expected dividend growth rate is constant, the Lucas asset pricing model implies that the fundamental price-dividend ratio (P_t^*/D_t) is constant, as shown in equation (A7) of the appendix. This means that the measure of deviation from fundamental price which we have used above (b_t^A) will attribute all of the variation in the price-dividend ratio to speculative behavior. There are many ways to relax this assumption. Here we consider an approach which allows for flexible variation over time in expected dividend growth. Based on Campbell and Shiller (1987), the key idea is that (in the absence of speculative behavior) stock market prices provide a way of capturing information about future dividend growth which is contained in the information set available to market participants. We define a second measure of apparent deviations from fundamental price as:

$$b_t^B = \frac{P_t - P_t^B}{P_t} \tag{29}$$

where

$$P_t^B = S_t' + \frac{1+r}{r} D_t \tag{30}$$

and the right hand side of the last equation uses the notation of Campbell and Shiller (1987)[19]. S_t' is an optimal linear forecast of future dividend changes based on a VAR representation which includes past prices and dividend changes.

Results for the second measure of apparent deviations form fundamental price are presented in Table 3.

The patterns are broadly similar to the first measure. The three likelihood ratio statistics allow us to reject the various "stylized fact" models at the 1% significance level in every case. We see that β_{Sb}, β_{Cb}, and β_{qb} are -0.008, -0.126, and -1.507, respectively (compared to -0.006, -0.111, and -1.560 for b_t^A) and that their significance levels are all roughly the same as before. Estimates of the other parameters are also similar to the estimates in Table 1. It therefore appears that the results are robust to allowing for linearly-predictable variation in dividend growth (and thus for variation in the fundamental price-dividend ratio).

Table 3. The Model of Speculative Behavior (Using b_t^B)

Parameters	Estimates
β_{S0}	1.006 (0.002)
β_{Sb}	-0.008 (0.006)
β_{C0}	0.972 (0.042)
β_{Cb}	-0.126 (0.079)
β_{q0}	2.103 (0.318)
β_{qb}	-1.507 (0.534)
σ_S	0.044 (0.002)
σ_C	0.172 (0.029)
Likelihood Ratio Tests	
Volatility Regimes	15.21 (0.004)
Mixture of Normals	15.12 (0.002)
Mean Reversion	12.68 (0.005)

Notes: See Table 1 for notes.

4. Switching Fundamentals

In this section, we examine the model of switches in fundamentals outlined in Section 2.3. One way to compare this model to the model of speculative behavior examined previously would be to estimate a more general model which nests both kinds of switching as special cases. In principle, this would allow each model to be tested against the more general alternative. However, it would require the estimation of a

4-state Markov switching regression subject to multiple nonlinear restrictions. Furthermore, testing for the absence of switching due to either of our two sources would then be equivalent to testing whether the number of discrete states could be reduced from 4 to 2. Inference in such cases would be complicated by the presence of numerous unidentified parameters under the null hypotheses. We chose not to pursue this avenue due to the difficultly of implementing it. Instead, we use a Monte Carlo experiment to compare the two models.

We begin by estimating the parameters of the switching process for dividend growth given in equations (15) and (16). We then choose reasonable values for the coefficient of relative risk aversion γ and the subjective discount rate β[20]. Next we generate data for dividends using the Markov switching model, derive the corresponding equilibrium stock prices using equations (17) and (A24) - (A27), and calculate the resulting returns[21]. This becomes artificial data with which we repeat the analysis of the preceding section; we estimate the switching regression and carry out the parametric tests. Repeating these steps with fresh draws of the random variables ε_t and S_t, we are able to estimate the distribution of the switching regression parameters and likelihood ratio statistics in a world with dividend switching but no speculative behavior.

Table 4: The Size and Variability of the Apparent Deviation From Fundamental Price in the Model of Switching Fundamentals

	Actual Data	2.5%	50%	97.5%
($\beta = 0.97, \gamma = -1.585$)				
Mean of $\lvert b_t^A \rvert$	0.2321	0.0048942	0.024275	0.051050
Standard Deviation of $\lvert b_t^A \rvert$	0.2019	0.017435	0.032532	0.032886
($\beta = 0.95, \gamma = -0.609$)				
Mean of $\lvert b_t^A \rvert$	0.2321	0.0027203	0.014700	0.030615
Standard Deviation of $\lvert b_t^A \rvert$	0.2019	0.011565	0.023080	0.025291

Notes: The entries in columns two, three, and four report the 2.5 percentile, median, and 97.5 percentile of the distribution of the mean and standard deviation of $\lvert b_t^A \rvert$ from Monte Carlo simulations of the model of switching fundamentals.

Table 5: The Model of Switching Fundamentals ($\beta = 0.97, \gamma = -1.585$)

Parameter Estimates	Actual Data	2.5%	50%	97.5%
β_{S0}	1.007	1.0021	1.0050	1.0070
β_{Sb}	-0.006	-.17521	-.14199	-.11095
β_{C0}	0.976	1.0956	1.1419	1.1773
β_{Cb}	-0.111	-2.4257	-2.2629	-2.0800
β_{q0}	2.098	3.0489	2.6491	2.4342
β_{qb}	-1.560	-7.4315	-9.2958	-14.312
σ_S	0.044	0.014178	.014948	.015707
σ_C	0.170	.003514	.012315	.021518
Likelihood Ratio Tests				
Volatility Regimes	16.29	56.30	154.49	282.48
Mixture of Normals	16.21	55.986	154.48	282.46
Mean Reversion	14.79	24.971	52.136	88.436

Notes: The entries in columns two, three, and four represent the probability distribution of the coefficient estimates and likelihood ratio statistics of the model of equations (19) - (21) under the null hypothesis of switching fundamentals, based on a Monte Carlo simulation with 1000 replications.

One interesting feature of the simulations is the size of the apparent deviation from fundamentals, which is presented in Table 4[22]. As in Section 3, this is constructed as the difference between the observed stock price and the "fundamental" price calculated under the (erroneous) assumption that there is no regime-switching in dividend growth. Therefore, the "apparent" deviation from fundamentals is due to misspecification rather than speculative behavior. In the actual data, the mean absolute value of b_t is about 0.232. The model of switching fundamentals produces much smaller and more tightly distributed values of b_t. The median over 1,000 simulations (with $\beta = 0.97$ and $\gamma = -1.585$) of the mean absolute value of b_t is about 0.024 and its 0.95 confidence interval lies below 0.052. The standard deviation of $|b_t|$ in the actual data is about 0.202, compared to the median standard deviation of about 0.033 and a 0.95 confidence interval lying below 0.033 for the simulations. Since b_t is defined as the actual price (P) minus the fundamental price (where the fundamental price is P^*, as defined in Section 2.1) divided by the actual price, Table 4 tells us that the stimulated prices stay much closer to the fundamental price than actual prices do. This implies that failing to account for regime switching in dividend growth can explain relatively little of the volatility of b_t^A.

The parameter estimates and likelihood ratio statistics from the Monte Carlo experiment with $\beta = 0.97$ and $\gamma = -1.585$ are shown in Table 5. The likelihood ratio statistics show that switching fundamentals are capable of accounting for the rejections of the null hypotheses of volatility regimes, a normal mixture, and mean reversion which we found in Section 4. This is an extremely interesting result, since it shows that switching in fundamentals does more than simply induce regime switching in returns; it also causes the apparent deviation from fundamental price to affect conditional returns.

The likelihood ratio statistics suggest that the model of switching fundamentals does not correspond very precisely to the actual data. The problem is that the rejections of the null hypotheses of volatility regimes, a normal mixture, and mean reversion are much stronger than what we find in the actual data. In each case, the likelihood ratio statistics from the actual data are in the extreme left tail of the distribution of likelihood ratio statistics generated by the model of switching fundamentals.

The parameter estimates from the simulations are also presented in Table 5 and are also somewhat different from those we obtain in the actual data. The most notable example is that the less-frequent state (which is labelled C in Table 5) is associated with highly positive returns (the median of β_{CO} is 1.142), whereas the actual data shows negative returns in the less frequent regime ($\beta_{CO} = 0.976$). On the other hand, the point estimates of β_{Cb} and β_{qb} are negative and $\beta_{Sb} > \beta_{Cb}$.

Table 6 reports Monte Carlo simulations using $\beta = 0.95$ and $\gamma = -0.609$. The parameter estimates for this case correspond somewhat more closely to the actual data. The infrequent regime (which is again labelled C in Table 6) is now associated with negative returns ($\beta_{CO} = 0.932$). The point estimates β_{Cb} and β_{qb} are again negative and $\beta_{Sb} > \beta_{Cb}$. As in the simulations for $\beta = 0.97$ and $\gamma = -1.585$, the likelihood ratio statistics are much larger than in the actual data.

The Monte Carlo simulations show that the model of switching fundamentals is able to reproduce many of the basic stylized facts of regime-switching found in the data, if not their exact magnitudes. It accounts for the rejection of the volatility regimes, mixture of normals, and mean reversion nulls; in fact, the model of switching fundamentals leads to much stronger rejections than we find in the actual data. It also gives slightly positive returns in the surviving state, balanced (when $\gamma > -1$) by large but infrequent losses in the collapsing state, a probability of collapse that increases with the absolute value of b_t, and a negative relation between b_t and returns in the collapsing regime.

In the Lucas model we presented in Section 2.3, there is no conceptual distinction between dividends, consumption and income. In performing our simulations, it would therefore be reasonable to calibrate the switching process for fundamental to data for any one of these three variables. We have considered only the results based on calibrating the model to dividends. This suffices to show that evidence of speculative behavior like that found in Section 3 can be generated by a model of switching

Table 6: The Model of Switching Fundamentals ($\beta = 0.95, \gamma = -0.603$)

Parameter Estimates	Actual Data	2.5%	50%	97.5%
β_{S0}	1.007	1.0030	1.0054	1.0072
β_{Sb}	-0.006	0.15473	0.20442	0.24982
β_{C0}	0.976	0.91088	0.93230	0.99372
β_{Cb}	-0.111	-1.9307	-1.7001	-0.78587
β_{q0}	2.098	3.0233	2.6026	2.0395
β_{qb}	-1.560	-7.4679	-12.956	-20.412
σ_S	0.044	0.014096	0.014908	0.015729
σ_C	0.170	0.004490	0.012987	0.046817
Likelihood Ratio Tests				
Volatility Regimes	16.29	43.153	123.57	232.04
Mixture of Normals	16.21	42.834	123.49	231.93
Mean Reversion	14.79	15.864	37.008	64.062

Notes: The entries in columns two, three, and four represent the probability distribution of the coefficient estimates and likelihood ratio statistics of the model of equations (19) - (21) under the null hypothesis of switching fundamentals, based on a Monte Carlo simulation with 1000 replications.

fundamentals. However, we will return to consider the possibility of regime-switching in other measures of fundamentals in Section 5.

5. Accounting for Historical Crashes

The evidence in Sections 3 and 4 suggests that either the model of speculative behavior or the model of switching fundamentals could account for the characteristics of U.S. stock market returns which are highlighted by the switching regression. In this section, we examine how well specific historical crashes are accounted for by each of the models. To do this, we use parameter estimates for each of the models to generate the probability of a stock market crash. For the model of speculative behavior, we focus

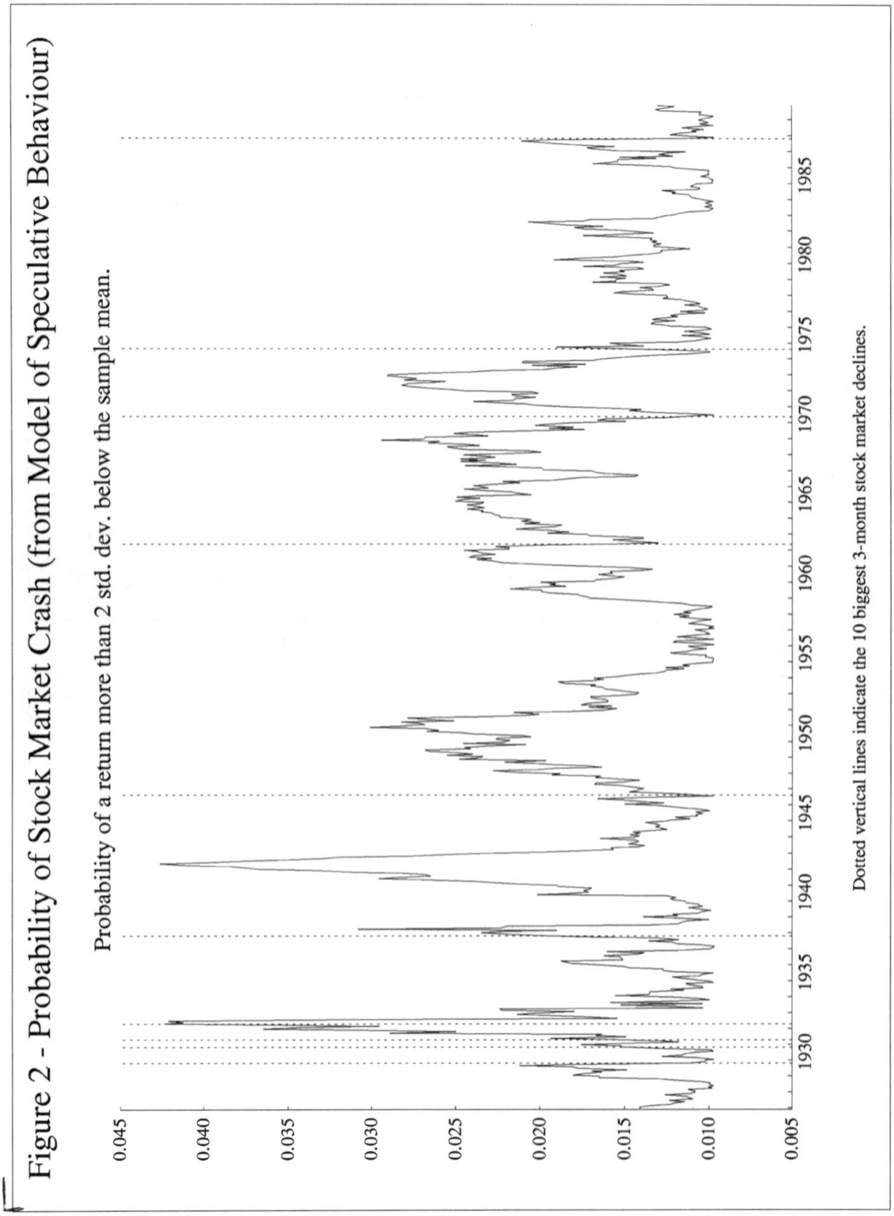

Figure 2 - Probability of Stock Market Crash (from Model of Speculative Behaviour)

Figure 3 - Probability of Stock Market Crash (from Model of Speculative Behaviour) Using Measure B of Fundamental Price

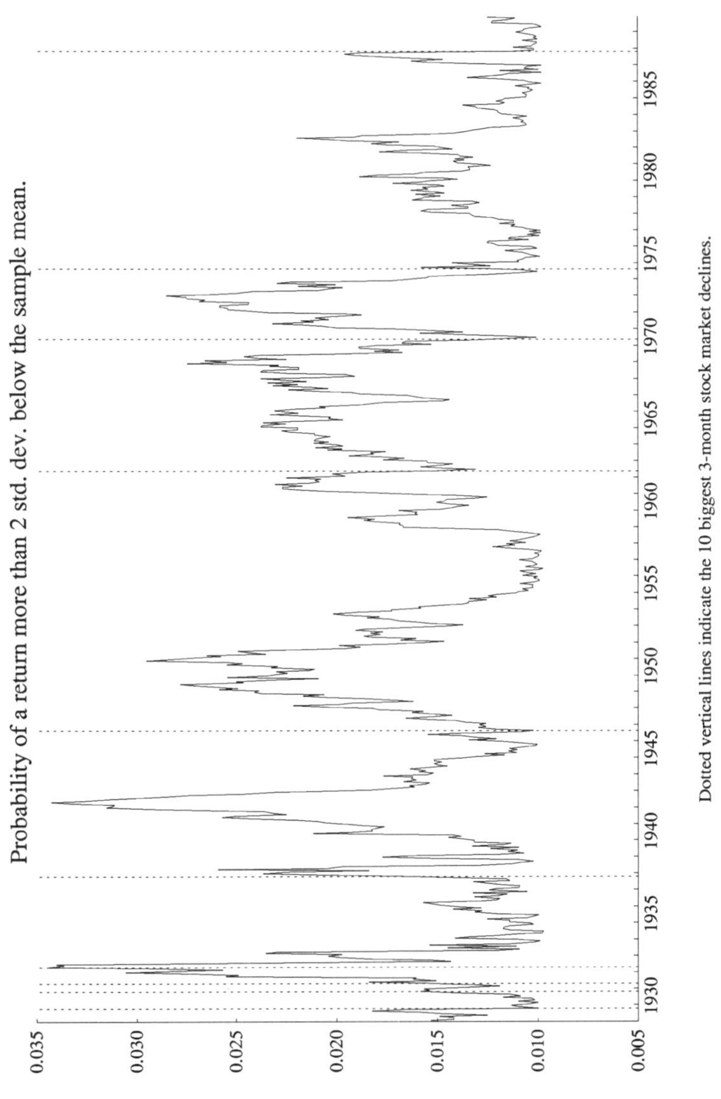

Measure B allows for variation in expected dividend growth based on Campbell and Shiller (1987)

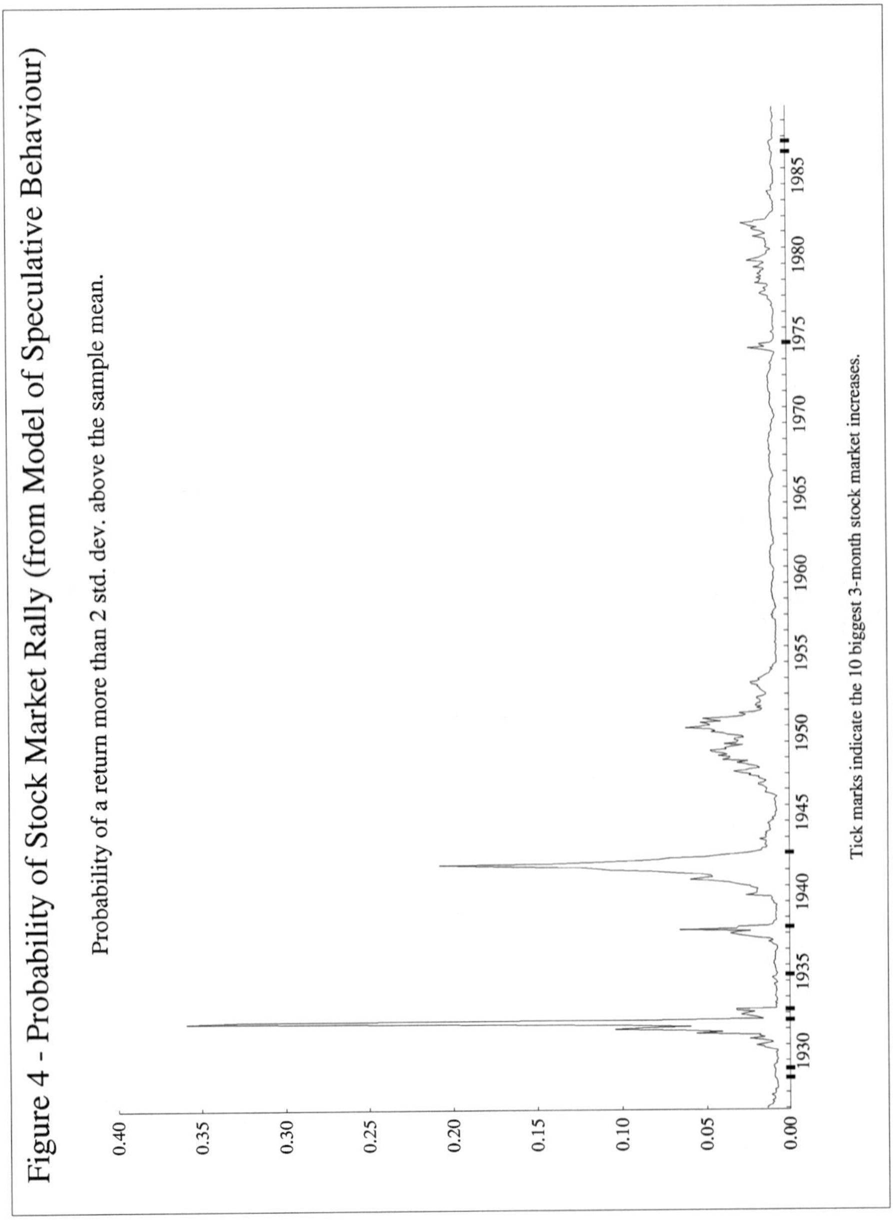

Figure 4 - Probability of Stock Market Rally (from Model of Speculative Behaviour)

Probability of a return more than 2 std. dev. above the sample mean.

Tick marks indicate the 10 biggest 3-month stock market increases.

Figure 5 - Real Per Capita Dividend Growth Probability of Being in Low-Growth State

Calculated from 1-sided filter.

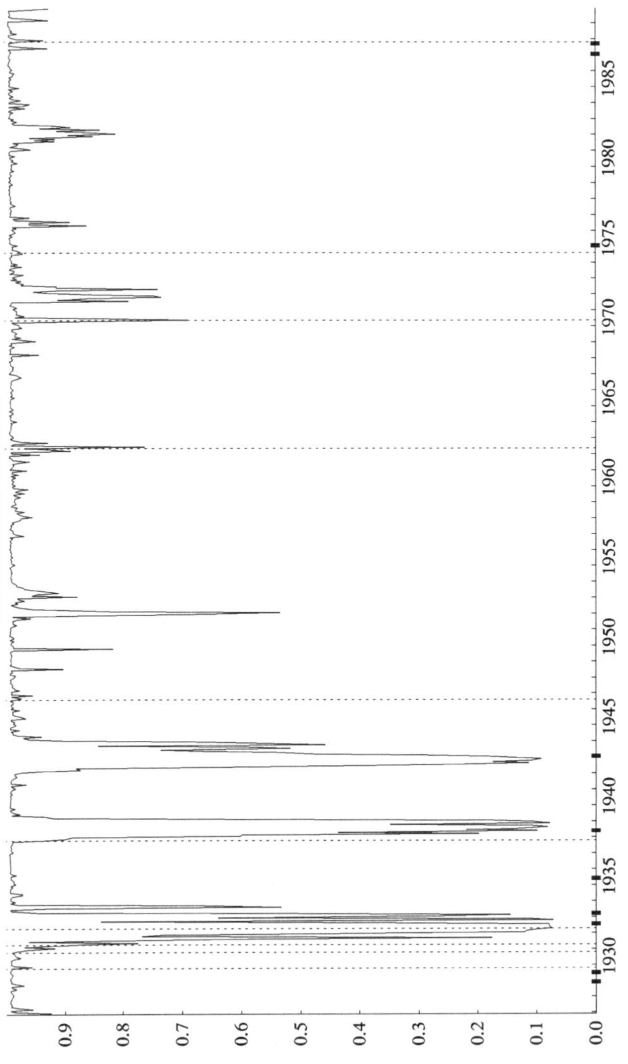

Dotted vertical lines indicate the 10 biggest 3-month stock market declines.
Tick marks indicate the 10 biggest 3-month stock market increases.

on the probability of a return which is two standard deviations below the mean return[23]. For the model of switching fundamentals, we focus on the probability of the low-growth regime.

5.1 Speculative Behavior

Figure 2 plots actual crashes versus the probability of a crash (generated from the empirical estimates of the model of speculative behavior)[24]. We begin by examining the two best-known crashes of this century. In both 1929 and 1987, there is an increase in the probability of a crash in advance of the actual crash. In both cases, the probability of a crash roughly doubles in the months preceding the crash. There are also substantial increases in the probability of a crash around the time of the 1930, 1931, 1932, 1937, 1946, 1962, and 1974 crashes[25]. The 1970 crash is an exception, coming at a time when the probability of collapse had recently decreased and was low. To check the robustness of the results, we present a graph based on b_t^B, which allows for variation in the expected dividend growth rate, in Figure 3. The pattern is similar: a rise in the probability of a crash precedes most of the largest crashes.

The model of speculative behavior also suggests that large undervaluations imply an increasing probability of a sharp movement towards fundamental price. Plots of the probability of a rally (Figure 4) show that it is less than 10% for all but two periods between 1926 and 1989. The first exception is 1932, when the probability rose very sharply to peak at about 0.35; this was followed immediately by one of the largest rallies in our sample. The second exception was in 1942, when the probability of a rally rose sharply to peak at about 0.20; this was followed (with a lag) by one of the largest rallies in our sample. The probability of a rally also rose before the 1938 rally. There was no increase in the probability of a rally in 1929 or 1987 because there was no apparent undervaluation at those times. (See Figure 1.)

5.2 Regime Switches in Fundamentals

In Figure 5, we examine how well switches in expected dividend growth account for actual stock market crashes. Figure 5 plots the ex ante probability of being in the high dividend growth state against actual market crashes. First, we note that the two most famous crashes (1929 and 1987) are marked by high and stable probabilities of high dividend growth. The 1946 and 1974 crashes are also marked by high and stable probabilities of being in the high dividend growth state. Therefore, the dividend switching model seems unable to explain these crashes. At the time of the 1930, 1931, 1937, 1962, and 1970 crashes, the probability of the high dividend growth state was falling rapidly, however. This is consistent with agents revising their expectations of future dividend growth and abruptly lowering stock prices[26].

Intuitively, the ex ante probability (which is based on Hamilton's (1989) one-sided filter) uses only the information on past dividend growth available to investors in period *t*. However, investors may have information which is not reflected in past dividend growth and which helps them to forecast future dividend growth. One way to account for this is to use all information on dividend growth in the entire sample to assign probabilities. The result can be thought of as an ex post probability and

Figure 6 - Real Per Capita Dividend Growth
Probability of Being in Low-Growth State

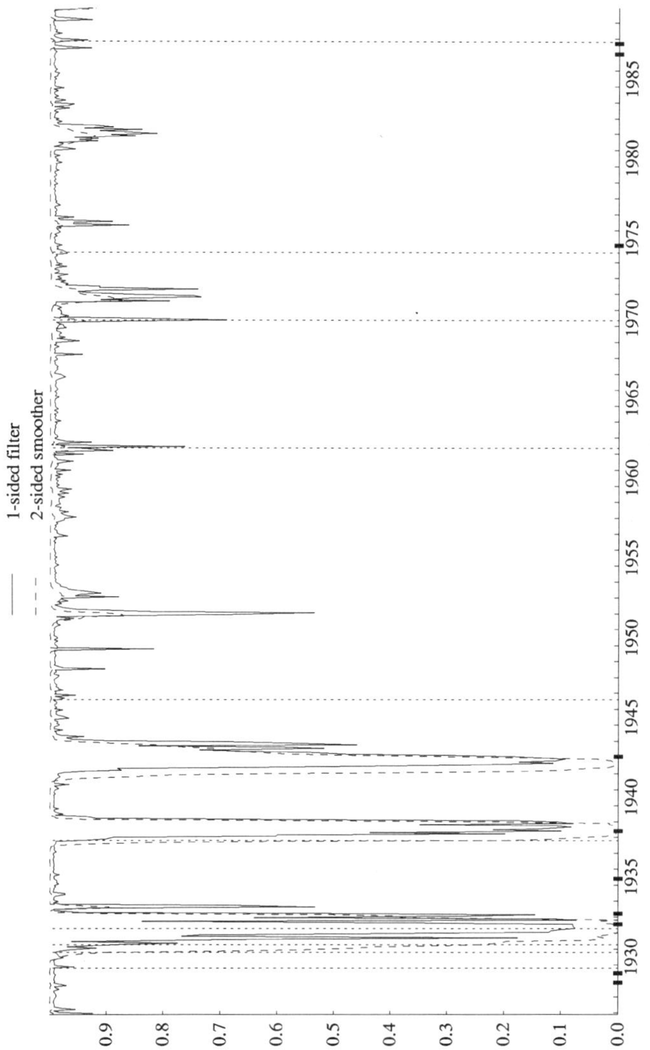

Dotted vertical lines indicate the 10 biggest 3-month stock market declines.
Tick marks indicate the 10 biggest 3-month stock market increases.

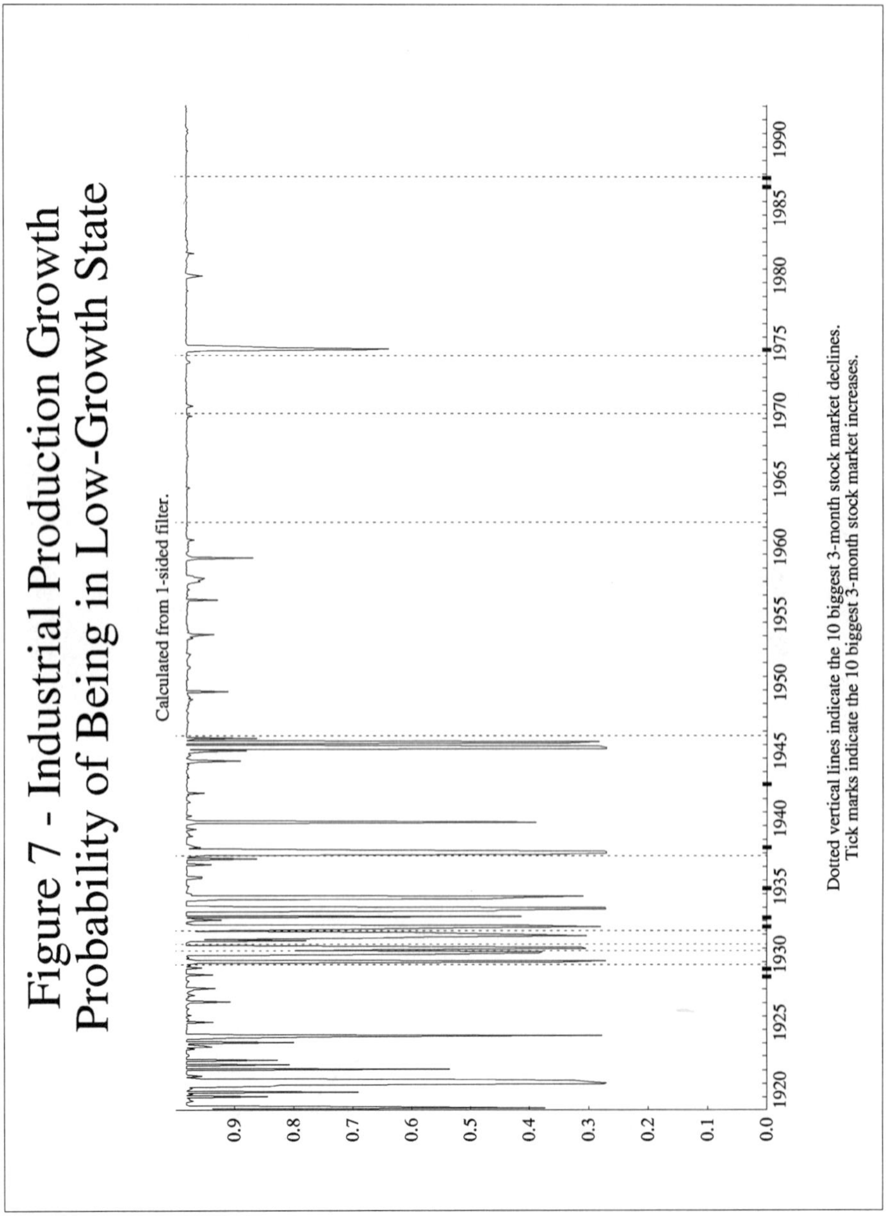

Figure 7 - Industrial Production Growth Probability of Being in Low-Growth State

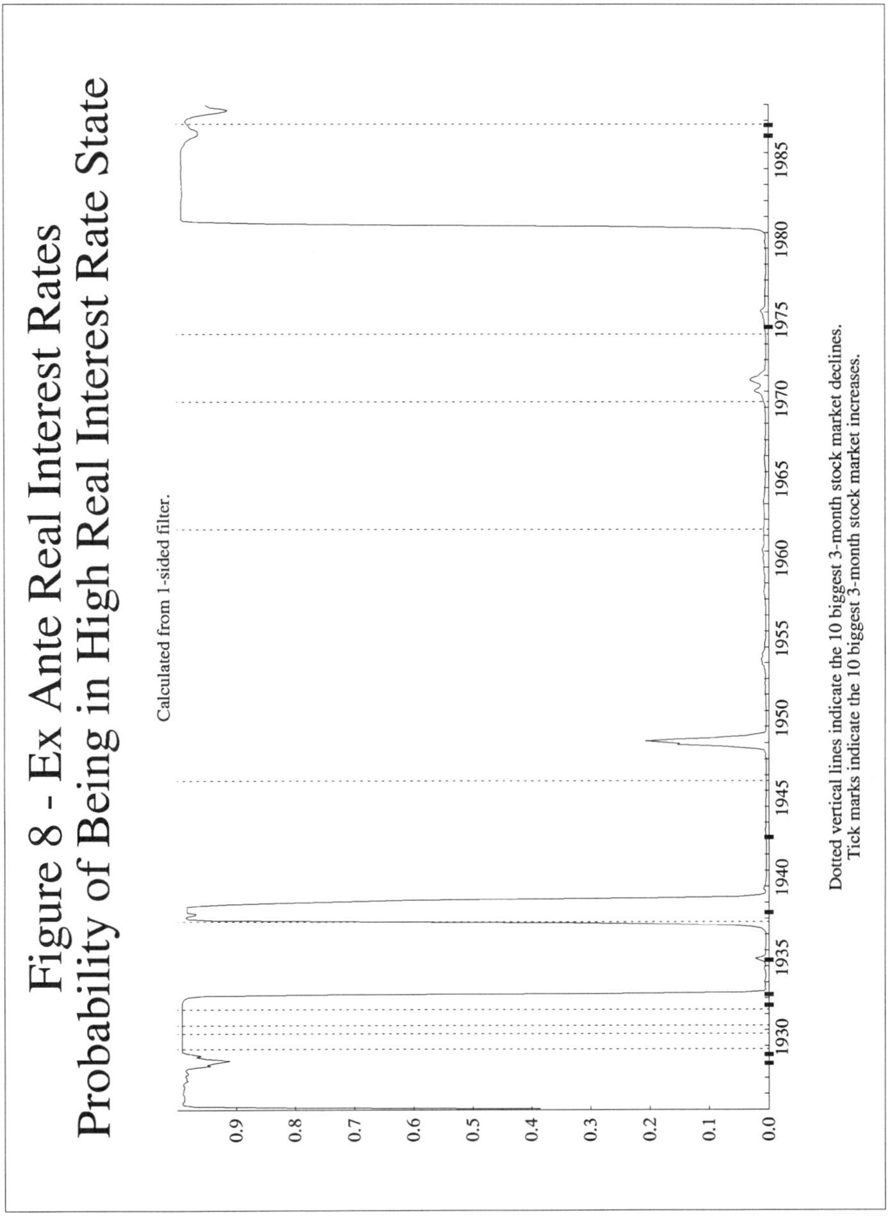

Figure 8 - Ex Ante Real Interest Rates
Probability of Being in High Real Interest Rate State

corresponds to Hamilton's (1989) two-sided smoother. In Figure 6, we plot both the ex ante and ex post probabilities. Endowing investors with more information makes only a small difference in the results, essentially pushing back by a few months the date at which the regime changes in the 1930's and early 1940's could have been perceived.

Intriguingly, the 1932, 1933, 1938, and 1943 rallies correspond to periods in which the probability of the high dividend growth state was rising. (See especially Figure 6.) The probability was stable, however, at the time of the 1929, 1935, 1974, and 1987 rallies.

It has often been argued that dividends are a poor measure of either corporate cash flow or the endowment[27]. We therefore consider two other variables which might capture changes in economic regime - industrial production and interest rates. Figure 7 presents the ex ante probability of being in the high growth state for industrial production. This probability is relatively low and variable in the early 1930's and 1937, but, with the exception of 1974, it does a poor job of explaining post-war crashes. For example, the probability of rapid growth in industrial production is rising in the months preceding the 1946 crash. Around the 1962, 1970, and 1987 crashes, the probability of rapid industrial growth is close to one and very stable. Switches in expected dividend growth do a better job of explaining stock market crashes than switches in expected industrial production growth. Figure 8 graphs the ex ante probability of high real interest rates[28]. The graph is quite different than Figures 5 and 7; it shows relatively rare but abrupt and decisive regime shifts. The figure does not offer much encouragement to the view that changes in the real interest rate regime explain stock market crashes. Only the 1937 crash corresponds to a rise in the probability of the high real interest rate regime.

6. Conclusion

Using the 1926-1989 monthly Center for Research in Securities Prices (CRSP) stock market data, our switching regressions find that there is nonlinear predictability of returns based on the degree of apparent market overvaluation. We are careful to show that this cannot be accounted for by previously documented stylized facts, such as regime shifts in volatility, the "leverage effect" or the linear predictability of returns.

The model of speculative behavior is consistent with this nonlinear relationship. We find that the regime which corresponds to the collapse of a speculative component is significantly more likely when the degree of overvaluation in the previous period is high. We also find a substantial difference in expected returns between the regime which corresponds to the survival of a speculative component and the regime which corresponds to a speculative collapse. In the regime where the speculative component survives, the typical return is positive. In the regime where the speculative component collapses, the typical return is -25.3% (on an annual basis). The difference in expected returns between the two regimes increases with the size of the apparent overvaluation. In a period where the apparent overvaluation is relatively large, such as September 1929, the expected return if the speculative component collapses is -44.6% (on an annual basis). The effects of apparent overvaluation are not restricted to the period that includes the Great Depression and World War II; in fact, the results are similar for each of three subperiods over which we estimate the model of speculative behavior. The results are also robust to changes in the measure of fundamental price that try to capture the effects of anticipated changes in dividend growth rates. However, since we use linear forecasts of dividend growth rates, we might conceivably fail to capture important nonlinearities in their behavior.

We therefore calibrate a Markov switching model of dividend growth to U.S. data for 1926-1989, simulate the resulting asset pricing model, and estimate the switching regression using the artificial data from the simulations. We find that switching fundamentals could account for much of the statistical evidence we find for speculative behavior; there is strong evidence of regime switching in the simulated returns and the degree of apparent "overvaluation" influences expected returns. However, we find that the degree and variability of the apparent market "overvaluations" are much smaller in the simulations than in the actual data.

The coefficient estimates and tests of parameter restrictions suggest that the model of speculative behavior and the model of switching fundamentals are substitutes in the sense that either model is capable of accounting for the switching characteristics of actual returns we have highlighted. A different picture emerges when we examine specific stock market crashes. We find that the probability of a crash from the model of speculative behavior rises in advance of several prominent stock market crashes (such as 1929 and 1987), but not in advance of all crashes. The probability of a crash calculated from the model of switching fundamentals fails to rise before the 1929 and 1987 crashes, although it does anticipate several major stock market crashes.

Taken together, the evidence suggests the model of speculative behavior and the model of switching fundamentals may be complements rather than substitutes. For example, the 1929 and 1987 crashes correspond well to the model of speculative behavior but not to switches in dividend growth, while other crashes correspond more closely to the model of switching fundamentals.

The results presented in this paper raise many interesting questions for both theorists and empirical researchers. If one is inclined to believe that speculative behavior is important, there are several areas in which an improved understanding would be helpful. First, why is it that actual prices sometimes begin to drift away from fundamental prices?[29] Second, are speculative episodes triggered by particular economic mechanisms, such as substantial increases in liquidity? If one is inclined to discount the importance of speculative behavior, then the empirical work presented in this paper poses other questions. Can a model based on economic fundamentals reproduce the magnitude and variability of the apparent "overvaluation" which we find in the actual data? How can a model based on economic fundamentals account for events like the 1929 and 1987 stock market crashes?

A variety of theoretical models have recently addressed the question of why stock markets might look as if they displayed speculative behavior, specifically why prices might move dramatically in the absence of important news[30]. An interesting question is whether theories along any of these lines can account for the features of the data highlighted here.

Acknowledgements

We would like to thank R. Barsky and O. Blanchard, as well as seminar participants at the American Economic Association, Bank of Canada, University of British Columbia, Carleton University, Econometric Society (Seattle and Uppsala), and Simon Fraser University for helpful comments. Professor Schaller acknowledges financial support from the Financial Research Foundation of Canada and the SSHRC. Any errors are our own. We are solely responsible for the views expressed here; no responsibility for them should be attributed to the Bank of Canada.

Notes

1. See Kindleberger (1989).
2. van Norden and Vigfusson (1996) compare these two testing approaches.
3. In controlled experiments where all investors receive the same dividend from a known probability distribution at the end of a fixed number of trading periods, there is a marked tendency for prices to rise relative to fundamental value and then crash. (See Smith, Suchanek, and Williams (1988); Van Boening, Williams, and LaMaster (1993) provide references to recent work.) Using surveys of market participants, Shiller, Kon-Ya, and Tsutsui (1991) find that during a speculative episode a large proportion of investors buy stocks because they think prices will continue to rise in the short term, even though they eventually expect prices to drop. Similarly, the experimental evidence suggests that capital gains expectations influence price changes.
4. There is an extensive literature noting restrictions on the admissibility of non-fundamental solutions, including Diba and Grossman (1988), Obstfeld and Rogoff (1983, 1986) and Tirole (1982, 1985). Blanchard and Fischer (1989, p. 238) argue that "[These restrictions] often rely on an extreme form of rationality and are not, for this reason, altogether convincing. Often bubbles are ruled out because they imply, with a very small probability and very far in the future, some violation of rationality, such as non-negativity of prices or the bubbles becoming larger than the economy. It is conceivable that the probability may be so small, or the future so distant, that it is simply ignored by market participants." Moreover, recent work by Allen and Gorton (1993), Allen, Gorton, and Postlewaite (1993), and Leach (1991) has shown that restrictions on non-fundamental solutions are not robust.
5. Note the use of the absolute value of b_t, since we allow the speculative component to be positive or negative.
6. Other empirical research has also considered the possibility that the speculative component may collapse only partially. See Evans (1991) and Hall and Sola (1993).
7. The assumptions that $u(0) = 0$ and $1 \geq u' \geq 0$ are added to ensure that the speculative component is expected to shrink in the collapsing state. To see this, draw a graph with $u(b_t)$ (which equals $E_t[B_{t+1}|C]/P_t$) on the vertical axis and b_t on the horizontal axis. The function $u(b_t)$ passes through the origin, since $u(0) = 0$. The 45° line represents a situation where $E_t[B_{t+1}|C]/P_t = B_t/P_t$; i.e., where a "collapsing" speculative component is the same size as the previous period's speculative component. Since $0 \leq u' \leq 1$, $u(b_t)$ always lies on or below the 45° line. Thus, these assumptions ensure that a collapsing speculative component is no larger than the speculative component in the previous period.
8. See Bonomo and Garcia (1994) for more on this point.
9. The derivation of the model may be found in the appendix. The high dividend growth state need not correspond to the high ρ state. In fact, it can be shown that

$$\left\{ \rho(0) \begin{matrix} > \\ = \\ < \end{matrix} \rho(1) \right\} \Leftrightarrow \left\{ \gamma \begin{matrix} > \\ = \\ < \end{matrix} -1 \right\}$$

where ρ(0) is defined to be the high-growth state.

10. For those familiar with Markov switching models, it may be useful to note that our model has a related stochastic structure. A two-state Markov model has two state-dependent probabilities: $q(t) = \Pr(S_t = 0|S_{t-1} = 0)$ and $p(t) = \Pr(S_t = 1|S_{t-1} = 1)$. The switching model presented here has one state-independent probability $q(t) = \Pr(S_t = 0)$. This is the special case of the Markov model where $q(t) = 1 - p(t)$; i.e., the probability of today's state is independent of the state yesterday. Evans and Lewis (1995) note that the standard Markov model is not compatible with the assumption that the expected rate of return is the same across the two states; our restriction is sufficient to ensure that this condition is satisfied.
11. This can also be interpreted as a test of asset pricing models, like that in Section 2.1, which imply that variations in returns should not be predictable and that deviations from fundamental price should therefore be irrelevant "noise."
12. Note that allowing $\sigma_s \neq \sigma_c$ simplifies the inference by allowing all parameters to be identified under the null.
13. In contrast to most of the empirical work in Cutler, Poterba, and Summers (1991), we use non-overlapping observations of one-period returns.
14. The population growth adjustment uses annual population data from 1924-1945 from Historical Statistics of the United States (1976) (series A29) and quarterly data from the BIS database (UBBA US01) from 1946 onwards. Monthly dates are linearly interpolated. Data from January 1960 onwards are divided by 1.0043 to correct for the inclusion of Alaska and Hawaii from that date onwards.
15. Under the hypothesis that the actual price corresponds to the fundamental price in the model in Section 2.1 ($P_t = P_t^*$), the mathematical expectation of P_t/D_t is ρ.
16. Expected returns are calculated using the point estimates in Table I. For example, the expected monthly return in the collapsing regime in September 1929 is 0.976 + (-0.111) (0.216).
17. See, for example, Fama and French (1988a) or Kim, Nelson, and Startz (1991). Evidence of mean reversion does not necessarily imply that asset prices deviate from fundamentals; see, for example, Brock and LeBaron (1990), Cecchetti, Lam, and Mark (1990), Fama (1991), Fama and French (1988a), or Jog and Schaller (1994).
18. The marginal significance level of the volatility regime test statistic is 0.052 in the period 1926-1954; in all other cases, the p-values are smaller.
19. We set $r=1/(\text{mean}[P]/\text{mean}[D]-1)$, so that the spread between price and a multiple of current dividends (S_t in the notation of Campbell and Shiller (1987)) is zero on average.
20. We consider values of γ on both sides of -1 because the model of switching fundamentals implies that the high dividend growth rate state will yield high stock prices when $\gamma>-1$ and low stock prices when $\gamma<-1$. We choose (β,γ) pairs so that the expected value of ρ matches the mean price-dividend ratio in our sample. The particular values we consider are ($\beta = 0.97$, $\gamma = -1.585$) and ($\beta = 0.95$, $\gamma = -0.609$).
21. The measure of apparent deviations from fundamental price (b_t^A) is calculated as before.
22. The results in the table are based upon the first definition of b_t, which we have referred to above as b_t^A.
23. In the switching regression model, the ex ante probability of observing a return equal to or less than some given value K is calculated according to the formula

$$\Pr(R_{t+1} \leq K) = \Phi\left(\frac{K - \beta_{S0} - \beta_{Sb} \cdot b_t}{\sigma_S}\right) \cdot \Phi\left(\beta_{q0} + \beta_{qb} \cdot |b_t|\right) + \Phi\left(\frac{K - \beta_{C0} - \beta_{Cb} \cdot b_t}{\sigma_C}\right) \cdot \Phi\left(-\beta_{q0} - \beta_{qb} \cdot |b_t|\right)$$

24. Actual crashes were defined by using monthly returns to calculate the 20 largest 3-month losses in our sample. Three-month rather than 1 month losses were used both to capture more gradual (but large) price declines, as well as to exclude transitory losses that are almost immediately offset by subsequent price increases. Only 10 distinct crashes are shown since half of the three-month losses either overlapped or were very close to each other.
25. A high probability of crash does not always coincide with one of the 10 biggest crashes, but this does not mean that it was not associated with a crash. For example, the probability of a crash was high in late 1941 and early 1942; in fact, the price declines in September-November 1941 and January-March 1942 were among the most severe in the sample but did not quite make the top 10 list.

26. In interpreting Figure 4, we have implicitly assumed that $\gamma > -1$ so the low-growth state is associated with lower stock prices. If $\gamma < -1$, crashes would be associated with transitions from low to high dividend growth states, rather than the reverse. We note that if we adopt this interpretation, then the dividend switching model does much worse in explaining stock market crashes, with none of the crashes corresponding to such a transition.
27. See, for example, Ackert and Smith (1993).
28. The monthly interest rate series we use is Moody's Industrial Bond Index Annual Yield; we created an ex post real interest rate series by deflating by the year over year changes in the CPI (all item urban consumer). We also examined an ex ante real interest rate series using the previous year's CPI to deflate; results were very similar.
29. Without any attempt to be exhaustive, two possibilities are learning and finance constraints. The key idea of the learning explanation is that agents may not have full information; the actual price may deviate from the full information fundamental price as agents attempt to draw inferences from the actions of the other agents (as they are reflected in prices, volume, and other variables). For theoretical and empirical work see, for example, Grossman (1989), and Timmerman (1993). The key idea of the finance constraint story is that the inability to borrow may depress cash flows relative to the unconstrained case, leading to lower asset prices; in a sense, a present value model still holds, but the resulting returns behavior may appear anomalous to the econometrician. For theoretical and empirical references, see Brock and LeBaron (1990), Jog and Schaller (1994), and Kiyotaki and Moore (1997).
30. See, for example, Bikhchandani, Hirshleifer, and Welch (1992), Bulow and Klemperer (1994), Gennotte and Leland (1990), and Romer (1993).

References

Ackert, L. and B. Smith (1993). "Stock Price Volatility, Ordinary Dividends, and Other Cash Flows to Shareholders," *Journal of Finance*, 48, 1147-1160.

Allen, F. and G. Gorton (1993). "Churning Bubbles," *Review of Economic Studies*, 60, 813-836.

Allen, F., S. Morris, and A. Postlewaite (1993). "Finite Bubbles with Short Sale Constraints and Asymmetric Information," *Journal of Economic Theory*, 61, 206-229.

Barsky, R. and J. B. DeLong (1993). "Why Does the Stock Market Fluctuate?" *Quarterly Journal of Economics*, 108, 291-312.

Bikhchandani, S., D. Hirshleifer, and I. Welch (1992). "A Theory of Fads, Fashion, Custom, and Cultural Change as Informational Cascades," *Journal of Political Economy*, 100, 992-1026.

Blanchard, O. J. (1979). "Speculative Bubbles, Crashes and Rational Expectations," *Economics Letters*, 3, 387-389.

Blanchard, O. J. and S. Fischer (1989). *Lectures on Macroeconomics* (1st edition) . Cambridge: MIT Press.

Blanchard, O. J. and M. Watson (1982). "Bubbles, Rational Expectations and Financial Markets," in P. Wachtel (ed.), *Crises in the Economic and Financial Structure*. Lexington: Lexington Books.

Bonomo, M. and R. Garcia (1994). "Can a Well-Fitted Equilibrium Asset-Pricing Model Produce Mean Reversion?" *Journal of Applied Econometrics*, 9, 19-29.

Brock, W., and B. LeBaron (1990). "Liquidity Constraints in Production-Based Asset-Pricing Models," in Hubbard, R. G. (ed.) *Asymmetric Information, Corporate Finance, and Investment*. Chicago: University of Chicago Press, 231-56.

Bulow, J. and P. Klemperer (1994). "Rational Frenzies and Crashes," *Journal of Political Economy*, 102, 1-23.

Campbell, J. Y. and R. J. Shiller (1987). "Cointegration and Tests of Present Value Models," *Journal of Political Economy*, 95, 1062-1088.

Cecchetti, S. G., P. S. Lam and N. C. Mark (1990). "Mean Reversion in Equilibrium Asset Prices," *American Economic Review*, 80, 398-418.

Cecchetti, S. G., P. S. Lam, and N. C. Mark (1993). "The Equity Premium and the Risk-Free Rate," *Journal of Monetary Economics*, 31, 21-45.

Cutler, D. M., J. M. Poterba and L. H. Summers (1991). "Speculative Dynamics," *Review of Economic Studies*, 58, 529-546.

Diba, B. T. and H. I. Grossman (1988). "Explosive Rational Bubbles in Stock Prices?" *American Economic Review*, 78, 520-528.

Donaldson, G. and M. Kamstra (1996). "A New Dividend Forecasting Procedure that Rejects Bubbles in Asset Prices: The Case of 1929's Stock Crash," *Review of Financial Studies*, 9, 333-383.

Evans, G. (1991). "Pitfalls in Testing for Explosive Bubbles in Asset Prices," *American Economic Review*, 81, 922-930.

Evans, M. D. D. and K. Lewis (1995). "Do Long-Term Swings in the Dollar Affect Estimates of the Risk Premia?" *Review of Financial Studies*, 8, 709-742.

Fama, E. F. (1991). "Efficient Capital Markets: II," *The Journal of Finance*, 46, 1575-1617.

Fama, E. F. and K. R. French (1988a). "Permanent and Temporary Components of Stock Prices," *Journal of Political Economy*, 96, 246-274.

Fama, E. F. and K. R. French (1988b). "Dividend Yields and Expected Stock Returns," *Journal of Financial Economics*, 22, 3-25.

Flood, R. P., and R. J. Hodrick (1990). "On Testing for Speculative Bubbles," *Journal of Economic Perspectives*, 4, 85-101.

Funke, M., S. Hall, and M. Sola (1994). "Rational Bubbles During Poland's Hyperinflation: Implications and Empirical Evidence," *European Economic Review*, 38, 1257-1276.

Gennotte, G. and H. Leland (1990). "Market Liquidity, Hedging, and Crashes," *American Economic Review*, 80, 615-32.

Goldfeld, S. M. and R. E. Quandt (1976). *Studies in Nonlinear Estimation*. Cambridge: Ballinger.

Grossman, S. (1989). *The Informational Role of Prices*, Cambridge: MIT Press.

Hall, S. and M. Sola (1993). "Testing for Collapsing Bubbles: An Endogenous Switching ADF Test," Discussion paper 15-93, *London Business School*.

Hamilton, J. D. (1989). "A New Approach to the Economic Analysis of Nonstationary Time Series and the Business Cycle," *Econometrica*, 57, 357-384.

Hamilton, J. D. and R. Susmel (1994). "Autoregressive Conditional Heteroskedasticity and Changes in Regime," *Journal of Econometrics*, 64, 307-33.

Hartley, M. J. (1978). "Comment on 'Estimating Mixtures of Normal Distributions and Switching Regressions' by Quandt and Ramsey," *Journal of the American Statistical Association*, 73, 738-741.

Jog, V. and H. Schaller (1994). "Finance Constraints and Asset Pricing: Evidence on Mean Reversion," *Journal of Empirical Finance*, 1, 193-209.

Kim, M. J., C. R. Nelson, and R. Startz (1991). "Mean Reversion in Stock Prices? A Reappraisal of the Empirical Evidence," *Review of Economic Studies*, 58, 515-528.

Kindleberger, C. P. (1989). *Manias, Panics and Crashes: A History of Financial Crises* (Revised Edition). New York: Basic Books, Inc.

Kiyotaki, N. and J. H. Moore (1997). "Credit Cycles," *Journal of Political Economy*, 105, 211-248.

Leach, J. (1991). "Rational Speculation," *Journal of Political Economy*, 99, 131-144.

Lucas, R. E. Jr. (1978). "Asset Prices in an Exchange Economy," *Econometrica*, 66, 429-445.

Obstfeld, M. and K. Rogoff (1983). "Speculative Hyperinflations in Maximizing Models: Can We Rule Them Out?" *Journal of Political Economy*, 91, 675-687.

Obstfeld, M. and K. Rogoff (1986). "Ruling Out Divergent Speculative Bubbles," *Journal of Monetary Economics*, 17, 349-362.

Poterba, J. M. and L. H. Summers (1988). "Mean Reversion in Stock Prices: Evidence and Implications," *Journal of Financial Economics*, 22, 27-59.

Romer, D. (1993). "Rational Asset Price Movements without News," *American Economic Review*, 1112-1130.

Schwert, G. W. (1989). "Business Cycles, Financial Crises, and Stock Volatility," *Carnegie-Rochester Conference Series on Public Policy*, 31, 83-126.

Shiller, R., F. Kon-Ya, and Y. Tsutsui (1991). "Investor Behavior in the October 1987 Stock Market Crash: The Case of Japan," *Journal of the Japanese and International Economies*, 5, 1-13.

Smith, V., G. Suchanek, and A. Williams (1988). "Bubbles, Crashes, and Endogenous Expectations in Experimental Spot Asset Markets," *Econometrica*, 56, 1119-1151.

Timmerman, A. (1993). "How Learning in Financial Markets Generates Excess Volatility and Predictability in Stock Prices," *Quarterly Journal of Economics*, 108, 1135-1145.

Tirole, J. (1982). "On the Possibility of Speculation under Rational Expectations," *Econometrica*, 50, 1163-1181.

Tirole, J. (1985). "Asset Bubbles and Overlapping Generations," *Econometrica*, 53, 1071-1100.

Turner, M., R. Startz, and C. R. Nelson (1989). "A Markov Model of Heteroscedasticity, Risk and Learning in the Stock Market," *Journal of Financial Economics*, 25, 3-22.

Van Boening, M., A. Williams, and S. LaMaster (1993). "Price Bubbles and Crashes in Experimental Call Markets," *Economics Letters*, 41, 179-185.

van Norden, S. (1996). "Regime Switching as a Test for Exchange Rate Bubbles," *Journal of Applied Econometrics*, 11, 219-251.

van Norden, S. and H. Schaller (1996). "Speculative Behavior, Regime-Switching, and Stock Market Crashes," working paper No. 96-13, Bank of Canada.

van Norden, S. and R. Vigfusson (1998). "Avoiding the Pitfalls: Can Regime-Switching Tests Reliably Detect Bubbles?" *Studies in Nonlinear Dynamics and Econometrics*, 3, 1-22.

Vigfusson, R. (1996). "Switching Between Chartists and Fundamentalists: A Markov Regime-Switching Approach," in Fenton, P. and J. Murray (eds.), *Exchange Rates and Monetary Policy*. Ottawa: Bank of Canada.

16 HIGHER-ORDER RESIDUAL ANALYSIS FOR SIMPLE BILINEAR AND THRESHOLD AUTOREGRESSIVE MODELS WITH THE *TR* TEST

Philip Rothman
East Carolina University

1. Introduction

Linear models have traditionally dominated the time series analysis of macroeconomic and financial data. Evidence of nonlinear behavior in the dynamic behavior of such data would imply that the conventional linear models are misspecified. As a result of such model misspecification, appropriately specified nonlinear forecasts would generally be superior to the optimal linear forecast.

Researchers indeed have achieved forecast improvement with simple nonlinear time series models. For example, Maravall (1983) used a bilinear model to forecast Spanish monetary data and reported a near 10% improvement in one-step ahead mean squared forecast error over several ARMA alternatives. de Gooijer (1989) also reported forecast improvement with bilinear models in a forecasting exercise with stock prices. More recently, Potter (1995), Montgomery, *et al.* (1998), Clements and Krolzig (1998), and Rothman (1998) used threshold autoregressive models to produce out-of-sample forecast improvement over linear alternatives for key U.S. macroeconomic indicators.

Time series tests have been developed which, based on Monte Carlo simulations, appear to have good power against simple bilinear and threshold models. These include, for example, Hinich's (1982) bispectrum test, the BDS test (for Brock, *et al.* (1987)), and White's (1989) neural network test. The Monte Carlo results on the estimated power for these test against simple bilinear and threshold autoregressive alternatives were reported in Ashley, *et al.* (1986) and Brock, *et al.* (1991), and Lee, *et al.* (1993).

While for some members of the respective alternative hypotheses these tests have good power, no simulation results reported so far suggest that any of these tests can serve as a direct guide to specification of an appropriate time series model. More specifically, these tests yield little information about the source of deviation from the respective null hypotheses. The information produced is essentially dichotomous, that is, nonlinear or nonlinear. This point has been stressed by other others; see, for example, Brock, *et al.* (1991).

The purpose of this chapter is to report Monte Carlo results on the estimated power of the Ramsey and Rothman (1996) time reversibility (*TR*) test statistics against ARMA residuals for simple bilinear and threshold autoregressive models.[1] The *TR* test statistics are shown to have proper size under the null and good power against these alternatives. More important, it is shown that distinct patterns of rejection appear in the *TR* test statistics for the two classes of models considered. This then offers a simple diagnostic which can be used in identification of these time series models. The Monte Carlo results presented supplement the analytical results for first-order random coefficient autoregressive and product autoregressive models obtained by Lawrance and Lewis (1987).

The paper proceeds as follows. The *TR* test is described in Section 2. In Section 3 a short discussion is given of bilinear and threshold autoregressive models. The small sample properties and estimated power of the *TR* test against ARMA residuals for simple bilinear and threshold alternatives are reported in Section 4. Section 5 concludes the paper.

2. The *TR* Test

If the probabilistic structure of a time series going forward in time is identical to that in reverse time, the series is time reversible. It is straightforward to show that a time series $\{X_t\}$ is time reversible if: (1) $\{X_t\}$ is independently and identically distributed; (2) $\{X_t\}$ is a stationary Gaussian process, but not necessarily independent.

Assume $\{X_t\}$ is a mean zero stationary process. The Ramsey and Rothman (1996) *TR* test statistic at lag k is constructed as follows:

$$\hat{\gamma}_{2,1}(k) = \hat{B}_{2,1}(k) - \hat{B}_{1,2}(k) \tag{1}$$

where $\hat{B}_{2,1}(k)$ and $\hat{B}_{1,2}(k)$ are method of moments estimators of the bicovariances $E[X_t^2 \cdot X_{t-k}]$ and $E[X_t \cdot X_{t-k}^2]$, respectively. By construction the *TR* test has no power against nonlinear processes for which these bicovariances are zero, such as ARCH and GARCH processes.[2]

Under the null hypothesis that $\{X_t\}$ is time reversible, the expected value of $\hat{\gamma}_{2,1}(k)$ is zero for all k. Under some mixing conditions one can easily show that $\hat{\gamma}_{2,1}(k)$ has an asymptotic normal distribution.

For $\{X_t\}$ independently and identically distributed, the exact expression for $Var[\hat{\gamma}_{2,1}(k)]$ is:

$$Var[\hat{\gamma}_{2,1}(k)] = 2(\mu_4\mu_2 - \mu_3)/(T-k) - 2\mu_2^3(T-2k)/(T-k)^2 \tag{2}$$

where T is the sample size of $\{X_t\}$.

To test the null hypothesis of time reversibility in the independently and identically distributed case, $\hat{\gamma}_{2,1}(k)$ is calculated directly on $\{X_t\}$. Given the asymptotic normality of $\hat{\gamma}_{2,1}(k)$, hypothesis testing is straightforward. Rothman (1992) provided Monte Carlo evidence which shows that $\hat{\gamma}_{2,1}(k)$ converges relatively quickly to its asymptotic normal distribution in several independently and identically distributed cases.

In the non-independently and identically distributed case, Ramsey and Rothman (1996) advocated a two stage approach. In the first stage, $\hat{\gamma}_{2,1}(k)$ is calculated on the raw data and hypothesis testing is carried on with standard errors generated by Monte Carlo simulation. In the second stage, $\hat{\gamma}_{2,1}(k)$ is computed on ARMA residuals. This paper employs the second stage of the *TR* test, in so far as rejection at this stage is more unambiguously indicative of what Ramsey and Rothman (1996) call "Type I" time irreversibility, i.e., irreversibility due to a nonlinear data generating process. Rothman (1990) showed that the variance of $\hat{\gamma}_{2,1}(k)$ is accurately approximated by equation (2) for the case in which $\{X_t\}$ consists of residuals from fitted Gaussian and non-Gaussian ARMA models. Convergence to the asymptotic normal distribution in this case is also relatively quick; some evidence on this is given below in Section 4.

3. Bilinear and Threshold Autoregressive Models

Systematic discussion of bilinear and threshold autoregressive models can be found in Subba Rao and Gabr (1984) and Tong (1983, 1990). They are both members of Priestley's (1980) general class of nonlinear models called "state dependent models."

The general discrete time univariate bilinear BL(p,r,m,k) model takes the form:

$$X_t = \sum_{j=1}^{p} \alpha_k X_{t-j} + \sum_{i=1}^{m}\sum_{j=1}^{k} \beta_{ij} X_{t-i}\varepsilon_{t-j} + \sum_{\ell=0}^{r} \gamma_\ell \varepsilon_{t-\ell} \tag{3}$$

where $\gamma_0 = 1$ and $\{\varepsilon_t\}$ is an independently and identically distributed zero mean sequence of random variables. If $\beta_{ij} = 0 \ \forall\ i,j$, then (3) reduces to a standard ARMA model.

Addition of the bilinear terms $\{X_{t-i}\varepsilon_{t-j}\}$ to the conventional ARMA model yields a parsimonious and powerful nonlinear time series model. Under suitable conditions, Brockett (1976) showed that the bilinear model can approximate arbitrarily closely any

Volterra series. This suggests that the bilinear model is a natural extension of the class of ARMA models.

In Section 4 the power properties of the *TR* test statistics are estimated against MA(1) residuals from simple diagonal bilinear BL(0,0,1,1) models of the form:

$$X_t = \beta X_{t-1}\varepsilon_{t-1} + \varepsilon_t, \quad \varepsilon_t \sim N(0,1) \tag{4}$$

for $\beta = 0.9, 0.8, ..., 0.1$. Tuan (1983) showed that the autocovariance functions of the processes $\{X_t\}$ satisfying (4) have the same form as that of an MA(1) process.

While bilinear models can capture a wide range of nonlinear behavior, there are certain nonlinear phenomena, such as limit cycles, which they cannot produce. Asymmetric limit cycle behavior can be captured by threshold autoregressive models. The fundamental idea behind threshold time series modelling is to use piecewise linear models in a systematic way to model nonlinear behavior. The self-exciting threshold autoregressive SETAR($k;d;p_1,...,p_k$) model with threshold $-\infty = r_0 < r_1 < \cdots < r_k = \infty$ and delay d is given, for $i = 1,...,k$, by:

$$X_t = \alpha_i + \varphi_i(B)X_{t-1} + \varepsilon_t, \quad \text{if } r_{t-1} < X_{t-d} \le r_i, \tag{5}$$

where α_i is a deterministic constant, B is the back shift operator, the ε_t are assumed to be independently and identically distributed with mean 0 and variance σ^2, and:

$$\varphi_i(B) = \varphi_{1i} + \varphi_{2i}B + \cdots + \varphi_{pi}B^{p_i-1}. \tag{6}$$

It is straightforward to show that any additive nonlinear piecewise continuous autoregressive model can be approximated arbitrarily closely by some SETAR model. Thus SETAR models are sufficiently general within the class of nonlinear autoregressive models.

In Section 4 the power properties of the *TR* test statistics are estimated against AR(1) residuals from SETAR(2;1;1,1) models of the form:

$$\begin{aligned} X_t &= \alpha X_{t-1} + \varepsilon_t, \quad \text{if } X_{t-1} \ge 1 \\ X_t &= -0.4X_{t-1} + \varepsilon_t, \quad \text{if } X_{t-1} < 1, \quad \varepsilon_t \sim N(0,1) \end{aligned} \tag{7}$$

for $\alpha = -0.9, -0.8, \ldots, -0.1$. Note that for $\alpha = -0.4$, (7) reduces to a Gaussian AR(1) model.

Table 1. Estimated Sized of *TR* Test Statistic at Lag k With 5 Per Cent Nominal Size.

	$T = 100$	$T = 250$	$T = 500$
	$\{X_t\}$ Gaussian AR(1) Residuals		
$k = 1$	0.102	0.056	0.050
$k = 2$	0.104	0.059	0.047
$k = 3$	0.112	0.057	0.052
$k = 4$	0.111	0.063	0.049
$k = 5$	0.113	0.060	0.054
	$\{X_t\}$ $\chi^2(1)$ AR(1) Residuals		
$k = 1$	0.166	0.083	0.080
$k = 2$	0.141	0.088	0.068
$k = 3$	0.130	0.083	0.055
$k = 4$	0.121	0.081	0.067
$k = 5$	0.131	0.061	0.071

Notes: These results are based on Monte Carlo simulations with 10,000 iterations for each sample size. In each iteration of the two cases studied, an AR(1) series of length T with, respectively, standard normal and $\chi^2(1)$ innovations, and an AR(1) coefficient equal to 0.9, was first generated. An AR(1) model was then fitted to the simulated series and the *TR* test statistic at lag k was calculated on the AR(1) residuals. Rejections were counted at a nominal size of 5%.

There are no exact analytical results for the autocovariance function of a general stationary SETAR model. Tong (1983) showed how the methods developed by Jones (1978) can be used to approximate the stationary distribution for SETAR models. Instead of calculating the Jones-type approximations, in this paper the estimated autocorrelation and partial autocorrelations functions for the models given by (7) were examined. The results suggested that the autocorrelation and partial autocorrelation structure of these models is the same as that of an AR(1) process.

4. Finite Sample and Power Properties of the *TR* Test

Rothman (1992) reported Monte Carlo results on the finite sampling properties of the *TR* test in several independently and identically distributed cases. Three distributions were utilized in those simulations: (1) the standard normal; (2) the chisquare with 1 degree of freedom; and (3) the *t*-distribution with three degrees of freedom. The results on the estimated sizes of $\hat{\gamma}_{2,1}(k)$ strongly suggest that convergence to the asymptotic normal distribution takes place by sample size 100 for all distributions considered, showing that this rate of convergence is not sensitive to asymmetry of the probability density function.

Simulation results on the size properties of $\hat{\gamma}_{2,1}(k)$ calculated on AR(1) residuals are presented in Table 1. In this table results are given for the AR(1) cases with the AR(1) coefficient set to 0.9 and for which: (a) the innovation sequences are distributed standard normal; and (b) the innovations are distributed chisquare with 1 degree of freedom

For the Gaussian case, at sample size 100 $\hat{\gamma}_{2,1}(k)$ appears to reject about twice as often as it should under the null hypothesis. The average estimated size across k is 0.11 while the nominal size is 0.05. The size properties improve by sample size 250. For this sample size $\hat{\gamma}_{2,1}(k)$ rejects roughly 6% of the time, slightly more often than should be rejected under the null. By sample size 500 $\hat{\gamma}_{2,1}(k)$ appears to have converged to its asymptotic normal distribution. Rothman (1990) showed that these results are independent of the value of the AR(1) coefficient (assuming the process is stationary). Thus, while the transformation to AR(1) residuals worsens the small sample properties of $\hat{\gamma}_{2,1}(k)$ relative to the independently and identically distributed case, by sample size 250 the rejection rate differs only slightly from what is should be under the null.

Convergence of $\hat{\gamma}_{2,1}(k)$ to the normal is slower for the non-Gaussian AR(1) case also reported in Table 1. With $\chi^2(1)$ distributed innovations, at sample size 100 $\hat{\gamma}_{2,1}(k)$ rejects about two and half times more often than should be the case under the null hypothesis. The estimated size of $\hat{\gamma}_{2,1}(k)$ is reduced by a little less than one half by sample size 250. At sample size 500, the average estimated size across k is roughly 0.06 while the nominal size is 0.05. Rothman (1990) obtained similar results for AR(1) models with $\chi^2(5)$ and standard exponential distributed innovations. Thus, by sample size 500 the rejection rate does not differ strongly from what is expected under the null hypothesis.

Table 2 presents Monte Carlo results on the estimated power for the bilinear models given by equation (4) at sample size 100, 250, and 500, respectively. As expected, the estimated probability of rejection increases as sample size increases. With the exception of lag $k=1$, at each sample size the estimated power is an increasing function of the parameter β. In so far as the coefficient β can be interpreted as an index of time irreversibility, the results show that the power of the test indeed increases as the degree of irreversibility increases.

Two distinct patterns emerge from the results. For high values of β, that is, for $\beta \geq 0.4$, significant rejections occur a lags $k=1$ and $k=2$ and then tend to decrease exponentially but smoothly as the lag length is increased; for $\beta = 0.5, 0.4$, the estimated power drops down to the nominal level beyond $k=2$. For the $\beta \leq 0.3$ cases, on the other hand, the estimated power is high only at lag $k=1$.

Table 2. Estimated Power of *TR* Test Statistic at Lag *k* Against MA(1) Residuals of BL(0,0,1,1) Model in Equation (4).

	Value of β								
	0.9	0.8	0.7	0.6	0.5	0.4	0.3	0.2	0.1
$T = 100$									
$k = 1$	0.428	0.369	0.359	0.412	0.539	0.651	0.611	0.400	0.150
$k = 2$	0.385	0.351	0.305	0.231	0.153	0.094	0.063	0.054	0.053
$k = 3$	0.152	0.109	0.080	0.055	0.048	0.041	0.047	0.052	0.053
$k = 4$	0.083	0.061	0.048	0.042	0.042	0.043	0.048	0.047	0.049
$k = 5$	0.053	0.048	0.034	0.037	0.039	0.045	0.046	0.050	0.050
$T = 250$									
$k = 1$	0.559	0.475	0.404	0.494	0.736	0.901	0.937	0.801	0.318
$k = 2$	0.587	0.586	0.539	0.427	0.280	0.143	0.071	0.048	0.049
$k = 3$	0.278	0.208	0.139	0.082	0.055	0.044	0.045	0.045	0.046
$k = 4$	0.156	0.105	0.067	0.047	0.043	0.045	0.048	0.044	0.048
$k = 5$	0.098	0.065	0.045	0.038	0.041	0.041	0.046	0.045	0.045
$T = 500$									
$k = 1$	0.671	0.575	0.455	0.553	0.843	0.981	0.996	0.981	0.569
$k = 2$	0.736	0.763	0.744	0.637	0.449	0.214	0.088	0.052	0.043
$k = 3$	0.424	0.328	0.217	0.116	0.063	0.045	0.047	0.046	0.045
$k = 4$	0.240	0.159	0.098	0.056	0.043	0.042	0.046	0.043	0.045
$k = 5$	0.140	0.086	0.055	0.040	0.041	0.043	0.046	0.048	0.047

Notes: Results of Monte Carlo simulations with 10,000 iterations. T is the sample size of the simulated series $\{X_t\}$ and k is the lag at which the *TR* test statistic is calculated. The nominal size of the test was set to 5%.

Table 3. Estimated Power of *TR* Test Statistic at Lag *k* for AR(1) Residuals of SETAR(2;1;1,1) Model in Equation (7).

	Value of α								
	-0.9	-0.8	-0.7	-0.6	-0.5	-0.4	-0.3	-0.2	-0.1
$T = 100$									
$k = 1$	0.223	0.160	0.114	0.084	0.061	0.051	0.060	0.075	0.095
$k = 2$	0.057	0.056	0.053	0.053	0.051	0.054	0.055	0.056	0.055
$k = 3$	0.059	0.053	0.052	0.049	0.053	0.052	0.49	0.049	0.049
$k = 4$	0.047	0.055	0.055	0.051	0.052	0.047	0.051	0.52	0.047
$k = 5$	0.049	0.051	0.056	0.049	0.048	0.047	0.049	0.052	0.047
$T = 250$									
$k = 1$	0.436	0.311	0.189	0.117	0.067	0.053	0.060	0.104	0.169
$k = 2$	0.061	0.057	0.051	0.048	0.046	0.048	0.051	0.054	0.054
$k = 3$	0.065	0.050	0.048	0.049	0.045	0.047	0.047	0.050	0.051
$k = 4$	0.047	0.050	0.046	0.051	0.048	0.048	0.048	0.045	0.046
$k = 5$	0.051	0.050	0.050	0.048	0.049	0.048	0.045	0.044	0.047
$T = 500$									
$k = 1$	0.706	0.500	0.328	0.163	0.078	0.047	0.072	0.160	0.303
$k = 2$	0.073	0.064	0.053	0.050	0.047	0.048	0.045	0.047	0.049
$k = 3$	0.069	0.057	0.048	0.047	0.049	0.047	0.044	0.049	0.047
$k = 4$	0.049	0.049	0.046	0.048	0.044	0.047	0.047	0.044	0.051
$k = 5$	0.048	0.048	0.045	0.042	0.046	0.046	0.044	0.047	0.047

Notes: Results of Monte Carlo simulations with 10,000 iterations. *T* is the sample size of the simulated series and *k* is the lag at which the *TR* test statistic is calculated. The nominal size of the test was set to 5%.

Table 3 shows Monte Carlo results for the SETAR(2;1;1,1) models given by equation (7) at sample size 100, 250, and 500, respectively. The following feature of this class of SETAR(2;1;1,1) models stands out in these tables. That is, significant rejections show up only at lag $k = 1$. For almost all sample sizes and almost all parameter values considered, the estimated probability of rejection at lag $k \neq 1$ is not far from the nominal value of 0.05.[3] In contrast, for the bilinear models studied significant rejections also show up at other values of the lag k.

As the quantity $|\alpha + 0.4|$ increases, at lag $k = 1$ the power of the *TR* test against the SETAR models unambiguously increases. When the value of this time irreversibility index equals zero, that is, for $\alpha = -0.4$ so that the model collapses to a Gaussian AR(1) model, the estimated probability of rejection at lag $k = 1$ equals the nominal size of 0.05 at all sample sizes.

5. Conclusions

Through Monte Carlo simulations the power of the *TR* test statistics was estimated for members of two classes of alternatives: BL(0,0,1,1) bilinear models and SETAR(2;1;1,1) threshold autoregressive models. The patterns of time irreversibility as revealed by the *TR* test statistics varies across the models studied. More specifically, the generic pattern for the SETAR models is very high power at lag $k = 1$ but with estimated probability of rejection at the nominal level at all other lags. For most of the bilinear models considered, the generic pattern was high power at both lags $k = 1$ and $k = 2$ but exponentially declining power at the lag k increased.

This then offers a simple diagnostic which can be used in identification of these time series models. By characterizing the time irreversibility of these stochastic processes through the *TR* test, systematic patterns are revealed which assist the time series analyst in carrying out Box-Jenkins-like identification. This higher-order residual analysis extends, in a simple way, the usual second-order analysis. Through such an approach, Rothman (1994) produced results which suggest that the well-known stock returns series analyzed by Scheinkman and LeBaron (1989) for the presence of chaotic dynamics, after standardization for ARCH effects, may actually be the output of a simple bilinear process. A bilinear model produced a reduction of more than 5% in the one-step ahead forecast mean squared error over the best ARMA alternative model.

Finally, it is important to recall that the *TR* test was not designed to have power against any particular member of the very large set of stochastic processes admitted under the alternative hypothesis. This is in contrast, for example, to the *LM* class of tests developed by Saikkonen and Luukkonen (1988). These authors introduced *LM* tests of pure autoregressive linearity against bilinear and threshold autoregressive alternatives. It is reasonable, then, to ask why a time series analyst should bother with *TR* testing when these specific alternatives are being considered. A good answer is that, while these *LM* tests indeed have reasonable power against the specific alternatives they were designed against, they also have power against many other alternatives. That is, rejection of the linearity null through use of these *LM* tests can spuriously lead to a conclusion of either bilinear or SETAR-type nonlinearity. As a result, it may prove useful in practice to supplement Saikkonen and Luukkonen (1988) *LM* test results with a battery of *TR* tests.

Acknowledgements

I wish to thank Buz Brock, Cliff Hurvich, James Ramsey, and Timo Teräsvirta for helpful comments. I also would like to recognize financial support provided by an East Carolina University Faculty Senate Project Expenses Grant.

Notes

1. In a companion paper, Rothman (1992), the estimated power of the *TR* test is compared to the simulated power of the BDS and Hinich bispectrum tests for simple threshold autoregressive models.
2. The zero bicovariance condition, however, does not imply that these time series processes are time reversible. It simply implies that the third-order *TR* test does not have power against them.
3. Rothman (1992) obtained similar results for different set of SETAR(2;1;1,1) models. In that paper, however, $\hat{\gamma}_{2,1}(k)$ was calculated directly on $\{X_t\}$ since the estimated autocorrelation functions for the models studied exhibited no evidence of serial correlation. Likewise, comparable results were reported by Chen, *et al.* (1998) for similar SETAR models.

References

Ashley, R. A, D. M. Patterson, and M. J. Hinich (1986). "A Diagnostic Test for Nonlinear Serial Dependence in Time Series Fitting Errors," *Journal of Time Series Analysis*, 7, 165-178.

Brock, W. A., W. D. Dechert, and J. A. Scheinkman (1987). "A Test for Independence Based on the Correlation Dimension," working paper, Department of Economics, University of Wisconsin. [Revised version, Brock, W. A., W. D. Dechert, J. A. Scheinkman, and B. LeBaron (1996), *Econometric Reviews*, 15, 197-235.]

Brock, W. A., D. A. Hsieh, and B. LeBaron (1991). *A Test of Nonlinear Dynamics, Chaos and Instability: Theory and Evidence*. Cambridge: The MIT Press.

Brockett, R. W. (1976). "Volterra Series and Geometric Control Theory," *Automatica*, 12, 167-176.

Chen, Y. -T., R. Y. Chou, and C. -M. Kuan (1998). "A New Time Reversibility Test Without Moment Restrictions," working paper, Department of Economics, National Taiwan University.

Clements, M. P. and H. -M. Krolzig (1998). "A Comparison of the Forecasting Performance of Markov-Switching and Threshold Autoregressive Models of US GNP," *Econometrics Journal*, 1, C47-C75.

Gooijer, J. de (1989). "Testing Non-Linearities in World Stock Market Prices," *Economics Letters*, 31, 31-35.

Hinich, M. J. (1982). "Testing for Gaussianity and Linearity of a Stationary Time Series," *Journal of Time Series Analysis*, 2, 169-176.

Jones, D. A. (1978). "Non-Linear Autoregressive Processes," *Proceedings of the Royal Society of London, Series A*, 360, 71-95.

Lawrance, A. J. and P. A. W. Lewis (1987). "Higher-Order Residual Analysis for Nonlinear Time Series with Autoregressive Correlation Structures," *International Statistical Review*, 55, 21-35.

Lee, T. H., H. White, and C. W. J. Granger (1993). "Testing for Neglected Nonlinearity in Time

Series Models: A Comparison of Neural Network Methods and Alternative Tests," *Journal of Econometrics*, 56, 269-290.

Maravall, A. (1983). "An Application of Nonlinear Time Series Forecasting," *Journal of Business and Economic Statistics*, 1, 66-74.

Montgomery, A. L, V. Zarnowitz, R. S. Tsay, and G. C. Tiao (1998). "Forecasting the U.S. Unemployment Rate," *Journal of the American Statistical Association*, 93, 478-493.

Potter, S. (1995). "A Nonlinear Approach to U.S. GNP," *Journal of Applied Econometrics*, 10, 109-125.

Priestley, M. B. (1980). "State-Dependent Models: A General Approach to Non-Linear Time Series Analysis," *Journal of Time Series Analysis*, 1, 47-71.

Ramsey, J. B. and P. Rothman (1996). "Time Irreversibility and Business Cycle Asymmetry," *Journal of Money, Credit, and Banking*, 28, 1-21.

Rothman, P. (1998). "Forecasting Asymmetric Unemployment Rates," *Review of Economics and Statistics*, 80, 164-168.

Rothman, P. (1994), "The Time Reversibility Test with Application to Financial Data," in Semmler, W. (ed.), *Business Cycles: Theory and Empirical Methods*. Boston: Kluwer Academic Press, 389-403.

Rothman, P. (1992). "The Comparative Power of the *TR* Test Against Simple Threshold Models," *Journal of Applied Econometrics*, 7, S187-S195.

Rothman, P. (1990). "Characterization of the Time Irreversibility of Economic Time Series," Ph.D. dissertation, Department of Economics, New York University.

Saikkonen, P. and R. Luukkonen (1988). "Lagrange Multiplier Tests for Testing Non-Linearities in Time Series Models," *Scandinavian Journal of Statistics*, 15, 55-68.

Scheinkman, J. A. and B. LeBaron (1989). "Nonlinear Dynamics and Stock Returns," *Journal of Business*, 62, 311-337.

Subba Rao, T. S. and M. M. Gabr (1984). *An Introduction to Bispectral Analysis. Lecture Notes in Statistics, Vol. 24*. New York: Springer-Verlag.

Tong, H. (1990). *Non-Linear Time Series: A Dynamical System Approach*. Oxford: Clarendon Press.

Tong, H. (1983). *Threshold Models in Non-Linear Time Series Analysis. Lecture Notes in Statistics, Vol. 11*. New York: Springer-Verlag.

Tuan, P. (1983). "Bilinear Markovian Representations and Bilinear Models," Technical Report No. 144, Department of Mathematics, University of Manchester Institute of Science and Technology.

White, H. (1989). "An Additional Hidden Unit Test for Neglected Nonlinearity in Multilayer Feedforward Networks," *Proceedings of the International Joint Conference on Neural Networks. Washington DC*. San Diego: SOS Printing, 451-455.

Index